Pest Control: Techniques and Management

Pest Control: Techniques and Management

Editor: Edwin Tan

www.callistoreference.com

Callisto Reference,
118-35 Queens Blvd., Suite 400,
Forest Hills, NY 11375, USA

Visit us on the World Wide Web at:
www.callistoreference.com

ISBN: 978-1-63239-781-2 (Hardback)

The publisher's policy is to use permanent paper from mills that operate a sustainable forestry policy. Furthermore, the publisher ensures that the text paper and cover boards used have met acceptable environmental accreditation standards.

Trademark Notice: Registered trademark of products or corporate names are used only for explanation and identification without intent to infringe.

Printed in the United States of America.

Cataloging-in-publication Data

Pest control : techniques and management / edited by Edwin Tan.
    p. cm.
Includes bibliographical references and index.
ISBN 978-1-63239-781-2
1. Pests--Control. 2. Pests--Integrated control. 3. Agricultural pests--Control. I. Tan, Edwin.
SB950 .P47 2017
632.9--dc23

Table of Contents

Preface

Regulation and management of pests in the agricultural setting is called pest control. It is considered hazardous not only to a person's health but also to the environment. This book will offer an insight into the umpteen numbers of measures that have been applied to combat pests across the world. It discusses the fundamentals as well as modern approaches of pest control. It strives to provide a fair idea about this field and to help develop a better understanding of the latest advances within this field. The topics included in this book are of utmost significance and bound to provide incredible information to readers and professionals.

Various studies have approached the subject by analyzing it with a single perspective, but the present book provides diverse methodologies and techniques to address this field. This book contains theories and applications needed for understanding the subject from different perspectives. The aim is to keep the readers informed about the progress in the field; therefore, the contributions were carefully examined to compile novel researches by specialists from across the globe.

Indeed, the job of the editor is the most crucial and challenging in compiling all chapters into a single book. In the end, I would extend my sincere thanks to the chapter authors for their profound work. I am also thankful for the support provided by my family and colleagues during the compilation of this book.

Editor

A New Light on the Evolution and Propagation of Prehistoric Grain Pests: The World's Oldest Maize Weevils Found in Jomon Potteries, Japan

Hiroki Obata[1]*, Aya Manabe[1], Naoko Nakamura[2], Tomokazu Onishi[3], Yasuko Senba[4]

1 Faculty of Letters, Kumamoto University, Kumamoto City, Kumamoto Prefecture, Japan, 2 Faculty of Law and Letters, Kagoshima University, Kagoshima City, Kagoshima Prefecture, Japan, 3 Faculty of Intercultural Studies, The International University of Kagoshima, Kagoshima City, Kagoshima Prefecture, Japan, 4 Project for Seed Impression Studies, Kumamoto University, Kumamoto City, Kumamoto Prefecture, Japan

Abstract

Three *Sitophilus* species (*S. granarius* L., *S. oryzae* L., and *S. zeamais* Mots.) are closely related based on DNA analysis of their endosymbionts. All are seed parasites of cereal crops and important economic pest species in stored grain. The *Sitophilus* species that currently exist, including these three species, are generally believed to be endemic to Asia's forested areas, suggesting that the first infestations of stored grain must have taken place near the forested mountains of southwestern Asia. Previous archaeological data and historical records suggest that the three species may have been diffused by the spread of Neolithic agriculture, but this hypothesis has only been established for granary weevils in European and southwestern Asian archaeological records. There was little archeological evidence for grain pests in East Asia before the discovery of maize weevil impressions in Jomon pottery in 2004 using the "impression replica" method. Our research on Jomon agriculture based on seed and insect impressions in pottery continued to seek additional evidence. In 2010, we discovered older weevil impressions in Jomon pottery dating to ca. 10 500 BP. These specimens are the oldest harmful insects in the world discovered at archaeological sites. Our results provide evidence of harmful insects living in the villages from the Earliest Jomon, when no cereals were cultivated. This suggests we must reconsider previous scenarios for the evolution and propagation of grain pest weevils, especially in eastern Asia. Although details of their biology or the foods they infested remain unclear, we hope future interdisciplinary collaborations among geneticists, entomologists, and archaeologists will provide the missing details.

Editor: Robert DeSalle, American Museum of Natural History, United States of America

Funding: This research was supported by Grant-in-Aid for Scientific Research on Areas A (subject number: 20242022) by Japan Society for Promotion of Science (http://www.jsps.go.jp/j-grantsinaid/index.html). The funders had no role in study design, data collection and analysis, decision to publish, or preparation of the manuscript.

Competing Interests: The authors have declared that no competing interests exist.

* E-mail: totori@kumamoto-u.ac.jp

Introduction

Granary (*Sitophilus granarius* L.), rice (*Sitophilus oryzae* L.), and maize (*Sitophilus zeamais* Mots.) weevils, known as "snout weevils", feed inside rice or barley grains during their larval stage and pupate inside the grains [1]. Rice and maize weevils are widespread in warm regions. In Europe and North America, they are replaced by temperate species such as the granary weevil [2]. The global distribution of these insects occurred relatively recently, and they have continued to spread as a result of worldwide cereal trading during the 20th century.

In archaeological records from Europe, the Mediterranean, and Asia, the oldest granary weevil discovery has been dated to ca. 7000 BP. The granary weevils recovered from many sites in these regions correspond to the diffusion of Neolithic agriculture or Roman cereal trading and transport of soldiers. In contrast, few prehistoric discoveries of the rice weevil and maize weevil had been reported in China and Japan by 2003. In general, the granary weevil and its two sister species are monophyletic species in genus *Sitophilus* and are believed to have originated in Asia [3].

However, many impressions of cereal weevils have been discovered in Jomon pottery in Japan since researchers first applied the "impression replica" method in 2004. This method involves making a model of the original form that created the cavity in a potsherd. Since 2004, the number of impression records has gradually increased. Until 2009, we had collected 37 samples, of which the oldest dated to ca. 4500 BP. However, our recent search for seed or insect impressions at archeological sites on Tanegashima Island in spring 2010 provided many new records, which suggest that the association between maize weevils and humans occurred in the Earliest Jomon (11 500 to 7300 BP), at ca. 10 500 BP. This discovery suggests that researchers must reconsider previous scenarios for the evolution and propagation of grain pest weevils, their adaptation to grain storage systems, and their spread along with prehistoric agriculture, especially in eastern Asia. Our new results suggest that eastern Asian grain pests, including the maize weevil and probably the rice weevil, evolved differently from the granary weevil in Europe.

The goal of this paper is to introduce our new discovery and discuss its preliminary significance for archaeological research in eastern Asia.

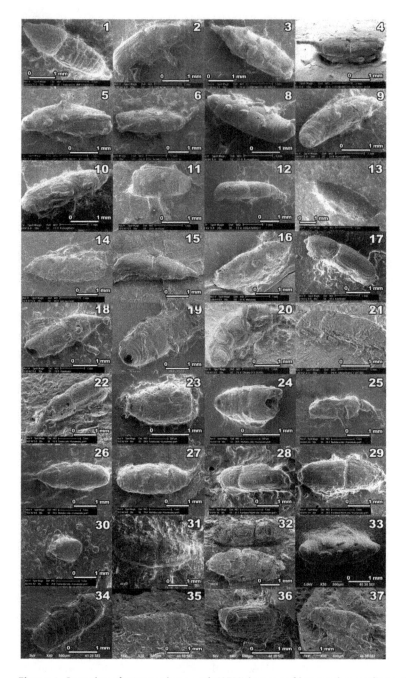

Figure 1. Scanning electron micrograph (SEM) images of impression replicas of maize weevils that had been discovered at Jomon sites by 2009. The 37 weevil impressions were discovered in Jomon pottery, mainly collected on Kyushu, dating to between ca. 4500 and ca. 3000 BP. Based on the diagnostic criteria described in the text, including the size of the specimens, these weevils appear to have been maize weevils (*Sitophilus zeamais*). Details of each impression replica are provided in Table 1.

Results and Discussion

Discovery of weevil impressions in Jomon potsherds

Greek and Roman records from ca. 2200 BP describe weevils infesting wheat. In China, the oldest record appears in the "Ěryǎ" dictionary, which was published between ca. 2500 to 2200 BP. By this time, weevils were already recognized as storage pests that infested rice. The first Japanese description of maize weevils appears in historical records from ca. 1000 BP [4,5].

Archaeological records from Europe and southwestern Asia reveal that granary weevils had spread along with the diffusion of agriculture and grain trading or transport between ca. 7000 BP and ca. 1900 BP [3]. For example, granary weevils were discovered in a funeral offering of barley to a dead king of the Egyptian Sixth Dynasty (ca. 4300 BP). In China, there is little archaeological evidence of prehistoric grain pests. The only example is of rice weevils infesting barley in a grave from the Han Dynasty (2118 BP) [4,5].

In Japanese archeological records, maize weevil impressions were first discovered in Jomon pottery in 2004. Before then, the oldest specimens (beetle carapaces) came from a ditch deposit at the Ikegamisone site (ca. 2000 BP). The beetle carapaces of maize

Table 1. Details of the maize weevil impressions that were obtained at Jomon sites by 2009. Numbers correspond to the photographs in Figure 1.

	Part of insect covered by the impression	Site	Sample name	Type of pottery	Shape/part	Phase during the Jomon	Estimated age (BP)
1	Dorsal view (missing rostrum, wings, and legs)	Ishinomoto (Kumamoto Pref.)	INM-47	Amagi	Deep bowl/rim	The end of the Late J.	ca. 3450
2	Side view (missing legs)	Ishinomoto (Kumamoto Pref.)	42-29629-1	Amagi	Deep bowl/rim	The end of the Late J.	ca. 3450
3	Ventral view (missing rostrum and legs)	Ishinomoto (Kumamoto Pref.)	45-1697-1	Goryo	Shallow bowl/rim	The latter half of the Late J.	ca. 3500
4	Dorsal view (missing legs)	Ishinomoto (Kumamoto Pref.)	39-SH01-2694	Koga	Shallow bowl/body	The end of the Late J.	ca. 3400
5	Ventral view (missing rostrum and legs)	Ishinomoto (Kumamoto Pref.)	47-SH35-31040-1	Amagi	Deep bowl/unknown	The end of the Late J.	ca. 3450
6	Ventral view (missing legs)	Ishinomoto (Kumamoto Pref.)	47-SX-07-b	Goryo	Shallow bowl/ unknown	Last half of the Late	ca. 3500
7	Unknown	Ishinomoto (Kumamoto Pref.)	Unknown	Unknown	Unknown	The Latest J.	ca. 3400 to 3000
8	Ventral view (missing rostrum and legs)	Kunugibaru (Kagoshima Pref.)	Kunugibaru-1	Unknown	Deep bowl/rim	The first half of the Latest J.	ca. 3200 to 3000
9	Side view (missing rostrum and legs)	Kunugibaru (Kagoshima Pref.)	Kunugibaru-2	Unknown	Deep bowl/rim	The first half of the Latest J.	ca. 3200 to 3000
10	Side view (missing rostrum and legs)	Kunugibaru (Kagoshima Pref.)	Kunugibaru-3	Unknown	Deep bowl/rim	The first half of the Latest J.	ca. 3200 to 3000
11	Ventral view (missing thorax and legs)	Ohnobaru (Nagasaki Pref.)	ONB1010	Tarozako	Deep bowl/base	The end of the Late J.	ca. 3600
12	Side view (missing legs)	Higataro (Nagasaki Pref.)	HIG115	Kurokawa?	Deep bowl/body	The first half of the Latest J.	ca. 3300
13	Side view (missing legs)	Higataro (Nagasaki Pref.)	10381-03	Kurokawa?	Deep bowl/body	The first half of the Latest J.	ca. 3300
14	Ventral view (missing rostrum and legs)	Gongenwaki (Nagasaki Pref.)	GGW-021	New Kurokasa	Bowl/body	The middle of the Latest J.	ca. 3000
15	Side view (missing rostrum and legs)	Ohnobaru (Nagasaki Pref.)	ONB1018	Tarozako?	Deep bowl/body	The latter half of the Late J.	ca. 3600
16	Ventral view (missing rostrum and legs)	Mimanda (Kumamoto Pref.)	MD0019	Tarozako	Bowl/base	The middle of the Late J.	ca. 3600
17	Side view (missing head and legs)	Kaminabe (Kumamoto Pref.)	KNB05	Amagi	Deep bowl/rim	The end of the Late J.	ca. 3450
18	Side view (missing head and legs)	Kaminabe (Kumamoto Pref.)	KNB32	Koga?	Deep bowl/body	The end of the Late J.	ca. 3400
19	Side view (missing legs)	Kaminabe (Kumamoto Pref.)	KNB34	Amagi?	Deep bowl/body	The end of the Late J.	ca. 3450
20	Side view (missing rostrum and legs)	Ohbaru D (Fukuoka Pref.)	Ohbaru-D-3 (9265)	Unknown	Deep bowl/unknown	The first half of the Latest J.	ca. 3200 to 3000
21	Side view (missing rostrum and legs)	Shigetome (Fukuoka Pref.)	Shigetome1 (8748)	Amagi	Deep bowl/body	The end of the Late J.	ca. 3450
22	Dorsal view (missing legs)	Toroku shell midden. (Kumamoto Pref.)	TR11	Kanezaki 3	Deep bowl/body	The first half of the Late J.	ca. 4000
23	Dorsal view (missing rostrum and trunk)	Toroku shell midden (Kumamoto Pref.)	TR21	Mimanda	Deep bowl/body	The latter half of the Late J.	ca. 3550
24	Ventral view (a trunk)	Nishibira shell midden (Kumamoto Pref.)	NB02	Nishibira	Bowl/body	The first half of the Late J.	ca. 3700
25	Side view (missing legs)	Nishibira shell midden (Kumamoto Pref.)	NB07	Nishibira?	Bowl/body	The first half of the Late J.	ca. 3700
26	Dorsal view (missing legs)	Nishibira shell midden .(Kumamoto Pref.)	NB08	Nishibira?	Bowl/body	The first half of the Late J.	ca. 3700
27	Ventral view (missing legs)	Nishibira shell midden (Kumamoto Pref.)	NB17	Goryou	Shallow bowl/rim	The latter half of the Late J.	ca. 3500
28	Dorsal view (missing rostrum and legs)	Kurokamimachi (Kumamoto Pref.)	KKN07	Unknown	Deep bowl/body	The latter half of the Late J.?	ca. 3600 to 3400

Table 1. Cont.

	Part of insect covered by the impression	Site	Sample name	Type of pottery	Shape/part	Phase during the Jomon	Estimated age (BP)
29	Dorsal view (missing rostrum and legs)	Kurokamimachi (Kumamoto Pref.)	KKN08	Unknown	Deep bowl/body	The latter half of the Late J.?	ca. 3600 to 3400
30	Dorsal view (a thorax)	Kaminabe (Kumamoto Pref.)	KNB22	Amagi?	Deep bowl/body	The end of the Late J.	ca. 3450
31	Dorsal view (missing rostrum and legs)	Nakaya (Yamanashi Pref.)	Nky01	Shimizutennouzan	Deep bowl/body	The first half of the Latest J.	ca. 3000
32	Side view (missing rostrum and legs)	Nakaya (Yamanashi Pref.)	Nky02	Shimizutennouzan	Deep bowl/body	The first half of the Latest J.	ca. 3000
33	Ventral view (missing rostrum and legs)	Mimanda (Kumamoto Pref.)	MMD2054	Unknown	Deep bowl/body	The latter half of the Late J.?	ca. 3600 to 3400
34	Ventral view (missing head and legs)	Ohnobaru (Nagasaki Pref.)	ONB1116	Unknown	Deep bowl/body	The latter half of the Late J.?	ca. 3600 to 3400
35	Ventral view (missing rostrum and legs)	Kakiuchi (Kagoshima Pref.)	KKU0019	Namiki-Nanpukuji	Deep bowl/body	The beginning of the Late J.	ca. 4500 to 4000
36	Dorsal view (abdomen)	Izumi shell midden (Kagoshima Pref.)	KZK0008	Izumi	Deep bowl/rim	The beginning of the Late J.	ca. 4300
37	Side view (missing rostrum and legs)	Nanbarauchibori (Kagoshima Pref.)	NBU0005	Ichiki	Deep bowl/rim	The beginning of the Late J.	ca. 4200

Japanese academic circles on archaeology divides the Jomon Period into six phases as followings. The Incipient Jomon (15000-11500BP), The Earliest Jomon (11500-7300BP), The Early Jomon (7300-5500BP), The Middle Jomon (5500-4500BP), The Late Jomon (4500-3300BP), The Latest Jomon (3300-3000BP).

weevils were also recovered at the Fujiwara Palace site (ca.1190 BP) and the Kiyosu Castle site (ca. 390 BP) [6]. In 2004, maize weevil impressions were discovered (ca. 3500 BP) in the Late Jomon (4500 to 3300 BP) pottery on Kyushu [7]. Another recent discovery revealed soybean and adzuki bean cultivation during the Early Jomon to the Latest Jomon [8,9]. These beans were also discovered in Jomon pottery as impressions.

Subsequently, many maize weevil impressions have been discovered at the Late Jomon sites, especially on Kyushu. In 2007, two maize weevil impressions were discovered at the Latest Jomon (3300 to 3000 BP) site in Yamanashi Prefecture [10]. By 2008, 32 weevil impressions had been found at 13 sites (Fig. 1, Table 1; samples 1 to 32). Most impressions appeared to be of maize weevils, based on their size and anatomical characteristics. These new discoveries suggest that maize weevils might have coexisted with humans as early as the Late Jomon period. The earliest impression was found inside a Kanezaki 3–type deep bowl dating to ca. 4000 BP. The number of specimens increases rapidly after the Nishibira pottery phase (ca. 3800 BP), with a peak during the last quarter of the Late Jomon [11]. Subsequently, impression discovery rates remain stable. We discovered three new impressions at early the Late Jomon sites in Kagoshima Prefecture in 2009. This finding dates the appearance of weevils 200 or 300 years earlier than previous research (Fig. 1, Table 1; samples 35-37).

Old hypothesis: Maize weevils document the origins of rice cultivation in Japan

Maize weevils feed on stored grain and on fruits such as peach or apple in North America, and also inhabit forests or grasslands in southern Japan, where they feed on flowers in the spring. The adult weevils feed on 37 families and 96 species of plants, but the larvae feed on only 11 families and 31 species [2]. The adult maize weevils from Kumamoto City that we tested preferred large grains such as rice, barley, and wheat. They did not feed on adzuki bean, hemp, or rice in the husk. The weevils also infested acorns, but

only those without seed coats, where they successfully matured [11]. However, during the Jomon period, acorns were generally stored with their seed coats to protect them from decay, so few or no maize weevils would have infested acorns during this period [12]. Although adult maize weevils feed on rice flour, they never oviposit there [13].

Two hypotheses have been proposed to explain the diffusion of rice cultivation into Japan. First, this form of agriculture may have diffused to Japan from the Shantong Peninsula through Korea ca. 4000 BP [11,14]; alternatively, it may have diffused from southern China through the Ryukyu Islands ca. 6700 BP [15]. The former hypothesis is accepted by most archaeologists. Evidence from archaeological sites dating to those times (charred seeds, seed impressions, and phytoliths) suggests the cultivation of cereals in the Poaceae (e.g., rice, barley), which were introduced to Kyushu from Korea. The second hypothesis is not supported by archeological or archaeobotanical evidence [16,17]. The earliest rice cultivation on the Ryukyu Islands occurred in ca. 1100 BP and was not introduced from southern China; instead, it was introduced at the time of human immigration from Kyushu into the region [11,14].

If the maize weevil depends on crop cultivation, then it should have appeared at roughly the same time as the expansion of agriculture, which is known from Japanese archaeobotanical records, and indeed, archaeological artifacts suggest increasingly strong cultural relationships between southern Korea and northwestern Kyushu at this time. Furthermore, rice phytoliths have been recovered from the Late Jomon pottery in northwestern Kyushu. The presence of maize weevils therefore suggests the existence of rice or barley cultivation during the Late to the Latest Jomon periods, and that they invaded Japan from Korea, accompanying rice cultivation [11].

New evidence: older maize weevil impressions

We have new evidence that contradicts the original hypothesis. In February 2010, we discovered the oldest impressions of maize

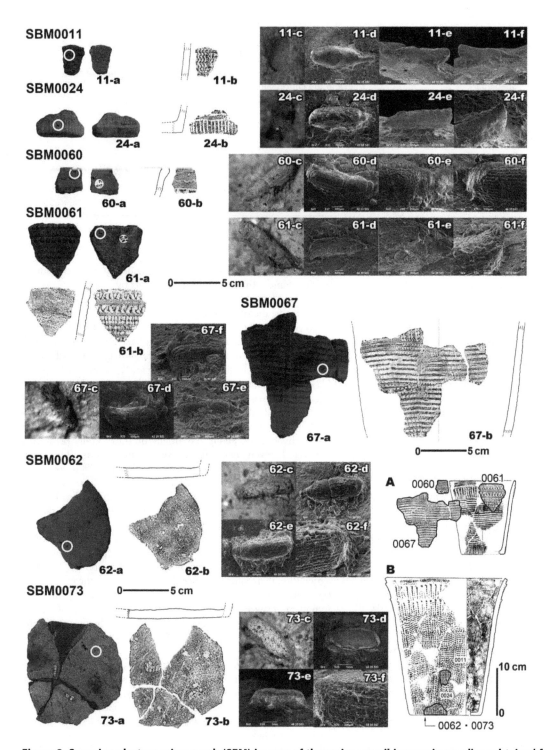

Figure 2. Scanning electron micrograph (SEM) images of the maize weevil impression replicas obtained from potsherds from the Sanbonmatsu Site. Maize weevil impressions on pottery (a), pottery section illustrations with rubbings (b), photos of cavities (c), and SEM images of the impression replicas (d,e,f). The white circles indicate the position of the cavities in the pottery. "A" and "B" show the positions of the potsherds with maize weevil impressions in pottery samples SBM0060, SBM0061, SBM0062, SBM0067, and SBM0073. "A "is from the Ohnakahara Site and "B" is from the Kakoinoharu Site in Kagoshima Prefecture. Details are provided in Table 2.

weevils, which we obtained from the Earliest Jomon potsherds dated to ca. 10 500 BP from the Sanbonmatsu (SBM) site in Kagoshima Prefecture. The site, which is on a terrace (50 m asl) on the eastern coast of Tanagashima Island, 40 km southeast of the Ohsumi Peninsula, was excavated in 2007 in a project organized by the Nishinoomote City Board of Education.

Researchers discovered cultural layers containing many artifacts, mainly from the Earliest Jomon period.

When we examined the potsherds to find seed and other impressions in February 2010, we found two fragments that contained maize weevil impressions. During our second examination, in April 2010, we found five fragments with maize weevil

Table 2. Details of the maize weevil impressions from the Sanbonmatsu (SBM) site.

No.	Pottery type	Shape/part	Part of the weevil in the impression	Length of impression replicas (mm)	Estimated length of original weevil (mm)
SBM0011	Yoshida	Deep bowl/body	Ventral view (missing legs)	3.69	3.98
SBM0024	Yoshida	Deep bowl/rim	Ventral view (missing rostrum and legs)	3.32	4.58
SBM0060	Yoshida	Deep bowl/rim	Dorsal view (missing rostrum and legs)	3.11	4.33
SBM0061	Yoshida	Deep bowl/rim	Ventral view (missing legs)	4.11	5.41
SBM0062	Yoshida	Deep bowl/bottom	Dorsal view (missing rostrum)	3.45	4.69
SBM0067	Yoshida	Deep bowl/body	Side view	3.17	4.39
SBM0073	Yoshida	Deep bowl/bottom	Dorsal view (missing rostrum)	3.58	4.83

Samples are illustrated in Figure 2.

impressions. These fragments came from Yoshida-type deep bowls and were ornamented with shells, a popular ornamentation during the first half of the Earliest Jomon period in southern Kyushu [18]. This cluster has ^{14}C dates from ca. 9240 BP to ca. 9330 BP, suggesting that the Yoshida pottery dates to ca. 10 500 cal BP [19].

Adult granary weevils have elongated punctations on their thorax and other body parts, whereas adult rice and maize weevils have round or irregular punctations [20]. Most of the replicas lacked legs and a rostrum, but had round or irregularly shaped punctations, similar to those of maize weevils (Fig. 2, Table 2).

However, two beetles are morphologically similar to maize weevils and should be excluded as possible explanations for the impressions: *Diocalandra* spp. and *Paracythopeus melancholicus* Roelofs. *Diocalandra* spp. are slenderer and longer than maize weevils [21]. The ratio of thorax to elytron length also differs among the three species: 0.898 for *Diocalandra* spp., 0.500 for *P. melancholicus*, and 0.757 for weevil impression replica SBM0024, which nearly equals the ratio of 0.776 for *S. zeamais* (Fig. 3). The side view of weevil impression SBM0024 is most similar to that of *S. zeamais* (Fig. 3). The diagnostic criterion that distinguishes *S. zeamais* from *P. melancholicus* is the elytron end, which is shorter than the abdomen in *S. zeamais* but covers the full length of abdomen in *P. melancholicus* (Fig. 3). These criteria can be seen clearly in the elytron end of the other weevil impression replicas (SBM0060, SMB0061, and SBM0062; Fig. 2), which suggest that the weevils from the Sanbonmatsu site were *S. zeamais*. These are therefore the oldest maize weevil relics in the world.

To confirm this identification, we obtained CT scans of the impressions. These revealed details of the insect's legs, rostrum end, and antennae that have never previously been seen in weevil impression replicas (Fig. 4). These findings demonstrate the superiority of the CT scans compared with the original replica method for correctly identifying insects.

Significance of maize weevils during the Earliest Jomon

Plarre [3] describes one accepted scenario for the evolution and propagation of grain pests, as follows. The currently existing *Sitophilus* species, including the three economically important grain pests discussed in this paper (*S. granarius*, *S. oryzae*, and *S. zeamais*), appear to be endemic to the forested areas of Asia. It therefore seems likely that the first infestation of stored grain occurred near the forested mountains of southwestern Asia. If these insects originally infested forest food sources such as acorns, the storage of these foods with cultivated grains would have provided the weevils with an alternative food source, and weevils capable of exploiting this resource would have had an advantage over other weevils, leading to co-evolution that produced weevils that were increas-

ingly dependent on stored grain. Subsequently, this coevolution would promote the spread of these insects by humans.

Researchers have not yet performed a phylogenetic analysis that would define the relationships among the various *Sitophilus* species. However, the high level of genetic similarity among the SOPE (*Sitophilus oryzae* primary endosymbiont) bacteria suggests that this endosymbiosis evolved only once, in a single original weevil species that subsequently evolved into *S. granarius*, *S. oryzae*, and *S. zeamais* [22]. Furthermore, dendrograms based on this DNA analysis suggest that these three *Sitophilus* species evolved from a common ancestor [23,24].

Our new discovery supports this hypothesis and contradicts earlier hypotheses about the origin of these grain pests. We discovered seven new maize weevil impressions. This high discovery rate (one weevil impression replica per 2000 potsherds) is similar to or greater than that (one among 3000 to 20 000 potsherds) from the Late Jomon sites on Kyushu, indicating that these weevils were already house pests in the Earliest Jomon. These weevils are therefore the oldest house pests in the world, as they are older than the oldest archaeological records in Europe, which date to ca. 7000 BP [3]. Even if the propagation of *S. granarius* in Europe occurred during the diffusion of Neolithic agriculture, there is no evidence that demonstrates the existence of cereal agriculture in Japan or its arrival at that time. If the Late Jomon maize weevils were associated with the spread of rice or barley cultivation into Japan, then what is the significance of the Earliest Jomon records? Were other cereals cultivated in Japan at that time?

The Earliest Jomon is nearly synchronous with the beginning of rice cultivation in the Lower Yangtze River Valley of China, which is believed to be the origin of Asian rice cultivation [25]. However, there is currently no archaeological evidence that connects the two regions at that time. And the oldest evidence of barley and wheat cultivation in eastern Asia is younger than ca. 4000 BP. Thus, the archaeological evidence strongly suggests there were no cultivated cereals that could have been infested with maize weevils in Japan at ca. 10 500 BP. This suggests that the existence of maize weevils in the Earliest Jomon was not associated with rice cultivation. Nevertheless, the high weevil density at that time suggests the weevils were closely related to the Jomon people and lived in Jomon houses, where they fed on stored foods.

We do not know what kind of wild plant food they fed on or infested. Judging from the archaeobotanical data in Japan, we propose three candidates: the seeds of bamboo (Bambusoideae), acorns (*Quercus* spp.), and chestnuts (*Castanea* spp.). Bamboo seeds can be stored for sufficiently long periods to permit weevil development and have sometimes been used as an emergency food

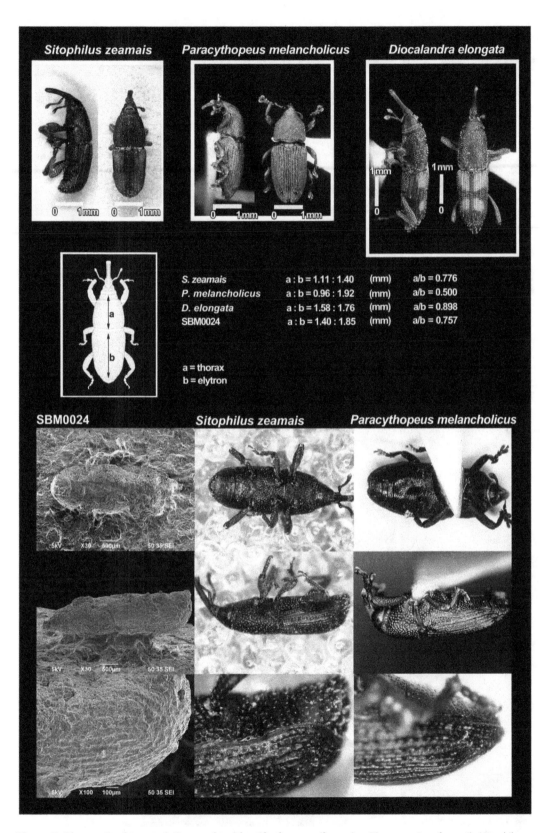

Figure 3. Diagnostic characteristics used to identify the weevil species. Three species of weevils (*Sitophilus zeamais*, *Diocalandra elongata*, and *Paracythopeus melancholicus*) are distinguishable by the ratio of thorax to elytron length. Another diagnostic criterion that distinguishes *S. zeamais* from *P. melancholicus* is the elytron end. Elytron does not cover the full abdomen in *S. zeamais* but extends the full length of the abdomen in *P. melancholicus* (bottom row of photographs).

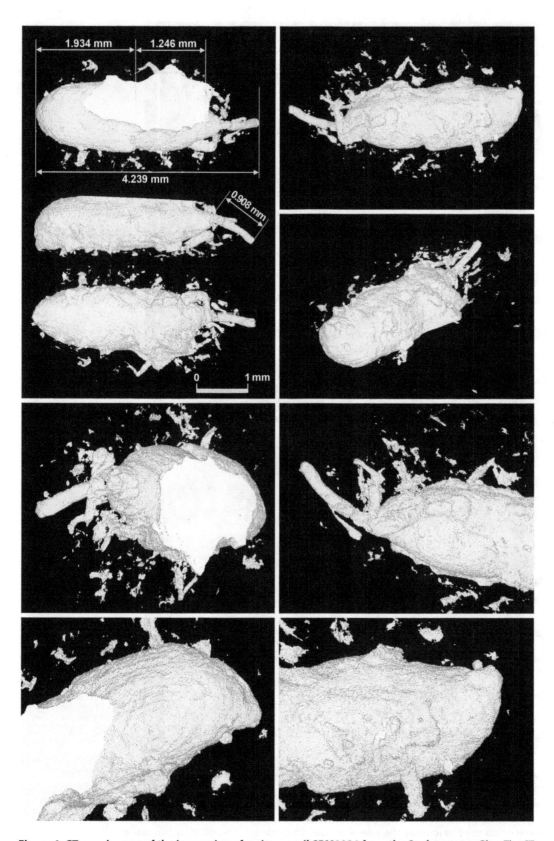

Figure 4. CT scan images of the impression of maize weevil SBM0024 from the Sanbonmatsu Site. The CT scan images show details of the insect's legs, rostrum end, and antennae that were previously unseen in the impression replicas. These findings demonstrate the method's superiority to the older impression replica method for correctly identifying insects. The lengths in this example are 0.908 mm (rostrum), 1.246 mm (thorax), and 1.934 mm (abdomen). In our experiment, the shrinkage of the mean lengths from the original maize weevils to the impressions they left in potsherd cavities were to 92.16% (rostrum), 91.63% (thorax), and 96.28% (abdomen) of the original length. Therefore, the original lengths would have been 0.985 mm (rostrum), 1.359 mm (thorax), and 2.008 mm (abdomen). These lengths are larger than those of modern reference specimens

that were reared in cleaned rice grains (mean lengths of rostrum, thorax, and abdomen were 0.903, 1.105, and 1.428 mm, respectively; $n = 20$) and were similar to the size of weevils reared in chestnuts (mean lengths of rostrum, thorax, and abdomen were 1.056, 1.338, and 1.882 mm, respectively; $n = 20$).

source during famines. Indeed, we found charred bamboo seeds at archaeological sites from the Late Jomon to the Ainu Culture period (from ca. 760 BP) on Hokkaido [26]. However, phytolith analysis suggests rapid decreases in bamboo flora in this region (Tanegashima Island) from 11 300 to 7300 BP because of the expansion of evergreen forest [27,28]. On the other hand, acorns and chestnuts are important stored foods and were popular in the Jomon Period [12]. According to the short introduction by Delobel and Grenier [29], cereal weevils are polyphagous, and they successfully reared them in both acorns and chestnuts. Rice weevils prefer larger mature wheat kernels to smaller immature ones [30], and the body size (weight) of the three cereal weevils depends on the size of the plant seeds they infest [29]. Our preliminary experiment showed that adult weevils reared in acorns or chestnuts were about 1.24 times the size of those reared in cleaned rice. Figure 4 and Table 2 shows that the maize weevils at the Sanbonmatsu Site are roughly the same size as those that were reared in acorns and chestnuts. This suggests that bamboo seeds, acorns, or chestnuts might have been the stored foods that became infested by maize weevils during in the Jomon Period. However, additional research is required to provide more data and more accurate data to support that hypothesis.

Do differences in the physiological characteristics and life cycle mean different origins?

Granary weevils and Japanese rice weevils cannot fly, unlike the maize weevil. Granary weevils and probably the Japanese rice weevil also depend strongly upon stored grain. Their complete larval development takes place inside the grains that they infest, and has not been observed in natural reservoirs [31,20]. In other words, they are fully synanthropic grain pests and their spread depends upon the transport of infested grain or a suitable substrate [3]. In contrast, maize weevils shelter in winter under fallen leaves or in the soil [2]. After awakening in the spring, they feed on nectar from flowers. Their home range is within 400 m from human villages with grain storage facilities; thus, dispersal over longer distances requires human assistance (e.g., via grain transport) [26]. The pattern of spending most of their life outdoors is thought to result from an ancient adaptation to surviving on wild plants that produce suitable seeds [32]. Consequently, the maize weevil is clearly different from the granary weevil in terms of its degree of dependence on stored grain and in its life cycle.

Plarre [3] has suggested that the separation of *Sitophilus* lineages that led to the current grain pests must have occurred much earlier than previously believed. Therefore, our evidence supports Plarre's hypothesis; we believe that weevil pests in the Earliest Jomon in Japan must have evolved from a single common Asian progenitor. Initially, they would not have infested cereal grains (which were not stored at that time) but would instead have infested various kinds of wild plant seeds that were stored by the Jomon people. Our new discovery supports this hypothesis, and should encourage additional research on the ecology and history of maize weevils. In particular, more information is needed about when they began infesting stored food. This will require additional maize weevil specimens from other time periods and regions. No fossils of maize weevils or their ancestors have been discovered, so the origin and history of the taxon remain unclear despite our new

evidence. In addition, explicit DNA analysis of modern grain weevils would provide important insights into their phylogenetic relationships.

Materials and Methods

In our previous research, we obtained weevil impression replicas from several sites [11]. Then and in the present study, we began examinations of all potsherds obtained from the Sanbonmatsu Site in Kagoshima Prefecture in 2010. In February and April 2010, we found seven fragments that contained maize weevil impressions.

Weevil models were made from impressions in the clay of the recovered potsherds using the "impression replica" method introduced and adopted by Japanese archaeological researchers in the late 1970s. In this method, researchers inject silicone into a cavity (impression) in the clay, producing a model (replica) of the original form that created the cavity. However, the wet clay used to create a pot contracts as it dries and is fired in a kiln, and this compresses insect specimens. To determine the magnitude of the compression, which can then be used to reconstruct the original size of the trapped insects, we conducted an experiment in which we measured the rostrum, thorax, and abdomen of maize weevils ($n = 20$), then trapped them in wet clay similar in composition to the clay that would have been used to create the potsherds we recovered. Once the clay was dry and had been fired in a kiln, we measured the dimensions of the impression replicas, and calculated the shrinkage ratio as the impression replica size divided by the original size.

To allow a comparison of the impression replicas with modern grain weevils, we obtained data on modern weevils from my collection in Kumamoto City. The maize weevil resembles the rice weevil, but adult rice weevils are 2.1 to 2.9 mm long (mean, 2.3 mm), versus 2.3 to 3.5 mm (mean, 2.8 mm) for adult maize weevils [33].

To understand the developmental biology of the weevils, which would provide clues to their potential anthropogenic food sources, we reared maize weevils on many different foods. In total, we reared the insects on eight foods: rice, wheat, barley, adzuki bean, soybean, acorn, chestnut, and hemp. We found that the weevils could survive on rice, wheat, barley, acorn, and chestnut seeds. For weevils that survived to adulthood in rice, acorn, and chestnut, we measured their total length ($n = 180$) to determine whether the size of the food source affected their maturation and growth.

Although impression replicas from potsherds are important sources of information, they are difficult to examine because of their small size. To test whether modern technology could improve our ability to extract information from the impression replicas, we obtained scanning electron microscope (SEM) micrographs of the impression replicas and CT scans of the impressions. This work was performed at the Archaeological Center of Kumamoto University and the X-Earth Center of Kumamoto University, respectively, following each lab's standard protocols for scans of small objects.

Acknowledgments

We thank Dr. Katsura Morimoto, an honorary professor at Kyushu University, for useful information about weevils in Japan, and Mr. Junichiro Okita, curator of the Board of Education, Nishinoomote City, for his assistance with our research. We also thank Prof. Yuzo Obara and Prof.

Jun Otani, Director and Vice-Director (respectively) of the X-Earth Center, Kumamoto University, for providing access to the CT scanner, and Yoichi Watanabe, graduate student in Engineering, Kumamoto University, for obtaining and processing the CT images.

References

1. Hayashi N (1999) Oviposition preference substances of the maize weevils. Environ Ent Behav Physiol Chem Ecol. pp 321–332.
2. Hara T (1971) Ecology and control of food pests. Tokyo: Korin Shoin. 529 p.
3. Plarre R (2010) An attempt to reconstruct the natural and cultural history of the granary weevil, *Sitophilus granarius* (Coleoptera: Curculionidae). Eur J Entomol 107: 1–11.
4. Yasue Y (1959) Old names of the maize weevils. New Insects 12(3): 25–28.
5. Yasue Y (1976) History of the maize weevils. Insectarium 13: 182–186.
6. Mori Y (2001) Urban insect assemblages from the pre-historical and historical sediments. House Household Insect Pests 23(1): 23–39.
7. Yamazaki S (2005) Agriculture in the western Japan, in Jomon. In: BakDongbeak, ed. Agriculture in Neolithic Korea and Japan. Changwuon: Institute for Study of Cultural Assets in Kyonnam, Society for Neolithic Study in Korea, Society for Study of Jomon in Kyushu. pp 33–55.
8. Nakayama S (2009) An archaeological study of utilization and cultivation of soybean (genus *Glycine*) in Jomon Period. Ancient Culture 61(3): 40–59.
9. Obata H (2010) Cultivation of adzuki bean and soybean in Jomon. Prehistory Archaeol 5: 239–272.
10. Nakayama S, Nagasawa H, Hosaka Y, Noshiro Y, Kushihara K, et al. (2008) Replica and SEM analysis of impressions on potsherds recovered from archaeological sites in Yamanashi Prefecture (2). Bull Yamanashi Museum 2: 1–10.
11. Obata H (2008) Origins of domesticated plants in Jomon period, Japan. Prehist Ancient Cultigens Far East 3: 43–93.
12. Obata H (2006) Utilization of acorns in Prehistoric Kyushu. Lowland sites and macro botanical remains in Kyuhu Jomon. Usa. Sakamoto Y, ed. Association for the Jomon Studies in Kyushu. pp 31–40.
13. Arakaki N, Takahashi F (1982) Oviposition preference of rice weevil, *Sitophilus zeamais* Mitschulsky (Coleoptera: Curculionidae), for unpolished and polished rice. Jpn J Appl Entomol Zool 26(3): 166–171.
14. Miyamoto K (2003) The emergence of agriculture on the Neolithic Korean Peninsula and Jomon agriculture. Cultura Antique 55(7): 1–16.
15. Sato Y (2002) History of rice cultivation in Japan. Tokyo: Kadokawa Shoten. 197 p.
16. Takamiya H (1999) Yanagida Kunio's ocean road hypothesis: evaluation based on the direct data—plant remains. Jpn J Ethnol 63(3): 283–301.
17. Udatsu T, ed (2004) Substantial study on the diffusion routes of rice cultivation in Jomon period. Miyazaki: Miyazaki University. 104 p.
18. Shinto K (2008) Shell-made ornamented potteries in South Kyushu, in the Earliest Jomon. Handb Jomon Pottery. Kobayashi T, ed. Tokyo: UM promotion. pp 186–193.
19. Onbe S (2009) Carbon 14 series of cylinder shaped pottery with shell-impressed patterns in south Kyushu. Newsletter of Jomon Studies in South Kyushu 20. Kagoshima. Association for Jomon Studies in South Kyushu. pp 114–153.
20. Yoshida T, Watanabe T, Sonda M (2001) Illustrations of pests of stored foods—identification and control. Tokyo: Zenkoku Noson Kyouiku Kyokai. 268 p.
21. Morimoto K (1980) Notes on the weevils of the genus *Diocalandra* injurious to the dried cane of bamboos in Japan. House Household Insect Pests 7/8: 65–67.
22. O'Meara B (2001) Bacterial symbiosis and plant host use evolution in Dryophthorinae (Coleoptera, Curculionidae): a phylogenetic study using parsimony and Bayesian analysis. Bachelor's thesis. Department of Biology: Harvard University. 66 p.
23. Lefever C, Charles H, Vallier A, Delobel B, Farrell B, et al. (2004) Endosymbiont phylogenesis in the Dryophthoridae weevils: Evidence for bacteria replacement. Mol Biol Evol 21: 965–973.
24. Conord C, Despres L, Vallier A, Balmand S, Miquel C, et al. (2008) Long-term evolutionary stability of bacterial endosymbiosis in the Curculionideae: additional evidence of symbiont replacement in the Dryophthoridae family. Mol Biol Evol 25: 859–868.
25. Fuller DQ (2007) Contrasting patterns in crop domestication and domestication rates: recent archaeobotanical insights from the Old World. Ann Bot 100: 903–924.
26. Yamada G, Tsubakisaka A (2009) About seeds of bamboo grass excavated from remains. Bull Hist Museum Hokkaido 37: 13–22.
27. Paleoenvironment Research Institute (2006) Phytolith analysis in Sankakuyama 1 site. Sankakuyama Sites 3: 252–266.
28. Sugiyama S (2001) Analysis of tephra and phytolith. Monthly J Earth 23(9): 645–650.
29. Delobel B, Grenier AM (1993) Effect of non-cereal food on cereal weevils and tamarind pod weevil (Coleoptera: Curculionidae). J Stored Prod Res 29(1): 7–14.
30. Campbell JF (2002) Influence of seed size on exploitation by the rice weevil, *Sitophilus oryzae*. J Insect Behav 15(3): 429–445.
31. Longstaff BC (1981) Biology of the grain pest species of the genus *Sitophilus* (Coleoptera: Curculionidae): a critical review. Prot Ecol 2: 83–130.
32. Yoshida T, Tamamura Y, Kawano K, Takahashi K, Takuma T, et al. (1956) Habit of maize weevils to visit flowers. Bull Fac Fine Arts Literature Miyazaki Univ 1(2): 137–178.
33. Yasutomi K, Umeya K (2000) An illustrated encyclopedia of house pests. Tokyo: Zenkoku Nouson Kyouiku Kyoukai. 310 p.

Author Contributions

Wrote the paper: HO. Performed lab work: AM. Took SEM Photos: AM. Co-Searcher of the findings: AM NN TO YS.

2

Genetical Genomics Identifies the Genetic Architecture for Growth and Weevil Resistance in Spruce

Ilga Porth[1]*, Richard White[2], Barry Jaquish[3], René Alfaro[4], Carol Ritland[1], Kermit Ritland[1]

1 Department of Forest Sciences, University of British Columbia, Vancouver, British Columbia, Canada, **2** Department of Statistics, University of British Columbia, Vancouver, British Columbia, Canada, **3** Kalamalka Forestry Centre, British Columbia Ministry of Forests, Lands and Natural Resource Operations, Vernon, British Columbia, Canada, **4** Pacific Forestry Centre, Canadian Forest Service, Victoria, British Columbia, Canada

Abstract

In plants, relationships between resistance to herbivorous insect pests and growth are typically controlled by complex interactions between genetically correlated traits. These relationships often result in tradeoffs in phenotypic expression. In this study we used genetical genomics to elucidate genetic relationships between tree growth and resistance to white pine terminal weevil (*Pissodes strobi* Peck.) in a pedigree population of interior spruce (*Picea glauca, P. engelmannii and their hybrids*) that was growing at Vernon, B.C. and segregating for weevil resistance. Genetical genomics uses genetic perturbations caused by allelic segregation in pedigrees to co-locate quantitative trait loci (QTLs) for gene expression and quantitative traits. Bark tissue of apical leaders from 188 trees was assayed for gene expression using a 21.8K spruce EST-spotted microarray; the same individuals were genotyped for 384 SNP markers for the genetic map. Many of the expression QTLs (eQTL) co-localized with resistance trait QTLs. For a composite resistance phenotype of six attack and oviposition traits, 149 positional candidate genes were identified. Resistance and growth QTLs also overlapped with eQTL hotspots along the genome suggesting that: 1) genetic pleiotropy of resistance and growth traits in interior spruce was substantial, and 2) master regulatory genes were important for weevil resistance in spruce. These results will enable future work on functional genetic studies of insect resistance in spruce, and provide valuable information about candidate genes for genetic improvement of spruce.

Editor: Pär K. Ingvarsson, University of Umeå, Sweden

Funding: This work was carried out with financial support from Genome British Columbia and Genome Canada. The funders had no role in study design, data collection and analysis, decision to publish, or preparation of the manuscript.

Competing Interests: The authors have declared that no competing interests exist.

* E-mail: porth@mail.ubc.ca

Introduction

Plants are sessile organisms that have evolved many resistance mechanisms to defend against insect pests. These resistance mechanisms are genetically complex and involve interactions between both host and pests [1,2]. Recently, a "cost-benefit" paradigm for resistance has emerged to enhance our understanding of these interactions [3–7]. This paradigm suggests that tradeoffs in the cost-benefit paradigm may be due to correlated selection (favored trait combinations) and spatio-temporal heterogeneity of the environment [8]. Theoretical approaches have also described the importance of resource allocation within biosynthetic pathways for the evolution of resistance [9].

Meta-analyses of plant-herbivore defenses suggest that trade-offs exist between constitutive and induced defenses. More competitive species tend to exhibit lower constitutive and higher induced resistance than less competitive species ([10], [11]). It has also been hypothesized that both constitutive and induced resistance are influenced by selection on traits that alter plant growth rates [12]. In spruce (*Picea* spp.), constitutive and induced defenses are thought to follow sequentially [13]. Strength and rapidity of traumatic resinosis have often been related to resistance. Nevertheless, Alfaro [13] suggested that in response to wounding, some resistant trees failed to produce the traumatic response and

some susceptible trees responded with an unexpectedly intensified response.

Phenotypic and genetic relationships between growth and resistance to white pine terminal weevil (*Pissodes strobi* Peck.) have been intensively studied in interior spruce (*Picea glauca* [Moench] Voss, *P. engelmannii* Parry and their hybrids); however, results have been inconsistent and seemingly contradictory. Kiss and Yanchuk [14] found a negative genetic correlation between mean family height and weevil damage in interior spruce, while King et al. [15] reported a positive phenotypic but a strong negative genetic correlation between attack level and leader height. Alfaro et al. [16] reported better developed bark resin canals in fast-growing trees, while Lieutier et al. [17] concluded that there is no relationship between tree growth and resistance in Norway spruce (*Picea abies*. ((L.) Karst.). Vandersar and Borden [18] suggested that weevils prefer faster growing trees, and more recently He and Alfaro [19] found a higher survival time for shorter trees. In Sitka spruce (*Picea sitchensis* (Bong.) Carr.), genetic resistance was most pronounced in families with average height growth [20]. This finding is interesting, since improved growth rate in spruce trees has lead to a higher predisposition to weevil attacks [21].

Trade-offs between correlated traits may be due to genetic and/or phenotypic variation. At the least, genetically correlated traits share quantitative trait loci (QTLs). However, pleiotropic genes, which control the hubs in such trade-offs, are difficult to distinguish from the confounding physically linked loci within a shared QTL. Moreover, biometric correlations and QTLs do not always concur because of the presence of obscuring antagonistic QTLs [22]. This failure to detect significant correlations may indicate the extent of independent variation between two traits and not necessarily the absence of a tradeoff. Other interacting factors can remain undetected [8]. In other species, pleiotropy and genetic correlations may be present. Examples include dehydration avoidance and flowering time in *Arabidopsis thaliana* [23], resistance and tolerance to herbivory in the common morning glory [24], and growth rate and flowering in *A. thaliana* [25]. While *A. thaliana* has facilitated research on tradeoffs for life-history traits in annual plants with short life cycles, research on long-lived forest trees promises new perspectives on molecular mechanisms of life-history control in non-model species [26].

In this paper, we present results on the use of genetical genomics to investigate a question of fundamental importance in plant genomics: How do genes underlying a pathway to pest resistance concertedly function? We investigated growth and insect resistance as a trait pair that defines the life history of interior spruce, a commercially valuable and ecologically important coniferous tree species. We show how genetical genomics can provide a fine-scale analysis of the genetic architecture in the study of pest resistance cost-benefit tradeoffs. Genetical genomics assays thousands of traits (gene expression levels) [27] and these "expression phenotypes" are subjected to standard QTL analysis. We use genetical genomics to infer the nature of resistance of interior spruce to the white pine weevil. We infer expression QTLs (eQTL) in segregating crosses of interior spruce, with variable resistance to white pine weevil. In this analysis, the positions of eQTLs indicate regions that harbor regulatory elements that control expression of genes in the same pathway. In the case of *cis*-regulation, the genomic location of the eQTL coincides with the physical location of the regulated gene, while *trans*-acting eQTLs identify regulatory elements for the gene elsewhere in the genome. The distribution of eQTLs may spread evenly on the genome or may appear in clusters or "hotspots", depending on the genetic architecture of these gene interactions [28]. At the QTL level, the action of a gene might suggest pleiotropy because multiple traits are affected. We search for pleiotropy based on common candidate genes between resistance and growth. Furthermore, a positive correlation between growth rate, and attack and oviposition rate (our resistance measure) might indicate a tradeoff between growth and defenses. This tradeoff could be due to the increased carbon cost required for higher defense chemical levels. We also search for master regulons that underlie *trans*-eQTL hotspots ("hubs") which tend to be at the center of gene expression networks (network eQTLs).

Results

Correlations between Phenotypic Estimates

The phenotypic response data consisted of tree height measurements and weevil attack and oviposition counts. Tree height measures were taken at the time of planting in 1995 and at three and five growing seasons thereafter. Leader length was measured in year five preceding the artificial augmentation of a local weevil population in October of the same year (hgt_1995, hgt_1997, hgt_1999, and ldr_99, respectively). Weevil attack rates counted in 2000 and 2001 were classified as successful top kills, failure to kill the leader, and no attack (atk_2000, atk_2001). In the same years, oviposition was assessed (egg_2000, egg_2001) and egg counts along the leaders were summarized into five discrete classes. The sum of weevil attacks and the sum of oviposition for

2000 and 2001 were also used as response traits (sum_atk, and sum_egg).

Forty-five pairwise phenotypic correlations were estimated for individual and sum traits (**Table 1**). In general, correlations between egg counts and attack classification were strong, between initial height (hgt 1995) and later growth correlation was weak, and between hgt_1999, ldr_99 and hgt_1997 correlations were strong. Correlations between growth (ldr 99 and hgt 1999) and attack (resistance) traits (atk_2000, egg_2000, sum_atk and sum egg) were generally positive (**Table 1**). Collectively, these results suggest a negative relationship between growth and the actual pest resistance (fewer attacks).

QTL Mapping

The spruce mapping population was genotyped for 384 SNP loci. These SNPs had been detected within expressed sequence tag (EST) contigs that represented assemblies of short expressed sequences with predicted open reading frames. These ESTs originated from the spruce Treenomix EST database (K. Ritland, personal communication). The genotypic information was used to estimate pairwise recombination rates between SNP loci and construct a framework genetic linkage map to localize quantitative trait loci (QTLs). Details about the genetic linkage map and the annotation of the mapped contigs can be found in the supplement material of [29]. The phenotypic variations that were obtained from four tree height, three weevil attack, three oviposition (see above), and extensive quantitative gene expression (21,840 transcripts) measures were mapped to the established genetic linkage map of the factorial progeny. The QTLs were mapped by using a likelihood function to assess the phenotype effect conditioned on the genotypic variation. A (e)QTL was significant with a LOD ≥3.84. In total, we identified 132,100 significant eQTLs (see **File S1** and legends in **Table S1**).

For each SNP locus along the genetic linkage map, we superimposed the mapped phenotypic trait QTLs (pQTLs) with the counts of significant eQTLs (**Figure 1**), and identified hubs of trait variation at several SNP loci that comprised of multiple pQTLs and an extensive accumulation of eQTLs. A goodness-of-fit test that assumed a uniform distribution was performed to test whether the observed frequencies of eQTLs along the linkage map differed significantly from the expected value. Then we used all detected eQTLs and all marker loci (see above) in a randomization procedure to assess the maximum number of eQTLs within eQTL clusters. According to this randomly generated data set, "eQTL hotspots" would be declared if the number of eQTLs at a given locus exceeded 630. However, we arbitrarily used a cutoff value of 786. This number was simply the value where the expected average value was exceeded by 50%. In this way, we focused on fewer hubs of *trans*-eQTLs that are associated with important regulators of quantitative trait variation. Fourteen loci coincided with eQTL hubs and with at least three pQTLs (**Figure 1**).

Hotspots of Phenotypic Trait Variation

From these 14 loci, seven loci were associated exclusively with resistance pQTLs, three loci with growth pQTLs, while four loci with pQTLs from individual traits of both growth and resistance, respectively (**Figure 1**). The composition of two *trans*-eQTL hotspots with extensive resistance pQTL overlap was analyzed in more detail (see below and **Figure 2**, and **Figure 3**; **Table S2**, and **Table S3**).

At eight map positions, at least four pQTLs overlapped with eQTL hotspots on five different linkage groups (LG): SNP44 on LG3 (termed Contig_2685_179 and annotated as unknown gene [29]; accumulation of eQTLs from 2040 transcripts, and

Table 1. Pearson's correlation coefficients for phenotypic (*above* diagonal) and for QTL correlations (*below* diagonal).

	ldr_99	hgt_1995	hgt_1997	hgt_1999	atk_2000	atk_2001	sum_atk	egg_2000	egg_2001	sum_egg
ldr_99		0.028	0.447	0.823	0.300	0.068	0.274	0.392	0.037	0.331
hgt_1995	0.035		0.357	0.142	−0.064	0.057	−0.009	−0.042	0.070	0.014
hgt_1997	0.034	0.095		0.723	0.089	0.163	0.181	0.093	0.235	0.231
hgt_1999	0.207	0.074	0.327		0.283	0.109	0.290	0.319	0.121	0.331
atk_2000	0.158	0.071	0.113	0.146		−0.050	0.719	0.782	−0.103	0.541
atk_2001	−0.042	−0.127	0.105	0.075	0.170		0.658	−0.001	0.799	0.540
sum_atk	0.218	0.141	0.169	0.056	0.560	0.178		0.589	0.478	0.784
egg_2000	−0.044	−0.106	0.184	0.255	0.465	0.282	0.348		−0.065	0.738
egg_2001	0.051	−0.121	0.133	0.111	0.198	0.634	0.032	0.250		0.625
sum_egg	0.177	0.131	0.352	0.183	0.357	0.283	0.309	0.279	0.166	

resistance pQTLs), SNP71 on LG4 (Contig_486_336, no hit [29];1341 transcripts, resistance pQTLs), SNP74 on LG4 (Contig_1623_510, unknown gene [29]; 1333 transcripts, resistance and growth pQTLs), SNP124 on LG6 (Contig_4096_434, caffeoyl-CoA 3-O-methyltransferase CCoAOMT [29];1307 transcripts, resistance pQTLs), SNP125 on LG6 (CCoAOMT_1_320, CCoAOMT [29]; 992 transcripts, resistance pQTLs), SNP210 on LG11 (Contig_1761_256, ubiquitin conjugating enzyme 2 [29]; 820 transcripts, resistance and growth pQTLs), SNP224 on LG11 (Contig_4216_318, predicted protein [29]; 1605 transcripts, growth pQTLs), and SNP252 on LG13 (Contig_1305_426, very weak similarity to cycloidea-like gene [29]; 1253 transcripts, resistance and growth pQTLs).

The two resistance trait associated eQTL hotspots on LG6 were localized at two annotated *bona fide* CCoAOMT genes that are separated by only 1.5 centiMorgan map distance, **Figure 1**. In a previous study that focused on spruce gene families from the phenylpropanoid pathway, we identified the same region enriched

for *trans*-eQTLs [29]. Both *CCoAOMT* genes also have *cis*-eQTLs that might represent promoter polymorphisms regulating the differential gene expression in those two loci. At the two different CCoAOMT loci a common set of pQTLs as well as eQTLs clustered (53% and 36% of their accumulated *trans*-eQTLs were generated by the same transcripts, respectively). We compared gene annotations to the 8366 unique *Arabidopsis* annotations from the entire microarray. For both loci the categories 'structural molecule activity', 'secondary metabolic process', 'ribosome', 'response to biotic stimulus', and 'cytosol' were overrepresented. A Fisher's exact test was employed to assess significance. Details about the overrepresented 'GOSlim Plant' categories within the two eQTL hotspots can be found in **Figure 2** & **Figure 3**. Five jasmonate-ZIM-domain protein (JAZ) genes (WS0105_K14, WS00918_B02, WS0063_E19, WS00918_P17, and WS00919_H21) contributed with eQTLs to this hotspot region, and expression variation for three JAZ genes mapped to both CCoAOMT loci (WS00918_B02, WS00918_P17, and

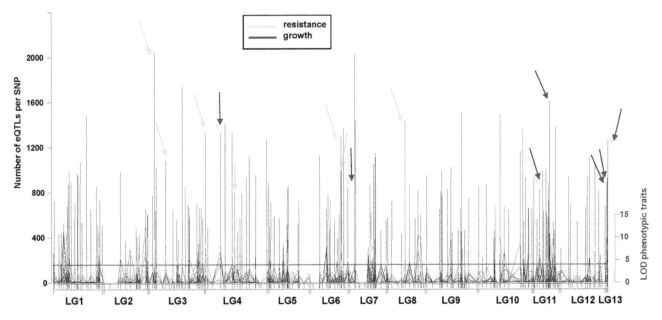

Figure 1. EQTL density map with overlapping positions of pQTLs at individual marker positions. Linkage groups (LG1-13) are displayed horizontally, black bars indicate SNP marker positions in linkage groups; arrows mark positions with at least three pQTLs (LOD ≥3.84, i.e. values above horizontal line) and eQTL numbers >786.

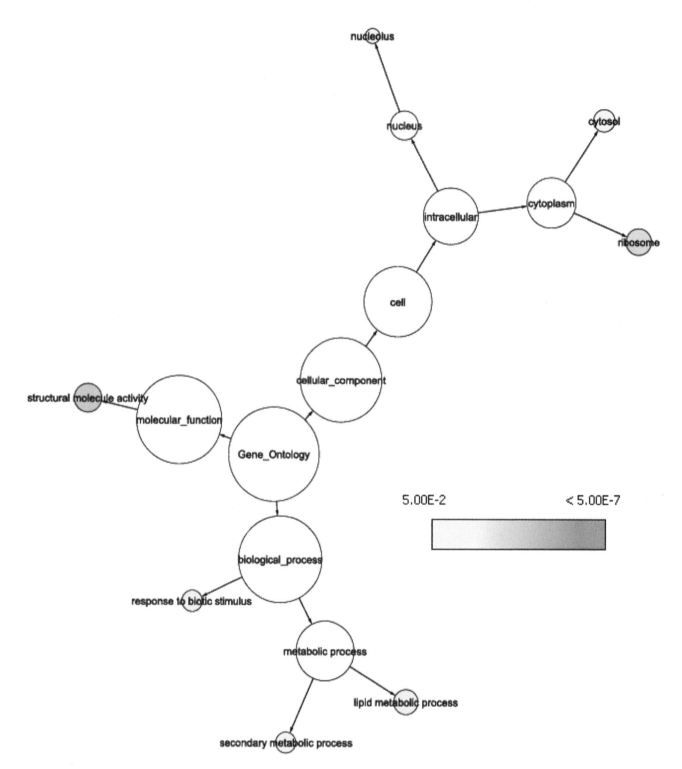

Figure 2. GO tree representation showing significantly (p ≤ 0.05) overrepresented GO categories at gene locus PiglCCoAOMT-1. Representing the *trans* eQTL-hotspot on LG6 (WS0031_301, SNP marker Contig_4096_434).

WS00919_H21). Three JAZ genes (WS00919_H21, WS0063_E19, WS00918_P17) were candidate genes directly associated with phenotypic variation for ht_1999, atk_2000 and atk_2001 traits, respectively (**Table S2**, **Table S3**, and **Table S4**). Transcripts from three putative carbonic anhydrase genes (WS00110_A15, WS00928_K21, and WS00936_A24) mapped *trans*-eQTLs to both hotspot locations.

Two carbonic anhydrase loci on LG4, SNP78 and SNP83 (contig_2079_440 and contig_103_602, [29]) are associated with extensive gene expression variation (**Figure S1** & **Figure S2**). At SNP78, three resistance traits mapped significant pQTLs. Seven overrepresented GO categories were in common between eQTL hotspots of carbonic anhydrase SNP83 and the CCoAOMT locus SNP125 ('biosynthetic process', 'cell wall', 'secondary metabolic

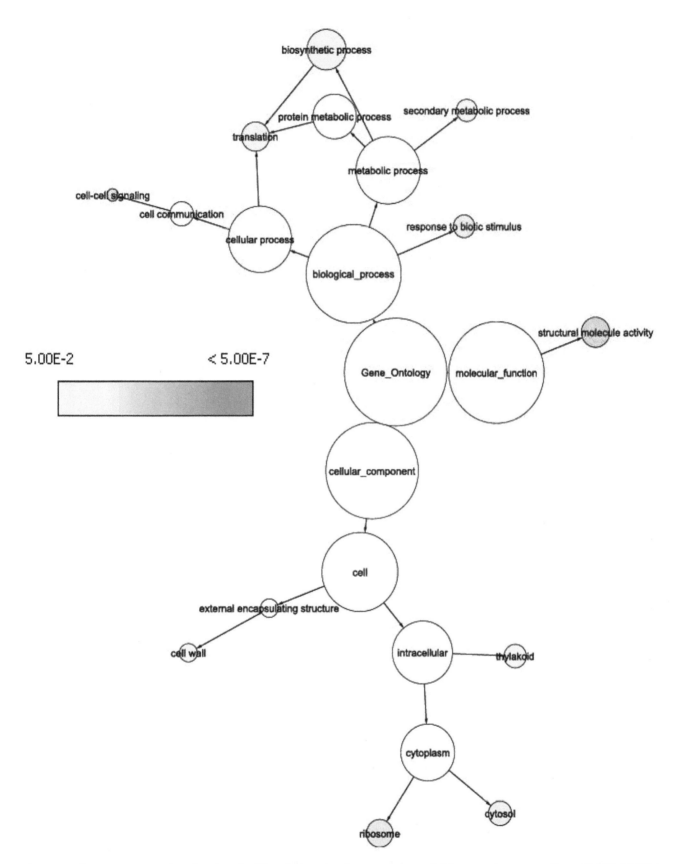

Figure 3. GO tree representation showing significantly (p ≤ 0.05) overrepresented GO categories at gene locus PiglCCoAOMT-2.
Representing the *trans* eQTL-hotspot on LG6 (WS0064_O09, SNP marker CCoAOMT_1_320).

process' and 'translation', e.g.), **Figure 3** and **Figure S2**. At locus SNP125, eQTLs from ethylene-responsive element binding factors (ERFs) linked to defensive gene expression were identified; specifically ERFs from group IX (ERF3, ERF4, ERF7).

The cluster of three phenotype associated eQTL hotspots on LG13 deserves further attention (**Figure 1**), yet we were unable to relate this group of markers to any of the 12 major linkage groups. This might be due to a limited number of segregating markers. At two SNP markers on LG13, SNP250 and SNP251 (WS0021_L13_301 and Contig_2062_390, [29]), related to sequences of glutamate decarboxylase (GAD) genes, significant growth variation and extensive expression variation co-localized.

On four spruce linkage groups (LG4, LG6, LG11 and LG13) hotspots of expression variation associated with QTLs from both growth and resistance traits (**Figure 1**). The pQTLs from resistance traits explained a higher % variation of the trait variation than pQTLs from growth traits. The largest pQTL at any of these eQTL hotspots was identified for sum_egg at SNP44 (on LG 3) and explained 10.3% of the trait variation. Also, for traits atk_2000, sum_atk, egg_2000 as well as for egg_2001 pQTLs explaining a higher portion of phenotypic variance (4.6–7.5%) mapped to this locus. Along the entire linkage map, most eQTLs mapped to this locus (2040 transcripts). At two loci, SNP74 (LG4) and SNP252 (LG13), the pQTLs from the same five traits (hgt_1997, hgt_1999, ldr_99, atk_2000 and egg_2000) were associated with the individual eQTL hotspots. On LG6, pQTLs from ldr_99, atk_2000 and sum_egg, while on LG11, pQTLs from hgt_1999, ldr_99, atk_2000, sum_atk and egg_2000 associated with the eQTL hotspot. At SNP210 related to ubiquitin conjugating enzyme 2 (LG11), the identified eQTL hotspot was associated with three resistance and two growth traits. In all four cases of eQTL hotspots where several pQTLs for growth and resistance traits collocated, the allelic effects of the pQTls for growth and resistance traits had the same sign. This indicates a positive correlation between the growth trait variation (as assessed by height data) and the resistance trait variation (from attack rates and oviposition data).

Positional Candidate Genes for Resistance and Growth Trait Variation

The combination of phenotype and gene expression datasets facilitates studying the genetic control of phenotypic traits of interest [30]. "Positional" candidate genes can be identified as genes for which transcript abundance correlates extensively with the quantitative phenotype.

Here, the positional candidate genes were identified by collocation of at least 40% of their eQTLs with pQTLs based on the criteria for identifying significant QTLs and running 10,000 randomizations (p ≤ 0.05), see Material and Methods. Thus, extensive co-segregation of transcript variation with the phenotypic trait of interest identifies positional candidate genes that are directly underlying the trait. We arbitrarily defined map intervals of 10 cM to measure collocation (on average 2 SNPs were binned into 10 cM). Thus, within a resolution window of 10 cM at a given SNP locus we determined the significance of co-localizations between the expression variation that was detected at a certain EST that was spotted on the microarray (gene spot) and the phenotypic trait variation (p ≤ 0.05). We screened 21,840 transcripts that represented distinct ESTs spotted on the microarray for co-localization with growth traits and for co-localization with resistance traits, respectively, and identified 1621 and 2002 distinct ESTs, respectively. These numbers comprised of the following trait associations: ldr_99 (217 gene spots), hgt_1995 (254), hgt_1997 (385), hgt_1999 (878); atk_2000 (346), atk_2001

(311), sum_atk (546), egg_2000 (361), egg_2001 (335), sum_egg (584), respectively (**Table S4**). Not unusual for conifers, many of those spruce genes had no *Arabidopsis* homologues. In total, 1191 gene spots from co-localizations with resistance traits gave hits with unique *Arabidopsis* entries (59%); 1000 gene spots from co-localizations with growth traits had unique annotations (62%).

We compared annotations independently for resistance and growth with the unique *Arabidopsis* annotations from the array. Twelve, and eight GOSlim Plants categories, respectively, were overrepresented among genes with expression variation significantly associated with individual phenotypic traits. The categories for resistance trait associations involved: 'response to stress' (64 compared to a total of 329 genes on the array), 'response to abiotic stimulus' (54/272), 'cellular component organization and biogenesis' (94/479) and 'cytosol' (32/145) as well as 'binding' (298/1908); **Figure 4** and **Table S5**. Growth trait categories were related to 'extracellular region' (8/32), 'signal transducer activity' (15/78), 'cell communication' (38/204), 'response to stress' (51/329) categories and the 'cell wall' category (19/63); **Figure 5** and **Table S5**.

In the gene lists a large number of kinases, phosphatases as well as transcription factors were identified; **Table S4**. Phosphorylation and dephosphorylation are important steps in various biosynthetic processes, and in signal transduction cascades within the organism. For resistance traits we counted 104, for growth traits 91 transcription (-related) factors/proteins. We identified 49 kinases associated with the growth traits, whereas 63 with the resistance traits. Phosphatases totaled to 22 candidate genes both for growth and for resistance traits, respectively.

Several multi-gene families were highly represented in both growth and resistance traits associations: among others GDSL-motif lipase/hydrolase, glycosyl hydrolase (GH), leucine-rich repeat (LRR) proteins, oxidoreductases, pentatricopeptide (PPR) repeat-containing proteins, disease resistance family proteins, DNAJ heat shock family proteins.

A number of auxin-related genes (including ABC transporters) co-localized with phenotypic trait variation: for resistance we found 19, for growth 15 co-localizing. A number of genes involved in embryo arrest/deficiency also co-localized with the phenotypes: 19 for resistance and 22 for growth traits. Jasmonic acid (JA)-forming and ethylene-forming/−responsive genes were also identified candidate genes based on collocations with the respective phenotypic traits.

The isoprenoid biosynthesis pathway generates many compounds relevant to plant defenses (terpenoids, tocopherol, e.g.) but also the precursors of 'plant hormones' like gibberellins (GA) and abscisic acid (ABA). Eight biosynthetic genes co-localized with resistance traits and six exclusively with the hgt_1999 trait. In addition, both for growth and resistance traits, four GA-regulated/ GA signaling-regulating, and two ABA-related proteins were identified as potential candidates.

The phenylpropanoid metabolic pathway provides various specialized metabolites important in plant development, polymeric lignin for structural support, anthocyanins for pigmentation, flavonoids with various protective functions, and antimicrobial phytoalexins [31]. Thirty-seven putative gene family members associated with individual resistance traits, while 25 with growth traits. For the traits related to egg counts (26 transcripts) and to height in 1999 (15 transcripts) the highest number of candidates from this pathway was identified.

Genes that are significant for both resistance and growth traits (p ≤ 0.05; 10,000 randomizations, see Materials and Methods) are summarized in **Table S6** (see **Table S4** for comprehensive results). In sum, 244 genes were identified. The majority of these

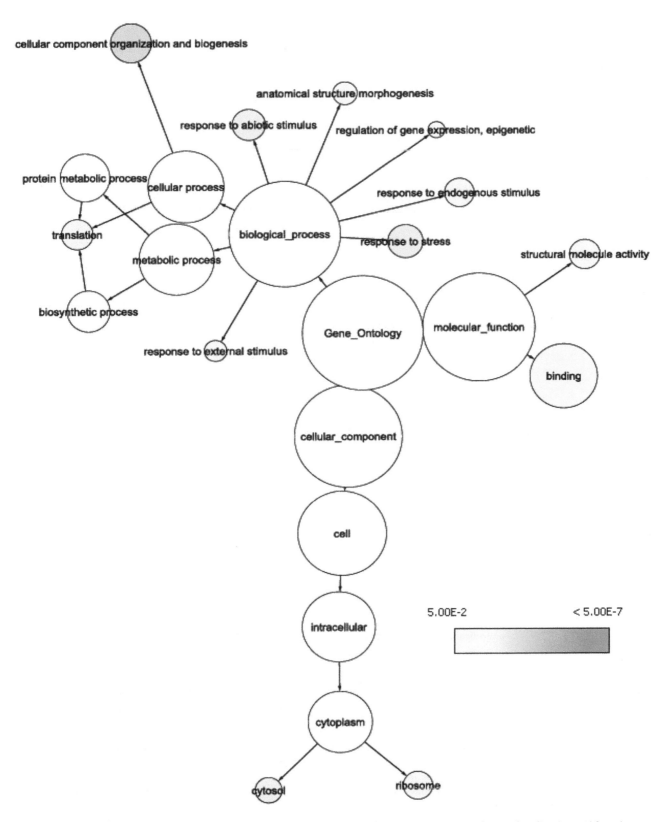

Figure 4. GO tree representation indicating significantly (p ≤ 0.05) overrepresented categories from colocalizations with resistance traits.

genes co-localized with Ht 1999, and also co-localized with atk_2000, egg_2000, respectively, resistance traits assessed in the next growth season. Since many identified genes have no angiosperm counterparts, they likely represent novel conifer-specific genes at the pivotal points of growth variation and defense. For one DNAJ heat shock family protein its eQTLs significantly

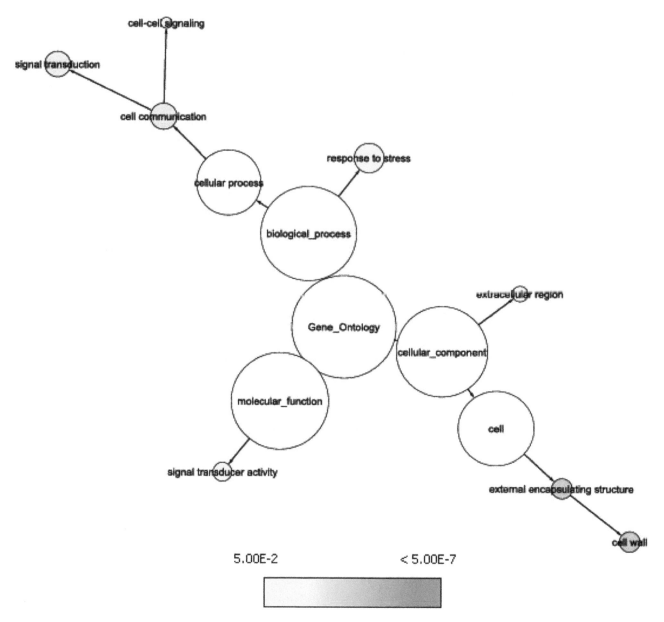

Figure 5. GO tree representation indicating significantly (p ≤ 0.05) overrepresented categories from colocalizations with growth traits.

co-localized with QTLs from as much as seven individual traits (**Table S6**). The functions of these chaperones are related to environmental challenges (involving stress tolerance) but are manifold due to the complexity of the whole gene family. Among the candidates in common between resistance and growth traits were genes involved in normal/optimal plant growth, stress signaling, defense, stress tolerance, glycine betaine synthesis, DNA repair, transcription regulation, post-transcriptional regulations, protein degradation as well as expansins with a proposed function related to cell wall architecture during rapid tissue expansion (**Table S6**).

Significant associations with all four individual growth and six individual resistance traits, respectively, allowed us to robustly identify 149, and 99 candidate genes for the composite 'resistance' and 'growth' phenotype, respectively (**Table S7** and **Table S8**). Gene identities with annotations are provided in **Table 2** and

Table 3. For about half of the gene spots identified as candidates no putative functions were unraveled by BLAST searches. Candidate genes for 'resistance' and 'growth' differ markedly in their functions. While collocations of expression with growth variation were predominantly found for gene products involved in (post-) transcription and post translational regulation (in total 18), collocations with resistance variation identified a larger number of biosynthetic proteins (17), signaling (20) and transporter/transport related molecules (13). We found that 19 genes are positional candidates for the composite 'resistance' phenotype, but are additionally associated with other growth trait(s); these genes are typically involved in signaling, transport and biosynthesis related processes. The 29 genes that are positional candidates for the composite 'growth' phenotype and at the same time associated with individual resistance traits have proposed functions in

Table 2. Display of 80 positional candidate genes (AGI annotated) for the composite resistance phenotype, p ≤ 0.1.

Gene id	P-value	E-value	AGI #	Annotation	Putative function
WS0262_L21	0.051	2.9E–45	AT2G38240	oxidoreductase, 2OG-Fe(II) oxygenase family protein	biosynthesis
WS00922_A20	0.061	1.4E–54	AT3G06810	acyl-CoA dehydrogenase-related	biosynthesis
WS01033_F16	0.065	5.30E–023	AT3G03780	AtMS2 (Arabidopsis thaliana methionine synthase 2)	biosynthesis
WS00818_F07	0.074	1.20E–073	AT2G43710	FAB2, SSI2 SSI2 (fatty acid biosynthesis 2); acyl-[acyl-carrier-protein] desaturase	biosynthesis
WS00812_J14	0.082	7.80E–052	AT3G04520	THA2 (THREONINE ALDOLASE 2)	biosynthesis
WS00113_D16	0.082	2.80E–111	AT5G08370	ATAGAL2 (ARABIDOPSIS THALIANA ALPHA-GALACTOSIDASE 2)	biosynthesis
WS0043_N05	0.084	2.3E–111	AT3G17390	SAMS3, MAT4, MTO3 MTO3 (S-adenosylmethionine synthase 3); methionine adenosyltransferase	biosynthesis
WS0073_B10	0.086	1.40E–027	AT1G14550	anionic peroxidase, putative	biosynthesis
WS0932_K15	0.086	7.5E–43	AT4G37970	mannitol dehydrogenase, putative	biosynthesis
WS01016_H01	0.089	5.50E–109	AT5G60540	EMB2407, ATPDX2, PDX2 ATPDX2/EMB2407/PDX2 (PYRIDOXINE BIOSYNTHESIS 2)	biosynthesis
WS00816_E04	0.092	5.4E–30	AT1G08250	prephenate dehydratase family protein	biosynthesis
WS00725_B17	0.093	2.80E–054	AT5G42800	TT3, M318, DFR DFR (DIHYDROFLAVONOL 4-REDUCTASE); dihydrokaempferol 4-reductase	biosynthesis
WS0045_J16	0.094	1.90E–031	AT5G04330	cytochrome P450, putative/ferulate-5-hydroxylase, putative	biosynthesis
WS0094_G24	0.096	1.50E–126	AT5G17990	PAT1, TRP1 TRP1 (TRYPTOPHAN BIOSYNTHESIS 1); anthranilate phosphoribosyltransferase	biosynthesis
WS0023_B12	0.097	5.50E–074	AT1G23800	ALDH2B, ALDH2B7 ALDH2B7 (Aldehyde dehydrogenase 2B7)	biosynthesis
WS0107_F01	0.097	3.1E–22	AT4G37970	mannitol dehydrogenase, putative	biosynthesis
WS0076_F23	0.098	4.9E–36	AT5G19730	pectinesterase family protein	cell wall
WS0105_N22	0.100	2.60E–138	AT1G77380	AAP3 (amino acid permease 3); amino acid permease	transport
WS00721_A21	0.100	1.6E–76	AT1G77120	ADH, ATADH, ADH1 ADH1 (ALCOHOL DEHYDROGENASE 1)	biosynthesis
WS0044_N09	0.083	7.2E–10	AT1G60390	BURP domain-containing protein/polygalacturonase, putative	cell wall, stress(?)
WS0086_K19	0.097	6.70E–046	AT1G27120	galactosyltransferase family protein	cell wall
WS0092_M11	0.064	6.3E–39	AT2G30410	KIS (KIESEL); unfolded protein binding	growth
WS0061_B17	0.073	1.40E–055	AT2G21530	forkhead-associated domain-containing protein	growth, development
WS01031_N10	0.079	4.10E–049	AT2G04030	EMB1956, CR88 CR88 (EMBRYO DEFECTIVE 1956); ATP binding	growth
WS0261_O16	0.088	6.40E–164	AT5G67270	ATEB1C (MICROTUBULE END BINDING PROTEIN 1); microtubule binding	growth
WS01021_K16	0.088	5.90E–196	AT2G33150	PED1, KAT2 PED1 (PEROXISOME DEFECTIVE 1); acetyl-CoA C-acyltransferase	growth
WS0261_G19	0.088	4.7E–34	AT5G53940	yippee family protein	growth, stress
WS00813_E05	0.060	5.80E–102	AT4G09670	oxidoreductase family protein	miscellaneous
WS0017_K15	0.066	5.80E–006	AT5G40470	similar to F-box family protein (FBL4) [Arabidopsis thaliana]	miscellaneous
WS00728_E10	0.076	1.0E–20	AT5G02450	60S ribosomal protein L36 (RPL36C)	miscellaneous
WS00921_L16	0.077	2.5E–19	AT1G27620	transferase family protein	miscellaneous
WS00912_K11	0.080	2.1E–47	AT2G19680	mitochondrial ATP synthase g subunit family protein	miscellaneous
WS00927_K21	0.082	2.10E–026	AT3G53850	similar to integral membrane protein, putative [Arabidopsis thaliana]	miscellaneous
WS00712_A12	0.091	8.40E–084	AT1G60420	DC1 domain-containing protein	miscellaneous
WS0032_G24	0.095	3.90E–057	AT2G37270	ATRPS5B ATRPS5B (RIBOSOMAL PROTEIN 5B); structural constituent of ribosome	miscellaneous
WS0057_N15	0.096	1.60E–102	AT1G10780	F-box family protein	miscellaneous
WS01032_N12	0.098	1.10E–053	AT1G44910	protein binding	miscellaneous
WS0039_A22	0.099	1.60E–116	AT5G53490	thylakoid lumenal 17.4 kDa protein, chloroplast	miscellaneous
WS00932_M11	0.061	5.10E–006	AT5G01020	protein kinase family protein	signaling
WS01011_I05	0.062	2.4E–23	AT1G73500	ATMKK9 ATMKK9 (Arabidopsis thaliana MAP kinase kinase 9)	signaling
WS00924_L15	0.071	1.20E–070	AT1G79110	protein binding/zinc ion binding	signaling
IS0011_J12	0.072	6.3E–28	AT5G53590	auxin-responsive family protein	signaling
WS0048_K17	0.074	1.70E–086	AT1G60490	ATVPS34 ATVPS34 (Arabidopsis thaliana vacuolar protein sorting 34); phosphatidylinositol 3–kinase	signaling
WS00824_D10	0.081	5.10E–041	AT5G14930	GENE101, SAG101 SAG101 (SENESCENCE–ASSOCIATED GENE 101); triacylglycerol lipase	signaling, stress

Table 2. Cont.

Gene id	P-value	E-value	AGI #	Annotation	Putative function
WS0016_M07	0.084	5.90E–030	AT5G14930	GENE101, SAG101 SAG101 (SENESCENCE–ASSOCIATED GENE 101); triacylglycerol lipase	signaling, stress
WS00946_N19	0.088	1.00E–079	AT2G38010	ceramidase family protein	signaling, stress
WS00922_M10	0.090	1.30E–043	AT2G32800	AP4.3A AP4.3A; ATP binding/protein kinase	signaling
WS00928_C13	0.091	1.4E–36	AT2G33040	ATP synthase gamma chain, mitochondrial (ATPC)	signaling
WS0063_H05	0.091	7.90E–075	AT4G15415	serine/threonine protein phosphatase 2A (PP2A) regulatory subunit B' (B'gamma)	signaling
WS0063_N04	0.093	3.20E–039	AT1G16670	protein kinase family protein	signaling
WS00924_P23	0.093	2.6E–45	AT5G26751	ATSK11 (Arabidopsis thaliana SHAGGY-related kinase 11); protein kinase	signaling
WS0261_J01	0.094	2.0E–85	AT1G73500	ATMKK9 (Arabidopsis thaliana MAP kinase kinase 9)	signaling
WS0097_P15	0.096	6.10E–024	AT3G12690	protein kinase, putative	signaling
WS0268_O17	0.096	6.80E–037	AT3G22190	IQD5 IQD5 (IQ-domain 5); calmodulin binding	signaling
WS00930_C14	0.096	5.80E–161	AT3G50960	similar to Thioredoxin domain 2 [Medicago truncatula]	signaling
WS0089_G23	0.097	2.6E–38	AT1G66410	ACAM-4, CAM4 CAM4 (CALMODULIN 4); calcium ion binding	signaling
WS00111_O11	0.099	1.80E–087	AT2G30020	protein phosphatase 2C, putative/PP2C, putative	signaling
WS01041_M05	0.099	5.40E–083	AT2G30020	protein phosphatase 2C, putative/PP2C, putative	signaling
WS00922_P23	0.069	2.80E–110	AT5G01230	FtsJ-like methyltransferase family protein	stress
WS01021_F15	0.092	3.00E–133	AT3G62550	universal stress protein (USP) family protein	stress
WS0263_B07	0.044	9.0E–21	AT4G23330	eukaryotic translation initiation factor-related	transcriptional, translational
WS0061_C09	0.061	6.90E–013	AT4G20970	basic helix-loop-helix (bHLH) family protein	transcriptional
WS00924_K23	0.074	8.00E–050	AT3G49430	SRP34A SRP34A (SER/ARG-RICH PROTEIN 34A); RNA binding	transcriptional
WS0078_K12	0.077	2.3E–11	AT1G10200	transcription factor LIM, putative	transcriptional
WS0087_O15	0.078	7.00E–051	AT5G47390	myb family transcription factor	transcriptional
WS00916_N06	0.083	1.2E–17	AT5G06550	similar to transcription factor jumonji (jmjC) domain-containing protein [Arabidopsis thaliana]	transcriptional
WS00826_M02	0.083	6.60E–065	AT1G13690	ATE1 (ATPase E1); nucleic acid binding	transcriptional
WS00825_H14	0.087	1.6E–32	AT1G29250	nucleic acid binding	transcriptional
WS0107_C03	0.062	5.00E–025	AT3G07490	AGD11 (ARF-GAP DOMAIN 11); calcium ion binding	transport
WS01013_E24	0.076	4.20E–103	AT2G35800	mitochondrial substrate carrier family protein	transport
WS00927_L04	0.076	1.60E–129	AT5G19760	dicarboxylate/tricarboxylate carrier (DTC)	transport
WS00939_B16	0.076	1.60E–158	AT2G20930	similar to intracellular transporter [Arabidopsis thaliana]	transport
WS0071_O18	0.081	1.90E–127	AT1G73030	SNF7 family protein	transport
WS0017_I09	0.085	4.80E–091	AT4G04860	DER2.2 Der1-like family protein/degradation in the ER-like family protein	transport, stress
WS00919_L11	0.085	1.00E–159	AT1G72280	AERO1 AERO1 (ARABIDOPSIS ENDOPLASMIC RETICULUM OXIDOREDUCTINS 1)	transport
WS0081_G16	0.088	2.40E–096	AT5G46630	clathrin adaptor complexes medium subunit family protein	transport
WS0105_B03	0.090	1.80E–069	AT5G12130	PDE149 PDE149 (PIGMENT DEFECTIVE 149)	transport
WS01012_D02	0.092	1.70E–088	AT4G02050	sugar transporter, putative	transport
WS00712_P20	0.095	8.10E–105	AT3G48420	haloacid dehalogenase-like hydrolase family protein	transport
WS0106_H16	0.100	3.80E–023	AT1G30690	SEC14 cytosolic factor family protein/phosphoglyceride transfer family protein	transport

transcriptional or (post-) translational control, growth and cell wall remodeling.

Correlations between Co-localization Estimates

We determined genetic (QTL) correlations based on co-localization estimates between gene expression and trait variation, **Table 1**. The correlation between the general 'growth' and 'resistance' trait based on associated expression variation of transcripts was significant ($R = 0.251$). However, the positional candidates for the general 'growth' and 'resistance' phenotypes were distinct. Co-localization estimates for hgt_1997, ldr_99 and hgt_1999, respectively, correlated with co-localization estimates for atk_2000, egg_2000, sum_atk as well as sum_egg traits. This means that a significant fraction of their eQTLs co-localize with both growth and resistance QTLs. Overall, 12% of the genes that were positional candidates for individual height growth traits were also positional candidates for individual resistance traits, and 15% of the genes that were positional candidates for

Table 3. Display of 49 positional candidate genes (AGI annotated) for the composite growth phenotype, p ≤ 0.1.

Gene id	P-value	E-value	AGI #	Annotation	Putative function
WS00924_B02	0.084	6.30E–101	AT2G24210	TPS10 (TERPENE SYNTHASE 10); myrcene/(E)-beta-ocimene synthase	biosynthesis
WS00923_K11	0.089	3.6E–83	AT5G03860	malate synthase, putative	biosynthesis
WS00924_G02	0.092	4.3E–85	AT4G35630	PSAT (phosphoserine aminotransferase); phosphoserine transaminase	biosynthesis
WS0079_D01	0.092	8.10E–160	AT3G54050	fructose-1,6-bisphosphatase, putative/D-fructose-1, 6-bisphosphate 1-phosphohydrolase, putative	biosynthesis
WS00945_C13	0.108	1.30E–165	AT1G79500	AtkdsA1 (Arabidopsis thaliana KDO-8-phosphate synthase A1); 3-deoxy-8-phosphooctulonate synthase	biosynthesis
WS00821_F22	0.058	1.60E–108	AT1G32860	glycosyl hydrolase family 17 protein	cell wall, remodeling(?)
WS00818_M19	0.095	2.50E–085	AT1G26770	ATEXPA10 (ARABIDOPSIS THALIANA EXPANSIN A10)	cell wall
WS0014_E13	0.108	7.40E–089	AT2G37870	protease inhibitor/seed storage/lipid transfer protein (LTP) family protein	cell wall, remodeling(?)
WS0087_G23	0.071	2.90E–058	AT3G06930	protein arginine N-methyltransferase family protein	plant development
WS00819_F17	0.095	1.10E–031	AT4G27745	Identical to Protein yippee-like At4g27740 [Arabidopsis Thaliana]	growth
WS00712_K23	0.100	6.6E–25	AT1G28480	glutaredoxin family protein	growth, development
WS00922_F02	0.104	1.70E–151	AT5G62390	ATBAG7 (ARABIDOPSIS THALIANA BCL-2-ASSOCIATED ATHANOGENE 7); calmodulin binding	growth arrest
WS0084_L12	0.104	2.30E–087	AT3G61780	EMB1703 (EMBRYO DEFECTIVE 1703)	growth arrest
WS00716_E11	0.106	3.00E–146	AT4G26850	VTC2 (VITAMIN C DEFECTIVE 2)	growth related
WS0264_I07	0.108	3.90E–080	AT1G60170	EMB1220 (EMBRYO DEFECTIVE 1220)	growth arrest
WS0083_N10	0.055	8.40E–095	AT2G18360	hydrolase, alpha/beta fold family protein	miscellaneous
WS00821_F12	0.081	2.4E–42	AT3G07480	electron carrier/iron ion binding	miscellaneous
WS01037_M20	0.085	1.1E–30	ATMG00810	similar to protein kinase family protein [Arabidopsis thaliana]	miscellaneous
WS00728_D14	0.092	3.7E–41	AT4G34670	40S ribosomal protein S3A (RPS3aB)	miscellaneous
WS01034_K20	0.095	6.70E–009	AT4G19380	alcohol oxidase-related	miscellaneous
WS00815_F18	0.100	3.00E–072	AT2G19750	40S ribosomal protein S30 (RPS30A)	miscellaneous
WS0041_I12	0.101	3.8E–08	AT1G12810	proline-rich family protein	miscellaneous
WS00926_B01	0.103	9.4E–57	AT4G18100	60S ribosomal protein L32 (RPL32A)	miscellaneous
WS0056_L17	0.104	1.20E–114	AT1G66530	arginyl-tRNA synthetase, putative/arginine–tRNA ligase, putative	miscellaneous
WS0097_I03	0.108	8.9E–08	AT5G54600	50S ribosomal protein L24, chloroplast (CL24)	miscellaneous
WS00819_E15	0.109	8.60E–053	AT4G38250	amino acid transporter family protein	miscellaneous
WS00112_E05	0.037	2.50E–062	AT2G22360	DNAJ heat shock family protein	posttranslational
WS01025_F14	0.067	1.60E–181	AT3G07780	protein binding/zinc ion binding	posttranslational
WS0011_I04	0.074	4.10E–010	AT3G54850	armadillo/beta-catenin repeat family protein/U-box domain-containing family protein	posttranslational
WS0047_F24	0.089	6.4E–24	AT3G06130	heavy-metal-associated domain-containing protein	posttranslational
WS00733_J11	0.090	9.6E–44	AT1G75690	chaperone protein dnaJ-related	posttranslational
WS0024_O12	0.093	5.30E–122	AT5G45390	NCLPP3, NCLPP4, CLPP4	CLPP4 (Clp protease proteolytic subunit 4)	posttranslational
WS01024_O16	0.103	4.9E–64	AT1G77460	C2 domain-containing protein/armadillo/beta-catenin repeat family protein	posttranslational
WS00815_B15	0.109	2.7E–19	AT1G01490	heavy-metal-associated domain-containing protein	posttranslational
WS0089_E10	0.102	6.50E–043	AT2G46225	ABI1L1 (ABI 1 LIKE 1)	signaling
WS00919_K24	0.104	9.20E–119	AT3G59520	rhomboid family protein	signaling
WS0104_D02	0.107	7.4E–16	AT1G61370	S-locus lectin protein kinase family protein	signaling
WS00928_J07	0.078	3.90E–152	AT1G17440	transcription initiation factor IID (TFIID) subunit A family protein	transcriptional/posttranscriptional
WS00912_K01	0.089	1.70E–019	AT2G41900	zinc finger (CCCH-type) family protein	transcriptional/posttranscriptional
WS00917_G03	0.091	3.90E–120	AT1G01350	zinc finger (CCCH-type/C3HC4-type RING finger) family protein	transcriptional/posttranscriptional

Table 3. Cont.

Gene id	P-value	E-value	AGI #	Annotation	Putative function
WS0097_H22	0.092	1.7E–30	AT4G25500	ATRSP35 (Arabidopsis thaliana arginine/serine-rich splicing factor 35)	transcriptional/posttranscriptional
WS00910_O08	0.096	7.00E–048	AT5G08390	similar to transducin family protein/WD-40 repeat family protein [Arabidopsis thaliana]	transcriptional/posttranscriptional
WS00815_A12	0.099	7.20E–120	AT1G20110	zinc finger (FYVE type) family protein	transcriptional/posttranscriptional
WS01031_K02	0.102	1.80E–139	AT3G26935	zinc finger (DHHC type) family protein	transcriptional/posttranscriptional
WS0261_F02	0.103	5.50E–089	AT3G10760	myb family transcription factor	transcriptional/posttranscriptional
WS00922_N21	0.106	1.0E–23	AT3G28917	MIF2 (MINI ZINC FINGER 2); DNA binding	transcriptional/posttranscriptional
WS0099_L07	0.106	2.70E–104	AT2G27110	FRS3 (FAR1-RELATED SEQUENCE 3); zinc ion binding	transcriptional/posttranscriptional
IS0014_L17	0.086	6.40E–100	AT2G21600	ATRER1B (Arabidopsis thaliana endoplasmatic reticulum retrieval protein 1B)	transport, vesicle trafficking
WS00917_J14	0.097	1.10E–020	AT1G33475	Identical to Probable VAMP-like protein At1g33485 [Arabidopsis Thaliana]	transport, vesicle trafficking

resistance traits were also positional candidates for height growth traits.

Discussion

Our work on spruce weevil resistance follows similar work on eucalyptus [32] and poplar ([33], [34]). Ours is the first study of expression QTLs for resistance in a conifer. Despite the enormous genome size of conifers (ca. 20 billion base pairs) studies on the transcriptome of conifers are just as manageable as those with angiosperms with much smaller genome sizes. Here we utilized a third generation microarray spotted with 21,840 spruce ESTs in combination with a multiplexed genotyping approach to examine expression QTLs involved with weevil resistance and height growth. This work represents an extended study of [29] that previously focused in detail on the phenylpropanoid pathway and related genes with respect to pest resistance in spruce.

In our experiment, we harvested plant material in spring at the optimal time point at which early seasonal growth and natural onset of weevil attacks coincided. Our microarray consisted of 70% cDNA from untreated tissue of many tissue types; no overrepresentation of specific metabolic pathways was attempted. The issue of cross-hybridizations in microarray experiments due to high nucleotide similarity [35], is reduced in genetical genomics because of the randomized genetic background, high sample size and the statistical procedures. Cross-species comparisons of QTL are also possible based on the white spruce/loblolly pine comparative mapping project (K. Ritland, pers. comm.). Linkage groups one to twelve (LG1-12) were assigned following [36] to facilitate comparisons within the family of *Pinaceae*.

The main focus of the present work was scanning the genome for transcripts whose abundance correlated with the quantitative phenotype in order to identify transcripts associated with the phenotype ([37]). We assigned groups of genes that significantly associated with individual resistance and growth traits, respectively, into functional, cellular component, and biological process GO categories. These genes were co-regulated and likely have combined functions in the studied phenotypes. Thus, the present work elucidates functional associations among genes and provides a comprehensive study to the evolution of transcription regulation in spruce. Overall we found that a significant fraction of eQTLs were in common between the general 'growth' and 'resistance' phenotypes. This result was based on expression variation from all

studied transcripts. This provides evidence for genetic pleiotropy of resistance and growth traits in interior spruce. In terms of directionality of gene expression with phenotypic trait variation, we found that the mean of correlations between transcript expression and quantitative traits was zero, however overall we found a wide distribution of correlations where some genes showed a clear positive, whereas others showed a clear negative correlation with traits (K. Ritland, personal communication).

Contribution from Single Gene QTL

From the myriad of candidate genes that were identified for the individual resistance traits (**Table S4**; **Figure 4**), our study also identified an array of single genes that were associated with both resistance and growth phenotype (**Table S6** and **Table S4**). The identification of shared candidates suggests that several of the general functionalities (notably, the signaling systems, [38]) important to normal plant development are also adopted for defense mechanisms. Among those 'pleiotropic' genes many had functionalities that prevalently involve signaling, transcription factor activity, functions in transcription/translation (including RNA editing), stress/stimulus response, as well as transport and cell wall functions. Transcription regulators have been suggested to be key targets for plant evolution ([39], [40]). Their high representation in our gene lists reinforces their broad importance to the sessile organism's potential to optimize growth under the given environmental conditions.

Multigene Family QTL Contributions

Large multi-gene families were represented in the associations with both growth and resistance traits. These involved GDSL-motif lipase/hydrolase, GHs, LRR proteins, oxidoreductases, PPR proteins, disease resistance proteins, and DNAJ heat shock family proteins. While GDSL, LRR, PPR and DNAJ proteins were equally represented among growth and resistance candidate genes, respectively, GH, oxidoreductases, and disease resistance proteins contributed twice as many resistance candidates than growth candidates. From this diverse group of disease resistance genes, two spruce sequences with similarity to f-family dirigent proteins that are implicated in constitutive resistance [41] were identified as candidates for weevil resistance. Several individual members were directly associated with both phenotypes and hence pleiotropic (one member in each case for GDSL-motif lipase/hydrolase, disease resistance proteins, and DNAJ heat shock family proteins; two each for GHs, and oxidoreductases; four each for LRR proteins, and PPR proteins). The spruce DNAJ heat shock protein co-localized with expression variation of seven individual traits (both resistance and growth), **Table S6**. Most of these candidates have previously been reported to be involved in defense reactions, some with proposed antifungal properties [42–47].

Dominant Themes among Gene Functions Associated with the Resistance Phenotype

The statistically overrepresented ontology categories among genes that were associated with resistance phenotypic traits revealed dominant themes in gene expression that involved response to all sorts of stimuli (biotic, abiotic, external, and endogenous), epigenetic gene expression, and translation (ribosome) for the *de novo* generation of gene products. Additionally, remodeling processes that involve the (re-) organization of cellular components, and anatomical structures by proteins that provide binding functions and other structural molecule activities are important components of defense (**Figure 4**). Interestingly, signaling is the most common molecular function and cellular process among growth phenotype associated genes (**Figure 5**). For genes associated with the composite weevil resistance phenotype, functions in signaling were also predominant (**Table 2**). There is evidence that signaling pathways in defense reactions are co-opted from normal developmental processes, see also above. Genes that have signaling functions and are negative regulators of ABA responsiveness [48] were identified (phosphatase 2C protein for resistance, while ABI-1-LIKE 1 for growth, **Table 2** and **Table 3**). It is assumed that these gene functions allow for the fine-tuning of stress responses. We could also identify an ATP synthase (**Table 2**). Recently, it was shown that the initiation of multiple defense elicitors in the host is triggered by herbivore proteolysis of a plant ATP synthase [49].

Biosynthesis is also an important function for genes associated with the composite resistance phenotype (**Table 2**). Two genes annotated as mannitol dehydrogenase were identified. Mannitol dehydrogenase counteracts the fungal suppression of the reactive oxygen species that are generated during host defenses [50]. Furthermore, several genes linked to the general phenylpropanoid pathway, ([32], [31], [51]) and to flavonoid and isoflavonoid biosynthesis [52] were positional candidates for the general resistance trait.

Significance of Phenylpropanoid vs. Terpenoid QTL for Constitutive Resistance

The observed eQTLs were the result of constitutive differences in gene expression. The importance of polyphenolics for constitutive defenses is reflected in a higher number of gene family members of the phenylpropanoid pathway (by homology to *A. thaliana* genes) whose eQTL co-localized with resistance QTLs. In this work, seven genes related to phenolics or flavonoid biosynthesis were positional candidates for the general resistance trait (**Table 2**). In contrast, no candidate gene for the resistance trait *per se* could be identified from the terpenoid pathway. Hence, we feel that this might reflect the higher importance of the phenolics over the terpenoid pathway in established resistance against this herbivore. We have previously suggested that monolignol formation may play an important role in defense reactions against the stem borer *Pissodes strobi* [29]. Based on in-depth analysis of genes involved in the shikimate pathway, monolignol biosynthesis and downstream condensation reactions as well as lignan formation with respect to weevil resistance, we

further conclude that gene family members that were duplicated in spruce may have acquired temporally and spatially diverse functions in defense [29].

Trans-eQTL Hotspots and their Significance

Certain phenotypes may be affected by gene expression regulators located within eQTL hotspots [30]. Although several previously conducted eQTL studies suggest that *cis*-eQTLs might have a greater effect on the phenotype than *trans*-eQTLs [53], *trans* eQTLs are important for our understanding of the complexity of phenotypes [54]. For example, by comparing the levels of *trans*-eQTLs for each gene the global regulatory hierarchy can be assessed [53]. While *cis*-eQTLs are physically linked to the causative locus of the phenotype, *trans*-eQTLs can identify many downstream genes and reveal unknown pathways. In our study, we were mainly limited to the detection of *trans*-eQTLs, since the majority of SNP loci on our genetic linkage map could not be annotated [29]. This limitation is due to the fact that we were working with a non-model species and in particular with a conifer of immense genome size for which the genome sequencing has yet to be completed (http://www.congenie.org/).

Several trait-associated SNPs that were enriched for *trans*-eQTLs were identified in our study. At seven map positions, hotspots of expression variation coincided with QTLs from multiple resistance traits (LG3, LG4, LG6 and LG8). At eight SNP positions at least four pQTLs overlapped with eQTL hotspots. On four spruce linkage groups (LG4, LG6, LG11 and LG13) hotspots of expression variation associated with QTLs from both growth and resistance traits (**Figure 1**). This indicates *gene expression regulators* [30].

For example, on LG13 the two SNP loci underlying extensive expression and growth variation (i.e. large number of mapped QTLs) are derived from GAD enzymes whose activity regulation is vital for normal plant development. This allows response to external stimuli [55]. In addition, the enzyme may also function in a host deterrence reaction towards herbivore attack [56]. The eQTL hotspot on LG11 was associated with three resistance traits and two growth traits. The SNP that is located within the eQTL hotspot region is a gene that plays an important role within the ubiquitin/proteasome system, regulating developmental processes in plants, but also involved in biotic defense responses [57]. SNP markers derived from two different contigs Contig_4096_434 and CCoAOMT_1_320, respectively, clustered on LG6 and represent two of the three lignin-forming CCoAOMT genes in spruce [29]. We identified *cis*-eQTLs for these genes as well as *trans*-eQTLs generated from a multitude of genes that mapped to the two loci. Both loci are also hotspots of weevil resistance QTLs. A GO analysis of the transcripts associated with the two *trans*-eQTL hotspots revealed significant over-representation of several molecular function, cellular component, and biological process GO categories. The fact that 36% and 53%, respectively, of the mapped *trans*-eQTLs were in common between CCoAOMT-1 and CCoAOMT-2 suggests extensive interactions between both CCoAOMT loci via eQTLs from a multitude of genes. This was also reflected by common GO categories that were overrepresented such as 'secondary metabolic process' and 'response to biotic stimulus' in both gene expression networks centered at these two SNPs (**Figure 2** & **Figure 3**). Thus, this demonstrates how epistasis between gene loci works at the transcriptional level by linking *cis*-eQTLs via *trans*-regulatory interactions. In the case of CCoAOMT-1 and CCoAOMT-2, three resistance pQTLs also contributed to this epistatic interaction. CCoAOMT-1 and CCoAOMT-2 represent both metabolic pathway–specific *trans*-eQTL hotspots [29] and based on the present study, they represent important global *trans*-eQTL hotspots that are of interest for pest resistance in spruce.

Jasmonate Signaling and its Central Role in Defense

We also identified JAZ genes that may play a role in the gene-gene interaction network between both CCoAOMT loci. JAZ genes were identified as central regulators of JA-mediated anti-insect defense [58]. Three JAZ genes were candidate genes directly associated with phenotypic trait variation. JA signaling is activated by repressor (i. e. JAZ) removal from the ubiquitin ligase complex [59]. Carbonic anhydrase genes are other important genes that have roles in jasmonate signaling and also in ethylene signaling [60], and in our study these genes mapped *trans*-eQTLs to both CCoAOMT hotspot locations. A previous study found that carbonic anhydrase genes were induced in spruce under stress treatments (budworm, weevil feeding, and mechanical wounding) [35]. At the second CCoAOMT locus on LG6, eQTLs from ERFs and specifically those from group IX [61] that represent known transcription repressors (ERF3, ERF4, ERF7) [62] were found.

In our study, transcriptional activators related to ethylene response (ERF-1, ERF-2, EIN5, [63]) as well as regulators for ethylene biosynthesis *per se* (RUB1, RUB1-conjugating enzyme, [64]) were found to co-localize exclusively with resistance traits. The hormonal cross-talks with JA (involving salicylic acid, ethylene, abscisic acid, and auxin) during growth and development as well as during adaptation to stress are highly complex [59,65]. In *Arabidopsis*, the major players in JA-mediated plant defense are tightly linked [66]. The differential regulation of certain components/steps in the pathway is expected to generate distinct responses to different stimuli (reproductive development, growth or defenses, [59,66]). Thus, establishing and maintaining defenses involves signaling systems that are co-opted from developmental processes [38].

Conclusions

Although genomics studies on forest trees have traditionally focused on wood attributes [67–70], the genomics of environmental challenges has recently gained importance [34,71]. Biotic stressors, herbivores and their accompanying pathogens pose an increased threat to tree populations, and knowledge of the genomic architecture can inform the management and breeding practices of conifers, as well as increase our general understanding of the evolution and adaptation of conifer species. Our result will add to this second vary important layer of genomics in forestry.

We have utilized expression QTL mapping to identify candidate genes. This will facilitate targeted association studies to further understand the genetic basis of host resistance to pests, the genomic basis of pest resistance. These results will enable both further functional studies to the nature of insect resistance in spruce, and provide valuable information about candidate genes for genetic improvement of spruce.

We identified several master regulators that underlie the genetic pleiotropy of pest resistance and developmental processes. Several candidate genes from the JA signaling pathway were identified for which we could show that central regulators of this pathway are contributing to extensive gene-gene interaction networks. Plant JA signaling provides a rapid response to various external stimuli [72] and is central to all biotic stress responses that directly influence the performance of the pest or contribute indirect defense responses to attract predators or herbivore parasitoids [73]. Importantly, this signaling pathway is not defense specific, but co-opted from normal developmental processes such as reproductive development [72]. In this way, the induction of defenses against herbivores or pathogens remains highly cost effective [38].

This work identified several pleiotropic genes as candidate genes whose proposed functions are important in stress response or disease resistance. In addition, our study revealed the presence of master genes which influence the global transcriptome. These genes are in "hotspots", sometimes linked to annotated loci which were in turn further annotated to developmental and defense associated processes. Since resistance and growth QTLs overlapped with eQTL hotspots along the genome, this suggests that: 1) genetic pleiotropy of resistance and growth traits in interior spruce was substantial, and 2) master regulatory genes were important for weevil resistance in spruce. Knowledge about the exact function of these master regulons in the conifer genome needs further investigation; however knock-out mutants for largely pleiotropic genes were shown to be largely lethal or exhibit highly deleterious phenotypes [53].

Materials and Methods

Interior Spruce Pedigree

Experimental interior spruce populations originated from a controlled cross progeny trial established in 1995 at Kalamalka Research Station in Vernon, BC, Canada [74]. The parental trees were selected from individuals previously ranked for weevil-resistance in open-pollinated progeny tests [14]. Out of twenty crosses segregating for weevil-resistance [74], four families with wide segregation for weevil resistance arranged as 2x2 factorial were harvested in May 2006 for gene expression profiling: cross 26 (♀PG87*♂PG165), cross 27 (♀PG87*♂PG117), cross 29 (♀PG21*♂PG165) and cross 32 (♀PG21*♂PG117). From 417 offspring of a 3x2 factorial (including the additional crosses 22 and 25), genomic DNA was isolated from flushing bud/needle tissue according to the cetyltrimethylammonium bromide (CTAB) method [75]. The studied trees represented individual genotypes that were planted in randomized plots within three replicate blocks in the field [74]. A detailed layout of the study site that shows the randomized location of plots for the QTL mapping families PG87*PG165 (cross 26), PG87*PG117 (cross 27), PG21*PG165 (cross29) and PG21*PG117 (cross32) within the replicate blocks can be found in [29].

A set of 384 SNPs were identified *in silico* from a collection of ESTs from the Treenomix EST database (K. Ritland pers.comm.) that were all derived from a single tree (PG 29). The genomic DNA was then genotyped for these SNPs using the multiplexed Illumina platform at the CMMT Genotyping and Gene Expression Core Facility, Centre for Molecular Medicine and Therapeutics, Vancouver, BC. Recombination rates were determined by joint likelihood [76] for each pair of loci and a consensus genetic map of 252 SNPs was constructed using JoinMap 3.0 [77], see [29] for details. Of all putative SNP loci, 73.4–76.0% were confirmed and included in the analysis; 394 individuals were true full-sibs. Those that could not be confirmed as full-sibs in the respective crosses (cross 26: 7%, cross 27: 10%, cross 29: 4%, and cross 32: 1% of the trees alive in 2006) were removed before phenotyping. The majority of spruce gene markers (i.e. the SNPs) that were used to build the framework map could not be annotated. This involved 67% of the ESTs when we used the TAIR7 database, while 54% of the ESTs when we used Viridiplantae databases [29].

Measures for Tree Height, Weevil Attack and Oviposition

The trial was screened for resistance to terminal leader weevil following the method described by [20]. In short, a population of weevils was raised in summer 1999 at the Canadian Forest Service (Pacific Forestry Centre), Victoria and released onto all test trees in fall 1999. Attack rates, egg counts and top kills were recorded in 2000–2004. Growth measurements included the initial tree height in 1995 (year one), and heights in years three, and five as well as leader length in year five preceding the artificial augmentation of the local weevil population in October of the same year (hgt_1995, hgt_1997, hgt_1999, and ldr_99, respectively). Attack rates in 2000 and 2001 (atk_2000, atk_2001) were classified as successful 'top kills', 'failure' to kill the leader and 'no attack' [74]. In addition, for the same years oviposition on the leaders was recorded (egg_2000 and egg_2001) by counting egg punctures into five discrete classes: 1 = 1–25, 2 = 26–50, 3 = 51–75, 4 = 76–100, 5 = 101 and more. Egg punctures contain egg covering fecal plugs and are easily distinguished from feeding punctures, which are not covered. The sums of weevil attacks and oviposition for 2000 and 2001 were also used as 'resistance' traits (sum_atk and sum_egg).

Tissue Collection, RNA Preparation, Microarray, Gene Expression Profiling

Tree material within a replicate block was sampled in a randomized fashion among the plots (i.e. crosses). Terminal leaders from trees in a block were collected in the mornings of May 16, 17 and 18, 2006, respectively. The weather was consistent among these days. Bark/phloem tissue was immediately harvested on site from cut leaders as described previously ([35]; [78]), flash frozen in liquid nitrogen, and stored at −80°C until processed. Total RNA from unattacked individuals was isolated following the protocol of [79] and quantified using NanoDrop® ND-1000 Spectrophotometer; RNA integrity was evaluated using the Agilent 2100 Bioanalyzer. The 21,840 spruce ESTs on the array involved elements from 12 different cDNA libraries, built from different tissues (bark, phloem, xylem), which were under different developmental stages, as well as wound/methyljasmonate treated (ca. 6,500 elements) and untreated (ca. 15,400). Complete details of cDNA microarray fabrication and quality control are described elsewhere (S. Ralph and co-workers, Gene Expression Omnibus database GEO: GPL5423 and http://www.treenomix.ca/). Labeling reactions, hybridizations, slide washes as well as scanning of slides were carried out as described in [35]. Fluorescent intensity data were extracted by using the ImaGene 6.0 software (Biodiscovery, El Segundo, USA). Signal intensity measurements were deposited in the Gene Expression Omnibus database under the accession number GSE22116.

Microarray Experimental Design and Pre-processing of Expression Data

Our experimental design is based on *a priori* known genotypes. Testing six genotyped crosses and using the previously collected phenotypic data (see above), we determined that genotype differences between most and least resistant progeny were highest in crosses 26, 27, 29 and 32. Since we used two-color microarrays, direct comparisons between Cy3-Cy5 labeled samples were required. A distant pair design for microarray analysis that maximized direct comparisons between different alleles at each locus was originally introduced by [80] and was modified in our study for outbred individuals. We estimated the genetic distance for possible probe-pairs genome-wide by using all segregating SNP loci (122 on average) and such we maximized the number of distant pairs in a given cross. A 25% improvement over random pairing was achieved. We also balanced the two dyes across the three replicate blocks (i.e. sampling on three different days), the different batches of microarray fabrication and different experimenters (see below). Our design resulted in 94 hybridizations profiling 48 individuals in cross 26, 36 in cross 27, and 50 in cross 29 as well as 54 individuals in cross 32 [29].

After quantification of the signal intensities in each array the local background was subtracted for each subgrid. Data were normalized to compensate for non-linearity of intensity distributions using the variance stabilizing normalization method [81]. We performed a single normalization of 188 columns of data. In this way each channel had a similar and array-independent overall expression level and variance. Signal intensities are deposited under the GEO accession number GSE22116. The linear model

$$h_i = \mu + dye + block + batch + person + \varepsilon_i$$

with μ as the overall mean was then fit to the normalized intensities of each gene i (h_i) in the Cy3 and Cy5 channels to account for technical effects within the experiment (gene-specific 'dye' effect, replicate 'block', microarray fabrication 'batch', experimenter 'person' are all fixed effects). The residuals were used in the subsequent QTL analysis. All of the above statistics was carried out using the R statistical package (www.r-project.org).

QTL Detection

A program was written in FORTRAN by K. Ritland for QTL mapping in the 3×2 factorial (for resistance and growth traits) and 2x2 factorial design (for gene expression traits). This program inferred QTL maps for each of the parents of the factorial. QTLs were mapped in the progeny by employing a likelihood function of the trait level (gene expression, other traits) conditioned on genotype of progeny, and compared to the likelihood of unconditioned genotype of progeny (no association of traits with progeny genotype) to give a log-odds (LOD) ratio. Due to the large number of gene expression traits, a single-marker model instead of an interval mapping approach was used, and QTLs were binned into 10 cM marker intervals, thus avoiding having two QTLs assigned to adjacent markers due to linkage of two markers to one QTL. We used R (www.r-project.org) to display QTL density maps. A QTL was significant at LOD ≥3.84 and had to be detected for a minimum of one parent in the factorial ([29], and **Table S1**). A goodness-of-fit test assuming a uniform distribution was performed to test whether the observed frequencies of eQTLs along the linkage map differed significantly from the expected value. Following the rejection of this null hypothesis ($\chi^2 = 96678$, df = 251, p-value <2.2e-16), we declared "eQTL hotspots" if the number of eQTLs at a given locus exceeded the expected average by 50%. These numbers are significantly above the maximum number within eQTL clusters (i.e. 630) from a randomly generated data set using all 132,100 detected eQTLs, 252 markers, and running 1,000 replicates. The positional candidate genes were identified by collocation of at least 40% of their eQTLs with phenotypic trait QTLs based on the criteria for identifying significant QTLs (see above) and running 10,000 randomizations (p ≤ 0.05).

Other Statistical Analyses

Phenotypic trait correlations were determined using SAS/STAT software, version 9.1.3 of the SAS system for Windows® (SAS Institute Inc., Cary, NC, USA). The cytoscape 2.5.1 plug-in BINGO [82] was used and a hypergeometric test was performed to determine statistically overrepresented Gene Ontology (GO) terms within the GOSlim Plants ontology for spruce genes with *Arabidopsis* homologs. In our case, this involved comparing the nearest Arabidopsis homologs for all genes that showed significant association of their expression variation with the previously assessed phenotypic trait variation (tree height, weevil attack, and oviposition) to all Arabidopsis homologs on the microarray.

Supporting Information

Figure S1 GO tree representation showing significantly (p ≤ 0.05) overrepresented GO categories within the *trans* eQTL-hotspot at the carbonic anhydrase gene locus contig_2079_440 (803 eQTLs) on LG4, for color code see Figures 2, 3, 4 and 5 in main text.

Figure S2 GO tree representation showing significantly (p ≤ 0.05) overrepresented GO categories within the *trans* eQTL-hotspot at the carbonic anhydrase gene locus contig_103_602 (1122 eQTLs) on LG4, for color code see Figures 2, 3, 4 and 5 in main text.

File S1 Comprehensive list of all 132,100 significant eQTLs (legends can be found in Table S1).

Table S1 Significant QTLs for gene expression (LOD ≥3.84); allele effect, and % phenotypic variation explained by QTL are given in File S1.

Table S2 Comprehensive list of eQTLs with annotations at locus Contig_4096_434 (see also Figure 2).

Table S3 Comprehensive list of eQTLs with annotations at locus CCoAOMT_1_320 (see also Figure 3).

Table S4 Comprehensive results for collocation estimations, with p-values.

Table S5 Statistically overrepresented Gene Ontology terms in the GOSlim Plant ontology for genes with expression variation co-localizing with resistance and growth traits, respectively, as presented in Figure 4 and Figure 5.

Table S6 Display of genes that are candidates for different resistance and growth traits (p≤0.05), for at least three phenotypic traits (ldr_99, hgt_1995, hgt_1997, hgt_1999, atk_2000, atk_2001, sum_atk, egg_2000, egg_2001, and sum_egg, respectively).

Table S7 Complete list of the identified 149 positional candidate genes for the general resistance trait, (p ≤ 0.1).

Table S8 Complete list of identified 99 positional candidate genes for the general growth trait, (p ≤ 0.1).

Acknowledgments

We acknowledge Charles Chen for help with R script. We thank Gillian Leung and Michelle Tang for technical support, Susan Findlay, Tristan Gillan, Jun Zhang, Cherdsak Liewlaksanyannawin, and Claire Cullis for help with sample collection.

Author Contributions

Conceived and designed the experiments: IP RW KR RA BJ CR. Performed the experiments: IP. Analyzed the data: IP RW KR. Wrote the paper: IP KR BJ. Contributed to project management: CR.

References

1. Dicke M (2000) Chemical ecology of host-plant selection by herbivorous arthropods: a multitrophic perspective. Biochemical Systematics and Ecology 28: 601–617.
2. Cornell HV, Hawkins BA (2003) Herbivore responses to plant secondary compounds: A test of phytochemical coevolution theory. American Naturalist 161: 507–522.
3. Strauss SY, Rudgers JA, Lau JA, Irwin RE (2002) Direct and ecological costs of resistance to herbivory. Trends in Ecology & Evolution 17: 278–285.
4. Zangerl AR, Arntz AM, Berenbaum MR (1997) Physiological price of an induced chemical defense: Photosynthesis, respiration, biosynthesis, and growth. Oecologia 109: 433–441.
5. Bergelson J, Purrington CB (1996) Surveying patterns in the cost of resistance in plants. American Naturalist 148: 536–558.
6. Mole S (1994) Tradeoffs and constraints in plant-herbivore defense theory - a life history perspective. Oikos 71: 3–12.
7. Herms DA, Mattson WJ (1992) The dilemma of plants - to grow or defend. Quarterly Review of Biology 67: 478–478.
8. Roff DA, Fairbairn DJ (2007) The evolution of trade-offs: where are we? Journal of Evolutionary Biology 20: 433–447.
9. Worley AC, Houle D, Barrett SCH (2003) Consequences of hierarchical allocation for the evolution of life-history traits. American Naturalist 161: 153–167.
10. Koricheva J, Nykanen H, Gianoli E (2004) Meta-analysis of trade-offs among plant antiherbivore defenses: Are plants jacks-of-all-trades, masters of all? American Naturalist 163: E64-E75.
11. Kempel A, Schaedler M, Chrobock T, Fischer M, van Kleunen M (2011) Tradeoffs associated with constitutive and induced plant resistance against herbivory. Proceedings of the National Academy of Sciences of the United States of America 108: 5685–5689.
12. Van Zandt PA (2007) Plant defense, growth, and habitat: A comparative assessment of constitutive and induced resistance. Ecology 88: 1984–1993.
13. Alfaro RI (1995) An induced defense reaction in white spruce to attack by the white-pine weevil, Pissodes strobi. Canadian Journal of Forest Research-Revue Canadienne De Recherche Forestiere 25: 1725–1730.
14. Kiss GK, Yanchuk AD (1991) Preliminary evaluation of genetic-variation of weevil resistance in interior spruce in British-Columbia. Canadian Journal of Forest Research-Revue Canadienne De Recherche Forestiere 21: 230–234.
15. King JN, Yanchuk AD, Kiss GK, Alfaro RI (1997) Genetic and phenotypic relationships between weevil (Pissodes strobi) resistance and height growth in spruce populations of British Columbia. Canadian Journal of Forest Research-Revue Canadienne De Recherche Forestiere 27: 732–739.
16. Alfaro RI, He FL, Tomlin E, Kiss G (1997) White spruce resistance to white pine weevil related to bark resin canal density. Canadian Journal of Botany-Revue Canadienne De Botanique 75: 568–573.
17. Lieutier F, Brignolas F, Sauvard D, Yart A, Galet C, et al. (2003) Intra- and inter-provenance variability in phloem phenols of Picea abies and relationship to a bark beetle-associated fungus. Tree Physiology 23: 247–256.
18. Vandersar TJD, Borden JH (1977) Visual orientation of Pissodes-strobi Peck (Coleoptera curculionidae) in relation to host selection behavior. Canadian Journal of Zoology-Revue Canadienne De Zoologie 55: 2042–2049.
19. He FL, Alfaro RI (2000) White pine weevil attack on white spruce: A survival time analysis. Ecological Applications 10: 225–232.
20. Alfaro RI, King JN, Brown RG, Buddingh SM (2008) Screening of Sitka spruce genotypes for resistance to the white pine weevil using artificial infestations. Forest Ecology and Management 255: 1749–1758.
21. vanAkker L, Alfaro RI, Brockley R (2004) Effects of fertilization on resin canal defences and incidence of Pissodes strobi attack in interior spruce. Canadian Journal of Forest Research-Revue Canadienne De Recherche Forestiere 34: 855–862.
22. Gardner KM, Latta RG (2007) Shared quantitative trait loci underlying the genetic correlation between continuous traits. Molecular Ecology 16: 4195–4209.
23. McKay JK, Richards JH, Mitchell-Olds T (2003) Genetics of drought adaptation in Arabidopsis thaliana: I. Pleiotropy contributes to genetic correlations among ecological traits. Molecular Ecology 12: 1137–1151.
24. Tiffin P, Rausher MD (1999) Genetic constraints and selection acting on tolerance to herbivory in the common morning glory Ipomoea purpurea. American Naturalist 154: 700–716.
25. MitchellOlds T (1996) Genetic constraints on life-history evolution: Quantitative-trait loci influencing growth and flowering in Arabidopsis thaliana. Evolution 50: 140–145.
26. Karrenberg S, Widmer A (2008) Ecologically relevant genetic variation from a non-Arabidopsis perspective. Current Opinion in Plant Biology 11: 156–162.
27. Rockman MV, Kruglyak L (2006) Genetics of global gene expression. Nature Reviews Genetics 7: 862–872.

28. Farrall M (2004) Quantitative genetic variation: a post-modern view. Human Molecular Genetics 13: R1–R7.
29. Porth I, Hamberger B, White R, Ritland K (2011) Defense mechanisms against herbivory in Picea: sequence evolution and expression regulation of gene family members in the phenylpropanoid pathway. BMC Genomics 12: 608.
30. Potokina E, Druka A, Luo ZW, Wise R, Waugh R, et al. (2008) Gene expression quantitative trait locus analysis of 16,000 barley genes reveals a complex pattern of genome-wide transcriptional regulation. Plant Journal 53: 90–101.
31. Ferrer JL, Austin MB, Stewart C, Noe JP (2008) Structure and function of enzymes involved in the biosynthesis of phenylpropanoids. Plant Physiology and Biochemistry 46: 356–370.
32. Kirst M, Myburg AA, De Leon JPG, Kirst ME, Scott J, et al. (2004) Coordinated genetic regulation of growth and lignin revealed by quantitative trait locus analysis of cDNA microarray data in an interspecific backcross of eucalyptus. Plant Physiology 135: 2368–2378.
33. Drost DR, Benedict CI, Berg A, Novaes E, Novaes CRDB, et al. (2010) Diversification in the genetic architecture of gene expression and transcriptional networks in organ differentiation of Populus. Proceedings of the National Academy of Sciences of the United States of America 107: 8492–8497.
34. Street NR, Skogstrom O, Sjodin A, Tucker J, Rodriguez-Acosta M, et al. (2006) The genetics and genomics of the drought response in Populus. Plant Journal 48: 321–341.
35. Ralph SG, Yueh H, Friedmann M, Aeschliman D, Zeznik JA, et al. (2006) Conifer defence against insects: microarray gene expression profiling of Sitka spruce (*Picea sitchensis*) induced by mechanical wounding or feeding by spruce budworms (*Choristoneura occidentalis*) or white pine weevils (*Pissodes strobi*) reveals large-scale changes of the host transcriptome. Plant Cell and Environment 29: 1545–1570.
36. Sewell MM, Sherman BK, Neale DB (1999) A consensus map for loblolly pine (Pinus taeda L.). I. Construction and integration of individual linkage maps from two outbred three-generation pedigrees. Genetics 151: 321–330.
37. Gibson G, Weir B (2005) The quantitative genetics of transcription. Trends in Genetics 21: 616–623.
38. Steppuhn A, Baldwin I (2008) Induced defenses and the cost-benefit paradigm. In: Schaller A, editor. Induced Plant Resistance to Herbivory: Springer. 61–83.
39. Doebley J, Lukens L (1998) Transcriptional regulators and the evolution of plant form. Plant Cell 10: 1075–1082.
40. Chen K, Rajewsky N (2007) The evolution of gene regulation by transcription factors and microRNAs. Nature Reviews Genetics 8: 93–103.
41. Ralph SG, Jancsik S, Bohlmann J (2007) Dirigent proteins in conifer defense II: Extended gene discovery, phylogeny, and constitutive and stress-induced gene expression in spruce (Picea spp.). Phytochemistry 68: 1975–1991.
42. Naranjo MA, Forment J, Roldan M, Serrano R, Vicente O (2006) Overexpression of Arabidopsis thaliana LTL1, a salt-induced gene encoding a GDSL-motif lipase, increases salt tolerance in yeast and transgenic plants. Plant Cell and Environment 29: 1890–1900.
43. Lipka V, Dittgen J, Bednarek P, Bhat R, Wiermer M, et al. (2005) Pre- and postinvasion defenses both contribute to nonhost resistance in Arabidopsis. Science 310: 1180–1183.
44. Shanmugam V (2005) Role of extracytoplasmic leucine rich repeat proteins in plant defence mechanisms. Microbiological Research 160: 83–94.
45. Saha D, Prasad AM, Srinivasan R (2007) Pentatricopeptide repeat proteins and their emerging roles in plants. Plant Physiology and Biochemistry 45: 521–534.
46. Belkhadir Y, Subramaniam R, Dangl JL (2004) Plant disease resistance protein signaling: NBS-LRR proteins and their partners. Current Opinion in Plant Biology 7: 391–399.
47. Miernyk JA (2001) The J-domain proteins of Arabidopsis thaliana: an unexpectedly large and diverse family of chaperones. Cell Stress & Chaperones 6: 209–218.
48. Gosti F, Beaudoin N, Serizet C, Webb AAR, Vartanian N, et al. (1999) ABI1 protein phosphatase 2C is a negative regulator of abscisic acid signaling. Plant Cell 11: 1897–1909.
49. Schmelz EA, LeClere S, Carroll MJ, Alborn HT, Teal PEA (2007) Cowpea chloroplastic ATP synthase is the source of multiple plant defense elicitors during insect herbivory. Plant Physiology 144: 793–805.
50. Jennings DB, Ehrenshaft M, Pharr DM, Williamson JD (1998) Roles for mannitol and mannitol dehydrogenase in active oxygen-mediated plant defense. Proceedings of the National Academy of Sciences of the United States of America 95: 15129–15133.
51. Koutaniemi S, Warinowski T, Karkonen A, Alatalo E, Fossdal CG, et al. (2007) Expression profiling of the lignin biosynthetic pathway in Norway spruce using EST sequencing and real-time RT-PCR. Plant Molecular Biology 65: 311–328.
52. Zhao JM, Last RL (1996) Coordinate regulation of the tryptophan biosynthetic pathway and indolic phytoalexin accumulation in Arabidopsis. Plant Cell 8: 2235–2244.
53. Kliebenstein D (2009) Quantitative Genomics: Analyzing Intraspecific Variation Using Global Gene Expression Polymorphisms or eQTLs. Annual Review of Plant Biology 60: 93–114.
54. Fehrmann RSN, Jansen RC, Veldink JH, Westra H-J, Arends D, et al. (2011) *Trans*-eQTLs Reveal That Independent Genetic Variants Associated with a Complex Phenotype Converge on Intermediate Genes, with a Major Role for the HLA. PLoS Genet 7: e1002197.
55. Baum G, LevYadun S, Fridmann Y, Arazi T, Katsnelson H, et al. (1996) Calmodulin binding to glutamate decarboxylase is required for regulation of glutamate and GABA metabolism and normal development in plants. Embo Journal 15: 2988–2996.
56. Bown AW, MacGregor KB, Shelp BJ (2006) Gamma-aminobutyrate: defense against invertebrate pests? Trends in Plant Science 11: 424–427.
57. Dreher K, Callis J (2007) Ubiquitin, hormones and biotic stress in plants. Annals of Botany 99: 787–822.
58. Chung HS, Koo AJK, Gao XL, Jayanty S, Thines B, et al. (2008) Regulation and function of Arabidopsis JASMONATE ZIM-domain genes in response to wounding and herbivory. Plant Physiology 146: 952–964.
59. Kazan K, Manners JM (2008) Jasmonate signaling: toward an integrated view. Plant Physiology 146: 1459–1468.
60. Ferreira FJ, Guo C, Coleman JR (2008) Reduction of plastid-localized carbonic anhydrase activity results in reduced Arabidopsis seedling survivorship. Plant Physiology 147: 585–594.
61. Nakano T, Suzuki K, Fujimura T, Shinshi H (2006) Genome-wide analysis of the ERF gene family in Arabidopsis and rice. Plant Physiology 140: 411–432.
62. Yang Z, Tian LN, Latoszek-Green M, Brown D, Wu KQ (2005) Arabidopsis ERF4 is a transcriptional repressor capable of modulating ethylene and abscisic acid responses. Plant Molecular Biology 58: 585–596.
63. Adams E, Devoto A, Turner J (2007) Analysis of a novel ethylene-induced COI1-dependent signalling pathway in Arabidopsis thaliana. In: al Re, editor. Advances in Plant Ethylene Research: Proceedings of the 7th International Symposium on the Plant Hormone Ethylene: Springer. 81–87.
64. Chae H, Kieber J (2005) Eto Brute? Role of ACS turnover in regulating ethylene biosynthesis. Trends in Plant Science 10.
65. Pauwels L, Goossens A (2011) The JAZ Proteins: A Crucial Interface in the Jasmonate Signaling Cascade. The Plant Cell Online 23: 3089–3100.
66. Xiao S, Dai LY, Liu FQ, Wang Z, Peng W, et al. (2004) COS1: An Arabidopsis Coronatine insensitive1 suppressor essential for regulation of jasmonate-mediated plant defense and senescence. Plant Cell 16: 1132–1142.
67. Gonzalez-Martinez SC, Wheeler NC, Ersoz E, Nelson CD, Neale DB (2007) Association genetics in Pinus taeda L. I. Wood property traits. Genetics 175: 399–409.
68. Pot D, McMillan L, Echt C, Le Provost G, Garnier-Gere P, et al. (2005) Nucleotide variation in genes involved in wood formation in two pine species. New Phytologist 167: 101–112.
69. Chagne D, Brown G, Lalanne C, Madur D, Pot D, et al. (2003) Comparative genome and QTL mapping between maritime and loblolly pines. Molecular Breeding 12: 185–195.
70. Sewell MM, Davis MF, Tuskan GA, Wheeler NC, Elam CC, et al. (2002) Identification of QTLs influencing wood property traits in loblolly pine (Pinus taeda L.). II. Chemical wood properties. Theoretical and Applied Genetics 104: 214–222.
71. Holliday JA, Ralph SG, White R, Bohlmann J, Aitken SN (2008) Global monitoring of autumn gene expression within and among phenotypically divergent populations of Sitka spruce (Picea sitchensis). New Phytologist 178: 103–122.
72. Howe GA, Jander G (2008) Plant immunity to insect herbivores. Annual Review of Plant Biology. Palo Alto: Annual Reviews. 41–66.
73. Thaler JS, Farag MA, Pare PW, Dicke M (2002) Jasmonate-deficient plants have reduced direct and indirect defences against herbivores. Ecology Letters 5: 764–774.
74. Alfaro RI, VanAkker L, Jaquish B, King J (2004) Weevil resistance of progeny derived from putatively resistant and susceptible interior spruce parents. Forest Ecology and Management 202: 369–377.
75. Doyle J, Doyle J (1990) Isolation of plant DNA from fresh tissue. Focus 12: 13–15.
76. Hu XS, Goodwillie C, Ritland KM (2004) Joining genetic linkage maps using a joint likelihood function. Theoretical and Applied Genetics 109: 996–1004.
77. Stam P (1993) Construction of integrated genetic-linkage maps by means of a new computer package - Joinmap. Plant Journal 3: 739–744.
78. Ralph S, Park JY, Bohlmann J, Mansfield SD (2006) Dirigent proteins in conifer defense: gene discovery, phylogeny, and differential wound- and insect-induced expression of a family of DIR and DIR-like genes in spruce (Picea spp.). Plant Molecular Biology 60: 21–40.
79. Kolosova N, Miller B, Ralph S, Ellis BE, Douglas C, et al. (2004) Isolation of high-quality RNA from gymnosperm and angiosperm trees. Biotechniques 36: 821–824.
80. Fu JY, Jansen RC (2006) Optimal design and analysis of genetic studies on gene expression. Genetics 172: 1993–1999.
81. Huber W, von Heydebreck A, Sultmann H, Poustka A, Vingron M (2002) Variance stabilization applied to microarray data calibration and to the quantification of differential expression. Bioinformatics 18: S96–S104.
82. Maere S, Heymans K, Kuiper M (2005) BiNGO: a cytoscape plugin to assess overrepresentation of gene ontology categories in biological networks. Bioinformatics 21: 3448–3449.

3

Mosquito Population Regulation and Larval Source Management in Heterogeneous Environments

David L. Smith[1,2,3]*, **T. Alex Perkins[3,4]**, **Lucy S. Tusting[5]**, **Thomas W. Scott[3,4]**, **Steven W. Lindsay[3,6]**

1 Department of Epidemiology, Johns Hopkins Bloomberg School of Public Health, Baltimore, Maryland, United States of America, **2** Malaria Research Institute, Johns Hopkins Bloomberg School of Public Health, Baltimore, Maryland, United States of America, **3** Fogarty International Center, NIH, Bethesda, Maryland, United States of America, **4** Department of Entomology, University of California, Davis, California, United States of America, **5** Department of Disease Control, London School of Hygiene and Tropical Medicine, London, United Kingdom, **6** School of Biological and Biomedical Sciences, Durham University, Durham, United Kingdom

Abstract

An important question for mosquito population dynamics, mosquito-borne pathogen transmission and vector control is how mosquito populations are regulated. Here we develop simple models with heterogeneity in egg laying patterns and in the responses of larval populations to crowding in aquatic habitats. We use the models to evaluate how such heterogeneity affects mosquito population regulation and the effects of larval source management (LSM). We revisit the notion of a carrying capacity and show how heterogeneity changes our understanding of density dependence and the outcome of LSM. Crowding in and productivity of aquatic habitats is highly uneven unless egg-laying distributions are fine-tuned to match the distribution of habitats' carrying capacities. LSM reduces mosquito population density linearly with coverage if adult mosquitoes avoid laying eggs in treated habitats, but quadratically if eggs are laid in treated habitats and the effort is therefore wasted (i.e., treating 50% of habitat reduces mosquito density by approximately 75%). Unsurprisingly, targeting (i.e. treating a subset of the most productive pools) gives much larger reductions for similar coverage, but with poor targeting, increasing coverage could increase adult mosquito population densities if eggs are laid in higher capacity habitats. Our analysis suggests that, in some contexts, LSM models that accounts for heterogeneity in production of adult mosquitoes provide theoretical support for pursuing mosquito-borne disease prevention through strategic and repeated application of modern larvicides.

Editor: Nikos Vasilakis, University of Texas Medical Branch, United States of America

Funding: DLS acknowledges funding from the Bloomberg Family Foundation. DLS and SWL acknowledge funding from National Institutes of Health/National Institute of Allergy and Infectious Diseases (U19AI089674). DLS, TAP, SWL, and TWS acknowledge funding from the Research and Policy for Infectious Disease Dynamics (RAPIDD) program of the Science and Technology Directorate, Department of Homeland Security, and the Fogarty International Center, National Institutes of Health.

* E-mail: dlsmith@jhsph.edu

Introduction

Dynamic models of malaria transmission have influenced strategic decisions about disease prevention from the time of Ronald Ross in the early 20th century, when larval source management (LSM) was the dominant form of vector control [1]. After early field deployment of DDT demonstrated that indoor residual spraying (IRS) was an extremely effective way to control malaria, George Macdonald's influential mathematical analysis showed that transmission was highly sensitive to adult mosquito mortality rates [2]. This analysis and emerging theory reinforced the prevailing notion at the time that DDT was a sufficient tool for malaria eradication [3,4], and IRS was implemented largely to the exclusion of LSM. The legacy of Macdonald's sixty-year old analysis can be seen in contemporary policy decisions by leading international organizations, including a recent evaluation by the World Health Organization (WHO) that was highly critical of larviciding in sub-Saharan Africa [5]. These recommendations, based largely on the Ross-Macdonald model that lacks dynamic mosquito populations and is ill suited to evaluate LSM, come despite evidence that LSM can achieve similar results and at a similar cost to ITNs [6]. Here, we re-examine the simple models

that have motivated such analyses, and we derive some basic lessons for mosquito population dynamic and control to guide policy for LSM.

Several mosquito population dynamic models have been developed that link adult and immature aquatic populations [7–11], and a few have explicitly considered LSM [12]. Most models of larval populations, whether simple or complex, make some assumption about density dependence and population regulation. Some have considered the complex structure that arises from having populations of eggs, four larval instars, and pupae [13]. Others have considered the dynamics of systems with predators or resource-based competition [14]. Complicated computer-simulation models have considered the effects of heterogeneity in rainfall and temperature, heterogeneous habitat geometries with variable responses to flushing, and desiccation [9,10,14–17]. Finally, a few models have considered how the distribution of larval habitat constrains egg laying and affects the adult mosquito population dynamics and pathogen transmission [18–20]. It remains unclear how heterogeneity affects the way mosquito populations are regulated and what variation in key processes means for LSM. Here, we present a simple theoretical framework that can be used to understand habitat heterogeneity, the local effects of density dependence, the factors that affect the outcome of LSM in dynamic, heterogeneous environments, and their total effects on pathogen transmission.

Methods

Many factors have been implicated in immature mosquito population dynamics, including egg laying, water temperature, resource limitations, predation, larval development rates, the ephemeral availability of mosquito habitat due to evaporation and desiccation, or filling or flushing habitat from the combination of rainfall and hydrology [21]. Here, we take a simpler approach that focuses narrowly population regulation when aquatic habitats are heterogeneous. The model may not be suitable for some purposes, such as simulating mosquito population dynamics when realistic lags for mosquito development are required (i.e. see [13]), but the models do provide insights into the regulation of mosquito populations, population dynamics in heterogeneous habitats and the effects of LSM. These lessons can, perhaps, serve as a theoretical basis for understanding more complicated models.

The Mosquito Population Dynamic Model

The following model considers the coupled dynamics of aquatic immature and terrestrial adult mosquito populations. We assume the population of larval mosquitoes is subdivided into N distinct aquatic habitats. Individual aquatic habitats are hereafter called "pools" to facilitate communication, even though this may not be the best description of many kinds of larval habitats.

Let $M(t)$ denote the population density of adult mosquitoes at time t, and let g denote the per-capita death rate. The number of larvae in each pool at any given time is denoted $L_i(t)$. Let f denote the mosquito blood feeding rate, v the number of eggs laid by a mosquito each egg laying cycle, and p_i the fraction of eggs laid in the ith pool. In the ith pool, mosquitoes are assumed to mature at rate α_i and die at the per-capita rate $\gamma_i + \psi_i L_i^{\sigma_i}$, where ψ_i represents a pool-specific increase in per-capita mortality in response to crowding. For $\sigma_i = 1$, which was assumed for most of our analysis, the relationship gives mean crowding, which is analogous to the classical first-order description of density dependence as described by the logistic growth equation. Under these assumptions, mosquito population dynamics are described by the following equations:

$$\dot{L}_i = fvp_iM - \left(\alpha_i + \gamma_i + \psi_i L_i^{\sigma_i}\right)L_i \tag{1}$$

$$\dot{M} = \sum_i \alpha_i L_i - gM \tag{2}$$

Homogeneous environments were defined by letting each pool have identical parameters and by distributing eggs evenly among the pools. Environments were made heterogeneous by varying parameters describing larval dynamics or egg laying (i.e. $\alpha_i, \gamma_i, \psi_i, \sigma_i,$ or p_i from Eq. 1) strategically to illustrate specific aspects of this system. We constructed completely heterogeneous environments by drawing random numbers for all larval dynamic parameters and for egg laying. Parameter names are summarized in Table S1 along with all the values used for the simulations. The population dynamics in these completely heterogeneous environments depends on some notion of the response to crowding, the

distribution of eggs laid, and the potential capacity for adult mosquito production of each pool.

Larval Source Management

LSM was simulated by assuming that control was applied either permanently or repeatedly to a subset of these pools, which were called "treated." This is done to make the analysis simpler and to illustrate properties of the models, even though there might be real constraints on the ability to completely and permanently nullify mosquito productivity. Coverage was defined as the proportion of aquatic habitats that were treated. In our simulations, LSM was assumed to prevent all larval development and eliminate productivity such that no adults emerged from treated pools. The analysis focused on the relationship between coverage and the "control effect size" on transmission, defined for LSM as the proportional decline in the adult mosquito densities when compared to the same system without control.

The control effect sizes of LSM were simulated under two different assumptions about changes in egg-laying behaviour of adult mosquitoes in response to LSM. First, mosquitoes could continue to lay eggs in the pools that had been treated, such as when modern non-repellent larvicides are applied to aquatic habitats. Second, mosquitoes could avoid treated pools and lay eggs elsewhere, either because larvicides in the water acted as repellents or because the habitat was modified or destroyed.

Control effect sizes of increased LSM coverage were examined for five classes of population dynamic simulations based on different assumptions about crowding and egg laying: (1) a homogeneous environment where all pools have identical parameters and eggs are laid evenly; (2) a simple extension of the homogeneous model in which a fraction of habitats in a homogeneous environment were simply non-productive, so that the fraction of eggs laid in productive habitats summed to less than 1; (3) LSM was applied in random order in a completely heterogeneous environment; (4) LSM was "targeted" by treating subsets of the most productive pools in a completely heterogeneous environment (this was done in a perfectly efficient order, such that as coverage increased, the pools with highest productivity were treated first); and (5) to show a contrast, LSM was then inefficiently targeted in a completely heterogeneous environment by treating subsets of the least productive pools.

Results

Mosquito Population Dynamics

The key dynamic feature of the equations describing mosquito population dynamics is that emerging adult mosquitoes become part of an adult mosquito population and that they distribute eggs among many independent pools (Eq. 2). Because of many factors affecting the distribution of eggs in habitats of differing qualities, including the patterns of blood feeding, different patterns emerge from examining the dynamics of completely heterogeneous systems compared with homogeneous systems.

The "carrying capacity" was defined as the equilibrium density of larvae in a system with only one pool or in a homogeneous system, and it is given by the formula:

$$K_i = \sqrt[\sigma_i]{\frac{fv\alpha_i g^{-1} - (\alpha_i + \gamma_i)}{\psi_i}}$$

In the homogeneous system, carrying capacity determines the maximum productivity of each pool, the emergence rate of adult mosquitoes $\Lambda_i = \alpha_i K_i$.

When egg-laying patterns are heterogeneous, larval density becomes decoupled from carrying capacity. The number of eggs laid and the mean crowding of each habitat affect larval densities. At the steady state, the number of eggs laid in a pool is $O_i = f v p_i \bar{M}$ and:

$$\bar{L}_i = \alpha_i \left(-\frac{\alpha_i + \gamma_i}{2\psi_i} + \sqrt{\left(\frac{\alpha_i + \gamma_i}{2\psi_i}\right)^2 + \frac{O_i}{\psi_i}} \right).$$

The number of adults emerging from any individual pool (*i.e.*, productivity), $\alpha_i \bar{L}_i$ depends on the number of eggs laid and the functional relationship that determines how larval mosquito mortality increases with crowding (Fig. 1a). (Different rules would likely follow from considering predation or other types of regulation that respond dynamically to larval population density.) These productivity curves show that carrying capacity is but one point along a continuum of adult output rates from a pool in relation to the egg input rate. Larval populations thin in response to crowding, so that the proportion of adults emerging from individual pools decreases with the number of eggs laid, but the number of adults emerging increases. Dynamics of individual pools linked by egg-laying females do not, therefore, generally conform to the rules of logistic growth. In heterogeneous environments, productivity and carrying capacity are therefore not generally given by the same quantity.

Some properties of the general system come from exploring a simple system of two pools with different carrying capacities. By varying the proportion of eggs laid in each pool, productivities of the individual pools and of the whole system were compared. In this system, the total productivity equals the total carrying capacity only when the proportion of eggs laid in each pool is equal to that pool's carrying capacity as a fraction of the total of all pool's capacities (*i.e.*, if $p_i = K_i / \sum_i K_i$, Fig. 1b).

Numerical simulations demonstrate that this rule holds in completely heterogeneous systems (Fig. 1c). In that case, total productivity is equal to the total carrying capacity only when the distribution of egg laying is fine-tuned to equal the relative distribution of carrying capacity. Unless the proportion of eggs laid is fine-tuned to match the carrying capacities, larval densities will be different than carrying capacity (Fig. 1c,d), often by a large margin. The net effect of this mismatch is unpredictable, but it will depend strongly upon the proportion of eggs laid in the most productive habitats. For the mathematical assumptions made in this model, productivity was lower than capacity in approximately one-third of the cases, but productivity often exceeded carrying capacity. In at least one case, productivity exceeded capacity by 270%.

Intuitively, the dynamic interplay of mobile adults, distributed aquatic habitats, and the response to crowding means that total productivity is strongly affected by the correlation between the distribution of eggs laid and the distribution of carrying capacities in aquatic habitats. The proportion of eggs that survive to become adults in any one pool is reduced as egg laying increases crowding, but the number of eggs being laid depends on the whole ensemble of aquatic habitats. Pools that receive the most eggs will tend to have population densities that exceed their carrying capacities, while those that have the fewest eggs are usually below capacity

(Fig. 1d). Crowding will be uneven and the effects of crowding in just a few pools dominate population regulation.

Productivity and carrying capacity should both be correlated with egg-laying (Fig. 2a–c), but the underlying functional relationship between eggs in and adults out is only revealed by plotting the ratio of egg-laying to carrying capacity against the ratio of productivity to carrying capacity (Fig. 2d). Though enlightening, this relationship may not have any practical use unless it is possible to measure carrying capacity directly, perhaps through surrogate measures such as a pool's surface area, volume, or key resource levels through bioassays [22]. The notion of a carrying capacity is, therefore, useful both conceptually and theoretically. Capacity is not, however, what is typically observed in individual pools or in populations at the steady state. Instead, productivity is determined by the carrying capacities of the individual pools, heterogeneity in egg laying proportions, and the mismatch between the two patterns.

Another important principle is that, in the absence of immigration from pools outside the study area, the dynamic feedback between egg-laying and aquatic population dynamics is subject to a threshold phenomenon governing mosquito persistence. A sufficient condition for mosquito persistence is $f v p_i \alpha_i > g(\alpha_i + \gamma_i)$. The mosquito population can, in theory, persist if at least one adult male and female mosquito is expected to emerge from a pool from an egg laid by a typical single adult mosquito originating from that pool (Analysis S1).

Larval Source Management

The control effect sizes of LSM depend strongly upon the adult female mosquito's egg-laying behaviour in response to LSM. The most important difference is whether mosquitoes continue to lay eggs in treated habitat. If mosquitoes avoid laying eggs in the treated habitats, then LSM simply reduces the amount of habitat available. The outcomes tend to be consistent with a common use of Macdonald's formula for R_0 with respect to LSM, which assumes linear responses. The dynamics of LSM with heterogeneous biting and targeting are more complicated, however, and different rules govern systems in which adults continue to lay eggs in treated pools. Some general aspects of LSM are best illustrated in homogeneous systems, but other aspects play out differently in heterogeneous systems.

LSM is more effective when mosquitoes continue to lay eggs in treated habitats. To illustrate why this behaviour changes the control dynamic, consider a simple heterogeneous system in which pools either have all the same carrying capacity, or they produce no adults at all. Holding the number of productive pools fixed, the total productivity of the system declines linearly with the number of unproductive habitats. The existence of unproductive aquatic habitats nearby can thus become "egg sinks" [23,24] and reduce the proportion of eggs laid in the productive pools (Fig. 3a).

To illustrate the relationship between coverage and control effect sizes, LSM was simulated in a homogeneous system with varying coverage levels and with both types of egg-laying responses. In these simulations, when adult mosquitoes continue to lay eggs in treated habitats, there are two effects of LSM. One effect, the reductions in the amount of productive habitat, is complemented by a second effect, an increase in the amount of habitat that serves as a sink for eggs. The two effects are multiplicative, so control effect sizes scale with LSM coverage in a way that is approximately quadratic: removing 50% of habitat reduces mosquito densities by approximately 75%, and removing 80% of the habitat reduces mosquito densities by approximately 96% (Figure 3b).

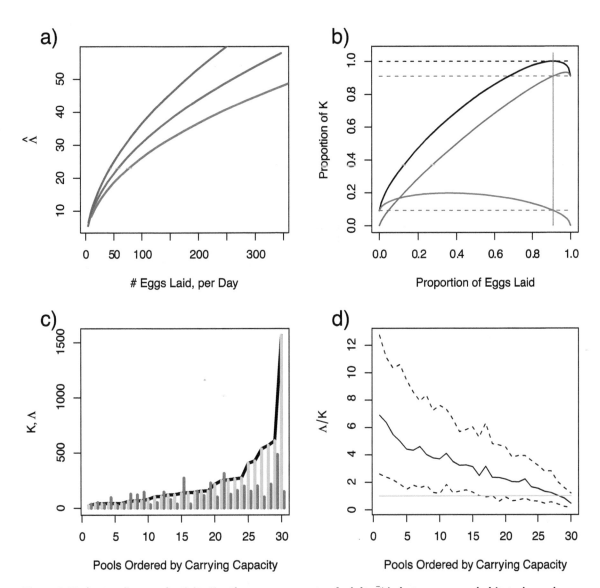

Figure 1. Understanding productivity (*i.e.*, the emergence rate of adults $\bar{\Lambda}$) in heterogeneous habitats depends upon understanding the relationship between egg laying, carrying capacity (K), and crowding. a) The functional relationship between the rate of egg-laying and productivity depends on the functional response to crowding. In this model, the relationship is sensitive to the power-law scaling relationship ($\sigma = 1$, blue; $\sigma = 1.1$, red; $\sigma = 0.9$, purple). Carrying capacity is given for a single value of egg laying rates, given at the steady state if that pool had existed in isolation. **b)** In a system with 2 pools linked by egg-laying, where the carrying capacity of pool 1 is approximately 90% of the total (dashed blue line) and pool 2 has the rest (dashed red line), the population totals overall (solid black) are generally below the maximum, unless egg laying is fine-tuned such that the proportion of eggs laid was equal to that pool's proportion of carrying capacity (vertical grey). **c)** A comparison of productivity (red) and carrying capacity (black line) for a typical set of heterogeneous aquatic habitats. Productivity equals carrying capacity when the distribution of eggs laid is finely tuned to match the distribution of carrying capacities (i.e. $p_i = K_i / \sum_i K_i$). **d)** The ratio of productivity to carrying capacity was computed for 100 sets of heterogeneous aquatic habitat. The green line plots the 1:1 ratio, when productivity equals carrying capacity. These distributions, plotted here as the median (solid line) and the 10th and 90th quantiles (dashed lines), shows the robust pattern that the habitats with the lowest productivity tend to be under capacity and the few highly productive habitats tend to be over capacity.

Similar results occur when habitat is heterogeneous, but the interpretation of "coverage" must be considered in a more nuanced way. In homogeneous systems, coverage describes reductions in capacity, productivity, and egg laying. In heterogeneous systems, however, the mismatch between productivity, capacity, and egg laying means that varying amounts of these three quantities remain as coverage increases. The control effect sizes of LSM in heterogeneous systems thus depend on both the adult egg-laying behaviour in response to LSM and the order that LSM is applied to the pools (Fig. 3c,d).

Not surprisingly, the control effect sizes of LSM would be substantially larger if LSM were targeted at the most productive

pools (Fig. 3c,d), and it would be substantially less efficient if not. The most efficient solution – targeting the most productive pools in rank order of their productivity from most to least – results in sharp increase in control effect sizes for even small coverage levels, regardless of adult mosquito egg-laying behaviour. The analysis here suggests that the control effect sizes are greater than log-linear, such that it is possible to reduce transmission by a hundred-fold with moderate coverage through targeted, repeated application of modern (i.e., non-repellent) larvicides and other modes of LSM that create an egg sink effect among the most productive pools.

Like homogeneous systems, the outcome of LSM in heterogeneous systems is also dependent on egg-laying behaviour of

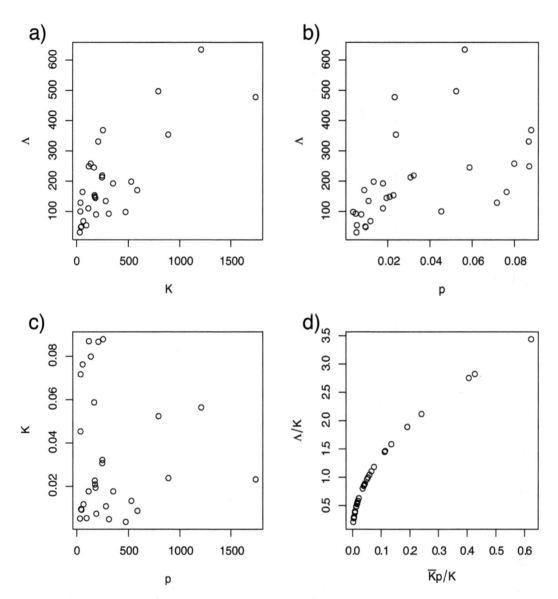

Figure 2. The scaling between egg-laying and productivity is only apparent after normalizing both productivity and egg laying by carrying capacity. In completely heterogeneous environments, there may be a poor correlation between **a**) carrying capacity and productivity; **b**) egg laying and productivity; and **c**) egg laying and carrying capacity. **d**) The crowding law governing density dependence is found by plotting the ratio of eggs laid to carrying capacity against the ratio of productivity to carrying capacity (i.e. $\{\Lambda_i/K_i, p_i/K_i\}$). The constant $\bar{K} = \sum_i K_i$ was used to scale the x-axis.

mosquitoes in treated pools. The effects of efficient targeting are similar between both types of egg-laying responses, but control effect sizes are always higher, all else equal, when mosquitoes continue to lay eggs in treated pools. Like the homogeneous systems, the "egg-sink" effect in heterogeneous systems complements the removal effect to further reduce population densities (Fig. 3c,d). The magnitude of the egg sink effect varies, however, because the mismatch between productivity and the fraction of eggs laid means that the egg-sink effect is only approximately linear with respect to LSM coverage.

Control effect sizes are, on the other hand, highly variable as a function of coverage when LSM is applied to pools in a random sequence. For the same random sequence, control effect sizes are always higher when eggs are laid in treated pools (Fig. 3c,d). Even with perfectly inefficient targeting, control effect sizes in relation to coverage are nearly linear when eggs are laid in treated pools.

The outcomes of LSM were surprising, however, for some random sequences and for inefficient targeting in the case when adult mosquitoes avoid treated pools and redistribute eggs elsewhere (Fig. 3c,d). In the case of perfectly inefficient targeting (*i.e.*, when a subset of the least productive pools is treated), eggs are redistributed from less to more productive pools. The effect is counteracted by a reduction in total carrying capacity. The net effects change with coverage and with the particular distribution of pools (Fig. 3c,d). Similarly, for a random sequence of pools, productivity can increase as coverage increases whenever the effect of redistributing eggs to more productive habitats is greater than the loss of capacity. The behavioural responses of mosquitoes thus make it possible for LSM to increase overall mosquito density by forcing adult mosquitoes to redistribute eggs in more productive pools when egg-laying under natural conditions is highly inefficient.

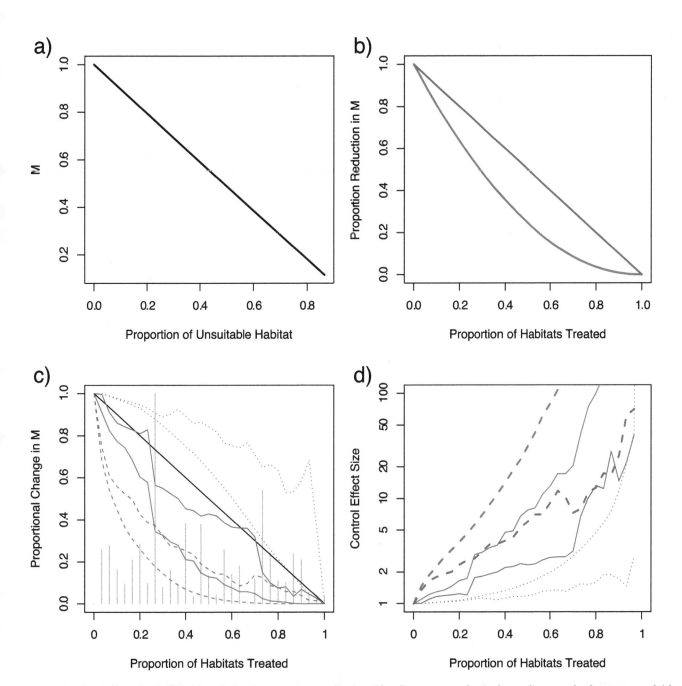

Figure 3. The "effect size" of LSM in relation to coverage tend to be either linear or quadratic depending on whether eggs are laid in "treated" habitats and how well LSM is targeted. a) Holding the total number of productive pools constant, adult mosquito population density declines as the number of unproductive pools increases and absorb eggs. **b**) The "egg sink" effect gives a non-linear effect to LSM if adult mosquitoes continue to lay eggs in the treated pools, so that treating 50% of the pools reduces adult density by 75%, and treating 75% of the pools reduces adult density by 95% (red). If adult mosquitoes do not lay eggs in the treated pools, however, then reductions in mosquito density are proportional to the % of habitat treated (blue). **c**) The change in adult mosquito density due to LSM in highly heterogeneous habitat as a function of the proportion of habitats treated depending on whether the adults lay eggs in treated pools (red) or avoid treated pools (blue), and depending on whether LSM was done in one particular random order (grey spikes), perfectly efficiently targeted (dashed lines), or perfectly inefficiently targeted (dotted lines). The black line represents a linear response with respect to coverage. **d**) For the same graphs as 3c, the effect sizes are plotted on a semi-log scale to highlight the benefits of LSM at high coverage. The best case for this system, with efficient targeting and egg-sink effects, predicts a hundred-fold (99%) reduction in mosquito density for 60% coverage. These benefits also get larger for higher coverage and show that there is enormous potential for LSM to reduce transmission through targeted repeated application of modern larvicides.

Discussion

We conclude that mosquito egg laying is an important factor for mosquito population dynamics and that it can have strong affects on the outcome of LSM. In our analyses, increasing coverage caused quadratic reductions in mosquito density if mosquitoes continued to lay eggs in treated pools. Therefore LSM has the potential to be a highly effective method of malaria control without extensive coverage. In particular, we predict that moderate coverage targeted at the most productive aquatic habitats can achieve reasonably large reductions in transmission. Some of these conclusions are inconsistent with statements from the recent WHO report, which was based on a Ross-Macdonald model perspective [5]. Moreover, recent evidence demonstrates that, in some circumstances, LSM has been effective in reducing clinical malaria outcomes [25–31], and for similar costs to those of IRS and long-lasting insecticidal nets (LLINs) [6,32–34]. Given all these caveats, generalizations about LSM are likely to depend heavily on the local context for pathogen transmission and operational constraints.

Our results highlight the lack of attention paid to heterogeneity in mosquito population dynamics and in considering the outcome of LSM. Habitat heterogeneity and local density dependence change the way that dynamics in mosquito population are regulated and could play a role in creating or controlling transmission hotspots [35,36]. The concept of carrying capacity in models with homogeneous habitat and logistic growth must be modified in light of the heterogeneous structure of aquatic habitats with local density dependence. Crowding could be highly uneven such that a few habitats would have very high larval densities while others would be scarcely populated. In general, adult mosquito population densities will differ from capacity unless there is fine-tuning in the relationship between the carrying capacities of aquatic habitats and the egg-laying patterns of adult mosquitoes. The rules governing mosquito population densities are more aptly described as a system in which crowding thins the aquatic mosquito populations.

When LSM is integrated into these models, egg-laying behaviour is identified, once again, as an important issue. The effects of LSM are approximately quadratic when mosquitoes continue to lay eggs in treated habitats, and these treated habitats function as egg sinks [23,24]. This effect is quadratic because LSM has two distinct linear effects that could be created separately by first removing productive habitat, and second replacing that habitat with oviposition traps that absorb just as many eggs. Using larvicides that repel mosquitoes has only the first effect, while using larvicides that do not repel mosquitoes has both effects. The product of these two linear effects creates a non-linear response (i.e. quadratic) much like the one that Macdonald identified in his oft-used analysis [2,4,5]. This analysis also raises an operationally relevant question about the repellent effects of modern larvicides at concentrations ordinarily used for field application [37–46]. Targeting of these systems [47] can lead to disproportionate efficiencies in the effectiveness of LSM, though practical advice about how to identify productive larval habitats for targeting remains a critical need. In places where the aquatic habitats are in the same places year after year, is possible for control programs to learn and adapt to local systems [47]. Factors that may seem to present technical limitations for LSM – such as the need to target the most productive habitats – can be turned into a long-term operational advantage. It may be possible, for example, to

accumulate knowledge about the local mosquito ecology and thereby improve the effectiveness of LSM over time.

The collective results of eleven decades of vector control have been mixed [1,48]. Overall, our study, along with many others [49], emphasizes the important role of various kinds of heterogeneity in transmission dynamics and control. Heterogeneities can have a strong influence on the ability to measure transmission or predict the outcomes of control programs. A general point to be made is that outcomes probably depend on R_0, but they also depend on specific aspects of human and vector behaviours in specific contexts. Interventions that have not been explicitly considered in the Ross-Macdonald model cannot be derived intuitively from the formula for R_0. Instead, they must be explicitly modelled and integrated into the underlying theory. Though some simple points can be made about the likely effects of LSM, simple mathematical models can often be misleading unless they identify the appropriate sources of heterogeneity. Application of the theory to LSM in this and other modelling studies [7,8,12] is increasingly based on information about local mosquito ecology and its relation to transmission. Given these concerns, even though analysis of mathematical models can help to inform policy, but empirical evidence should perhaps play a stronger role in evaluation of policy and in making policy recommendations. The consideration of LSM, LLINs, IRS, spatial repellents, attractive sugar toxic baits, genetic strategies, oviposition traps, and other new vector control tools designed to reduce transmission of a pathogen by mosquitoes all lead to the realization that there will be some advantages and some disadvantages for each approach, and that intervention success will vary by the transmission context and the efficiency with which programs are implemented. In the context of increasingly widespread insecticide and drugs resistance, limitations in delivery and coverage, and national and international funding constraints, what is urgently needed is programmatic flexibility. Success in ever changing environments will depend on the capacity to select from a suite of options a package of interventions that is best suited for local to national and regional vector-borne diseases prevention goals [50,51].

Author Contributions

Conceived and designed the experiments: DLS TAP LST TWS SWL. Performed the experiments: DLS. Analyzed the data: DLS TAP. Wrote the paper: DLS TAP LST TWS SWL. Developed the model, created the graphics wrote the first draft: DLS. Analyzed the model: DLS and TAP.

References

1. Ross R (1907) The prevention of malaria in British possessions, Egypt, and parts of America Lancet: 879–887.
2. Macdonald G (1952) The analysis of the sporozoite rate. Trop Dis Bull 49: 569–586.
3. Macdonald G (1956) Theory of the eradication of malaria. Bull World Health Org 15: 369–387.
4. Macdonald G (1956) Epidemiological basis of malaria control. Bull World Health Org 15: 613–626.
5. W.H.O. (2012) The role of larviciding for malaria control in sub-Saharan Africa. Interim Position Statement. Geneva: World Health Organization.
6. Worrall E, Fillinger U (2011) Large-scale use of mosquito larval source management for malaria control in Africa: a cost analysis. Malar J 10: 338.
7. Dye C (1984) Models for the Population Dynamics of the Yellow Fever Mosquito, Aedes aegypti. J Animal Ecology 53: 247–268.
8. Yakob L, Yan G (2010) A network population model of the dynamics and control of African malaria vectors. Trans R Soc Trop Med Hyg 104: 669–675.
9. Magori K, Legros M, Puente ME, Focks DA, Scott TW, et al. (2009) Skeeter Buster: a stochastic, spatially explicit modeling tool for studying Aedes aegypti population replacement and population suppression strategies. PLoS Negl Trop Dis 3: e508.
10. Focks DA, Daniels E, Haile DG, Keesling JE (1995) A simulation model of the epidemiology of urban dengue fever: literature analysis, model development, preliminary validation, and samples of simulation results. Am J Trop Med Hyg 53: 489–506.
11. Focks DA, Haile DG, Daniels E, Mount GA (1993) Dynamic life table model for Aedes aegypti (Diptera: Culicidae): analysis of the literature and model development. J Med Entomol 30: 1003–1017.

12. White MT, Griffin JT, Churcher TS, Ferguson NM, Basanez MG, et al. (2011) Modelling the impact of vector control interventions on Anopheles gambiae population dynamics. Parasit Vectors 4: 153.

13. Hancock PA, Godfray HCJ (2007) Application of the lumped age-class technique to studying the dynamics of malaria-mosquito-human interactions. Malaria Journal 6: 98.

14. Depinay JM, Mbogo CM, Killeen G, Knols B, Beier J, et al. (2004) A simulation model of African Anopheles ecology and population dynamics for the analysis of malaria transmission. Malar J 3: 29.

15. Bomblies A, Duchemin J, Eltahir E (2008) Hydrology of malaria: Model development and application to a Sahelian village. Water Resources Research 44: W12445.

16. Bomblies A, Duchemin J-B, Eltahir EAB (2009) A mechanistic approach for accurate simulation of village scale malaria transmission. Malaria Journal 8: 223.

17. Focks DA, Haile DG, Daniels E, Mount GA (1993) Dynamic life table model for Aedes aegypti (diptera: Culicidae): simulation results and validation. J Med Entomol 30: 1018–1028.

18. Le Menach A, McKenzie FE, Flahault A, Smith DL (2005) The unexpected importance of mosquito oviposition behaviour for malaria: Non-productive larval habitats can be sources for malaria transmission. Malaria Journal 4: 23.

19. Gu WD, Regens JL, Beier JC, Novak RJ (2006) Source reduction of mosquito larval habitats has unexpected consequences on malaria transmission. Proceedings of the National Academy of Sciences of the United States of America 103: 17560–17563.

20. Gu W, Novak RJ (2009) Agent-based modelling of mosquito foraging behaviour for malaria control. Trans R Soc Trop Med Hyg 103: 1105–1112.

21. Reiner RC, Jr., Perkins TA, Barker CM, Niu T, Chaves LF, et al. (2013) A systematic review of mathematical models of mosquito-borne pathogen transmission: 1970–2010. J R Soc Interface 10: 20120921.

22. Arrivillaga J, Barrera R (2004) Food as a limiting factor for Aedes aegypti in water-storage containers. J Vector Ecol 29: 11–20.

23. Wong J, Morrison AC, Stoddard ST, Astete H, Chu YY, et al. (2012) Linking oviposition site choice to offspring fitness in Aedes aegypti: consequences for targeted larval control of dengue vectors. PLoS Negl Trop Dis 6: e1632.

24. Wong J, Stoddard ST, Astete H, Morrison AC, Scott TW (2011) Oviposition site selection by the dengue vector Aedes aegypti and its implications for dengue control. PLoS Negl Trop Dis 5: e1015.

25. Yapabandara AM, Curtis CF, Wickramasinghe MB, Fernando WP (2001) Control of malaria vectors with the insect growth regulator pyriproxyfen in a gem-mining area in Sri Lanka. Acta Trop 80: 265–276.

26. Yapabandara AM, Curtis CF (2004) Control of vectors and incidence of malaria in an irrigated settlement scheme in Sri Lanka by using the insect growth regulator pyriproxyfen. J Am Mosq Control Assoc 20: 395–400.

27. Sharma SK, Tyagi PK, Upadhyay AK, Haque MA, Adak T, et al. (2008) Building small dams can decrease malaria: a comparative study from Sundargarh District, Orissa, India. Acta Trop 107: 174–178.

28. Geissbuhler Y, Kannady K, Chaki PP, Emidi B, Govella NJ, et al. (2009) Microbial larvicide application by a large-scale, community-based program reduces malaria infection prevalence in urban Dar es Salaam, Tanzania. PLoS One 4: e5107.

29. Fillinger U, Ndenga B, Githeko A, Lindsay SW (2009) Integrated malaria vector control with microbial larvicides and insecticide-treated nets in western Kenya: a controlled trial. Bull World Health Organ 87: 655–665.

30. Fillinger U, Kannady K, William G, Vanek MJ, Dongus S, et al. (2008) A tool box for operational mosquito larval control: preliminary results and early lessons from the Urban Malaria Control Programme in Dar es Salaam, Tanzania. Malar J 7: 20.

31. Castro MC, Tsuruta A, Kanamori S, Kannady K, Mkude S (2009) Community-based environmental management for malaria control: evidence from a small-scale intervention in Dar es Salaam, Tanzania. Malar J 8: 57.

32. Fillinger U, Lindsay SW (2011) Larval source management for malaria control in Africa: myths and reality. Malar J 10: 353.

33. Lengeler C (2004) Insecticide-treated bed nets and curtains for preventing malaria. Cochrane Database Syst Rev: CD000363.

34. Pluess B, Tanser FC, Lengeler C, Sharp BL (2010) Indoor residual spraying for preventing malaria. Cochrane Database Syst Rev: CD006657.

35. Bousema T, Griffin JT, Sauerwein RW, Smith DL, Churcher TS, et al. (2012) Hitting hotspots: spatial targeting of malaria for control and elimination. PLoS Med 9: e1001165.

36. Chaves LF, Hamer GL, Walker ED, Brown WM, Ruiz MO, et al. (2011) Climatic variability and landscape heterogeneity impact urban mosquito diversity and vector abundance and infection. Ecosphere 2.

37. Xue RD, Ali A, Crainich V, Barnard D (2007) Oviposition deterrence and larvicidal activity of three formulations of piperidine repellent (AI3-37220) against field populations of Stegomyia albopicta. J Am Mosq Control Assoc 23: 283–287.

38. Xue RD, Barnard DR, Ali A (2003) Laboratory evaluation of 18 repellent compounds as oviposition deterrents of Aedes albopictus and as larvicides of Aedes aegypti, Anopheles quadrimaculatus, and Culex quinquefasciatus. J Am Mosq Control Assoc 19: 397–403.

39. Ansari MA, Mittal PK, Razdan RK, Sreehari U (2005) Larvicidal and mosquito repellent activities of Pine (Pinus longifolia, Family: Pinaceae) oil. Journal of Vector Borne Diseases 42: 95–99.

40. Karunamoorthi K, Ramanujam S, Rathinasamy R (2008) Evaluation of leaf extracts of Vitex negundo L. (Family: Verbenaceae) against larvae of Culex tritaeniorhynchus and repellent activity on adult vector mosquitoes. Parasitology Research 103: 545–550.

41. Pandey SK, Upadhyay S, Tripathi AK (2009) Insecticidal and repellent activities of thymol from the essential oil of Trachyspermum ammi (Linn) Sprague seeds against Anopheles stephensi. Parasitology Research 105: 507–512.

42. Rajkumar S, Jebanesan A (2009) Larvicidal and oviposition activity of Cassia obtusifolia Linn (Family: Leguminosae) leaf extract against malarial vector, Anopheles stephensi Liston (Diptera: Culicidae). Parasitol Res 104: 337–340.

43. Carroll JF, Tabanca N, Kramer M, Elejalde NM, Wedge DE, et al. (2011) Essential oils of Cupressus funebris, Juniperus communis, and J. chinensis (Cupressaceae) as repellents against ticks (Acari: Ixodidae) and mosquitoes (Diptera: Culicidae) and as toxicants against mosquitoes. Journal of Vector Ecology 36: 258–268.

44. Govindarajan M (2011) Larvicidal and repellent properties of some essential oils against Culex tritaeniorhynchus Giles and Anopheles subpictus Grassi (Diptera: Culicidae). Asian Pacific Journal of Tropical Medicine 4: 106–111.

45. Govindarajan M, Mathivanan T, Elumalai K, Krishnappa K, Anandan A (2011) Mosquito larvicidal, ovicidal, and repellent properties of botanical extracts against Anopheles stephensi, Aedes aegypti, and Culex quinquefasciatus (Diptera: Culicidae). Parasitology Research 109: 353–367.

46. Kumar SV, Mani P, Bastin JTMM, Kumar RA, Ravikumar G (2011) Larvicidal, oviposition deterrent and repellent activity of annona squamosa extracts against hazardous mosquito vectors. International Journal of Pharmacy and Technology 3: 3143–3155.

47. Fillinger U, Sombroek H, Majambere S, van Loon E, Takken W, et al. (2009) Identifying the most productive breeding sites for malaria mosquitoes in The Gambia. Malar J 8: 62.

48. Najera JA, Gonzalez-Silva M, Alonso PL (2011) Some lessons for the future from the Global Malaria Eradication Programme (1955–1969). PLoS Med 8: e1000412.

49. Woolhouse ME, Dye C, Etard JF, Smith T, Charlwood JD, et al. (1997) Heterogeneities in the transmission of infectious agents: implications for the design of control programs. Proc Natl Acad Sci U S A 94: 338–342.

50. Eisen L, Beaty BJ, Morrison AC, Scott TW (2009) Proactive vector control strategies and improved monitoring and evaluation practices for dengue prevention. J Med Entomol 46: 1245–1255.

51. Scott TW, Morrison AC (2008) Longitudinal field studies will guide a paradigm shift in dengue prevention. In: Vector-borne Diseases. Understanding the Environmental, Human Health, and Ecological Connections. The National Academies Press: Washington, DC. 132–149.

4

Explaining Andean Potato Weevils in Relation to Local and Landscape Features: A Facilitated Ecoinformatics Approach

Soroush Parsa[1,4]*, Raúl Ccanto[2], Edgar Olivera[2], María Scurrah[2], Jesús Alcázar[3], Jay A. Rosenheim[4]

1 CIAT (Centro Internacional de Agricultura Tropical), Cali, Colombia, 2 Grupo Yanapai, Miraflores, Lima, Peru, 3 CIP (Centro Internacional de la Papa), Lima, Peru, 4 Department of Entomology, University of California Davis, Davis, California, United States of America

Abstract

Background: Pest impact on an agricultural field is jointly influenced by local and landscape features. Rarely, however, are these features studied together. The present study applies a "facilitated ecoinformatics" approach to jointly screen many local and landscape features of suspected importance to Andean potato weevils (*Premnotrypes* spp.), the most serious pests of potatoes in the high Andes.

Methodology/Principal Findings: We generated a comprehensive list of predictors of weevil damage, including both local and landscape features deemed important by farmers and researchers. To test their importance, we assembled an observational dataset measuring these features across 138 randomly-selected potato fields in Huancavelica, Peru. Data for local features were generated primarily by participating farmers who were trained to maintain records of their management operations. An information theoretic approach to modeling the data resulted in 131,071 models, the best of which explained 40.2–46.4% of the observed variance in infestations. The best model considering both local and landscape features strongly outperformed the best models considering them in isolation. Multi-model inferences confirmed many, but not all of the expected patterns, and suggested gaps in local knowledge for Andean potato weevils. The most important predictors were the field's perimeter-to-area ratio, the number of nearby potato storage units, the amount of potatoes planted in close proximity to the field, and the number of insecticide treatments made early in the season.

Conclusions/Significance: Results underscored the need to refine the timing of insecticide applications and to explore adjustments in potato hilling as potential control tactics for Andean weevils. We believe our study illustrates the potential of ecoinformatics research to help streamline IPM learning in agricultural learning collaboratives.

Editor: Lee A. Newsom, The Pennsylvania State University, United States of America

Funding: This project was funded by the UC Davis University Outreach and International Programs (UOIP) and the McKnight Foundation Collaborative Crop Research Program (CCRP). The funders had no role in study design, data collection and analysis, decision to publish, or preparation of the manuscript. There are no current external funding sources for this study.

Competing Interests: The authors have declared that no competing interests exist. CIAT is not a private company, but an international non-profit research center, belonging to the CGIAR.

* E-mail: s.parsa@cgiar.org

Introduction

Modern advances in information science, statistics, and computing power are creating unprecedented opportunities to advance agricultural science. An underexploited opportunity exists to address research questions using data routinely generated by farmers and agricultural consultants through their record-keeping activities [1–5]. This research approach falls under the umbrella of ecoinformatics, an emerging field in ecology that is thought to hold particular promise for integrated pest management (IPM) research [5]. There is no uniform definition for "ecoinformatics;" but its scope covers the management, integration and analysis of diverse streams of data to answer complex questions in ecology. Here, we apply an ecoinformatics approach to the joint study of local and landscape factors explaining infestations of a key potato pest in the Andes, the Andean potato weevil (*Premnotrypes* spp.).

It is widely recognized that pest impact on an agricultural field is jointly influenced by the features of that field (i.e. local features) and by features of the area surrounding it (i.e. landscape features).

Seldom, however, are these local and landscape features studied jointly (but see [6–8]). Perhaps because they are easier to manipulate, local-level processes like host-plant resistance and chemical control are most often studied experimentally. In contrast, landscape-level processes, like the spillover of natural enemies from unmanaged areas into agricultural areas, are almost exclusively studied observationally [9]. The lack of an integrated approach results in a dearth of knowledge on the relative importance of local versus landscape-level influences and the relative payoffs of managing each.

The Andean potato weevils, a complex of tuber-boring herbivores dominated by the genus *Premnotrypes*, are the most important pests of potatoes in the high Andes [10]. They are native to the Andes, feed only on potatoes, and complete only one generation per year under traditional (rain-fed) potato agriculture [10]. Andean potato weevils appear to have reached pest status only in the past century, in response to the intensification of Andean farming systems [11]. Despite much searching, no weevil-resistant potato cultivars and only a few modestly-suppressive natural enemies (all of them generalists) have been found to date [12], [13]. An effective integrated pest management (IPM) program has been developed for these weevils [14], but adoption rates are poor due to its high labor requirements (Ortiz, personal communication).

Under rain-fed agriculture, the Andean potato weevil life cycle begins with the onset of rains (October–November), when adults emerge from their overwintering sites in soils of previous-season potato fields and potato storage facilities, and disperse by walking to find germinating potato plants [10], [15]. Most weevils disperse to find newly-planted potato fields. The remaining minority stays within previous-season fields to feed and reproduce on potato plants that re-emerge spontaneously (i.e. volunteer potatoes). A potato field's proximity to previous-season fields and storage facilities (overwintering sites) is positively correlated with infestation, while its proximity to potato fields planted during same season is negatively correlated with infestation [15]. When potatoes are reached, female adults lay eggs on the base of the plants. Upon hatching, neonate larvae dig into the soil and burrow into tubers where they feed until completing their larval life cycle. Towards the end of the potato-growing season (April–May), mature larvae begin to abandon tubers to pupate in the soil. Larvae that mature before harvest pupate in the soil of potato fields. The remaining larvae are transferred to and pupate in the dirt floors of potato storage facilities. Because volunteer potatoes are harvested and consumed by farmers early in the season (February–March), larvae within them seldom complete their life cycle (S. Parsa, personal observation).

Our objective was to jointly screen and compare the explanatory value of many local and landscape factors thought to influence Andean potato weevil populations. This task gave us an opportunity to systematically unearth and validate much knowledge on this pest reported only in the "gray" literature. We hoped this analysis would help us propose a shorter set of "priority" variables to be examined more closely by future studies seeking to refine the Andean potato weevil IPM program. We build upon a previous study that examined the explanatory importance of landscape factors [15]. This article builds on the previous analysis by evaluating the role of local factors, with data partly generated by farmers using record-keeping forms produced and distributed by our research team. We call our approach "facilitated ecoinformatics," because farmers were included in the generation of hypotheses and accompanied in field monitoring and record-keeping activities needed to test them.

Materials and Methods

Ethical considerations

No specific permits were required for the described field study. Participatory work involved only adults and it was non-experimental, anonymous and voluntary. Due to high illiteracy rates, informed consent was obtained verbally. The study was designed in consultations with faculty at the University of California in Davis and it was approved by the McKnight Foundation Collaborative Crop Research Program and by regional indigenous authorities of the Chopcca Nation. No participant-identifiable data was recorded. The principles expressed in the Declaration of Helsinki were followed. The individual pictured in this manuscript has given written informed consent (as outlined in the PLoS consent form) to appear in the published photo.

Study Site

The study was conducted from November 2008 to May 2009 in four adjacent farming villages in the department of Huancavelica, in Peru (74°45′W, 12°46′S). The villages belong to the Chopcca indigenous nation, characterized by a traditional, subsistence-based agriculture. Potato (*Solanum* spp.) provides the main means of subsistence in the area, but it is complemented by barley (*Hordeum vulgare*), oats (*Avena sativa*), fava beans (*Vicia faba*), pearl lupine (*Lupinus mutabilis*) and minor quantities of other Andean crops. Fields are cropped once per year, with a cycle that invariably begins with potato, typically-followed by three years of non-potato cropping, and three years of fallow. Agriculture is rain-fed, with a yearly growing season that spans from October to May. As is typical for other Andean farming systems, the area is mountainous (3,500 to 4,200 meters), cold (6–12°C mean temperature) and semi-humid (500–1,000 mm/year rainfall) [16]. Farmers recognize Andean potato weevils (Fig. 1) and potato flea beetles (*Epitrix* spp.) as their most important potato pests. The dominant weevil species in the area are *P. suturicallus* and *P. piercei*. Both have similar life cycles and behavior [10]; therefore our analyses do not distinguish between the two. The potato tuber moths *Symmestrischema tangolias* and *Phthorimaea operculella* do not reach economically-important levels in the area.

Sample of potato fields

The potato fields in this study belonged to 138 farmers randomly-selected from a roster of 643 total farmers (Fig. 2, Top). Only the cultivars Yungay (improved, *S. tuberosum*) and Larga (landrace, *S. chaucha*) were considered for the study, because they were the most abundant in the Chopcca nation and they are widely distributed in the Peruvian Andes. When a farmer had fields of both cultivars, one field was randomly-selected for the study. Data collection was coordinated by four local farmers who were extensively trained and had collaborated with us as research associates for more than a planting season. We refer to these facilitators as "community knowledge workers." Each knowledge worker was randomly assigned to assist 20 farmers and was asked to choose roughly 15 more, both from our list of randomly-selected

Figure 1. Heavy infestations by Andean potato weevils (*Premnotrypes* spp.) on improved potato cultivar Yungay (*S. tuberosum*). Photo credit: Soroush Parsa.

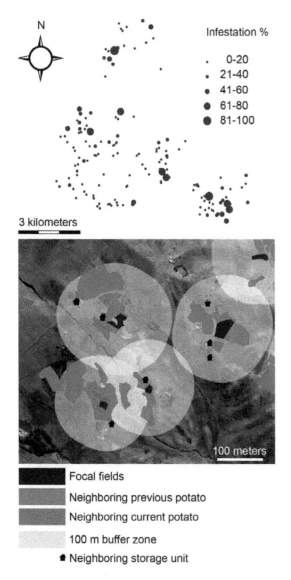

Figure 2. Landscape in the study area in Huancavelica, Peru. (Top) Distribution of study fields showing percentage infestation by Andean potato weevils (*Premnotrypes* spp.). (Bottom) Representative study fields showing landscape predictors of percentage infestations by Andean potato weevils.

farmers. This flexibility allowed us to improve farmer participation rates. No farmer approached refused to participate in our study.

Response variable

To account for edge effects on weevil distribution [15], fields were divided into edge and center sections. The outer 3 meters of the field was considered "edge," provided it was not adjacent to (1) a barrier putatively inhibiting weevil immigration (e.g., a wall, a stream or a ditch >1 m deep) or (2) to another potato field. The remaining area was considered "center." We measured edge and center areas using geographic positioning system (GPS) receivers (GPSMap 76CSx, Garmin Ltd., Olathe, KS) and geographic information system (GIS) software (ArcGIS, ESRI, Redlands, CA). We then sampled 20 evenly-distributed plants from each stratum (i.e., 40 per field) and inspected tubers to score for presence/absence of larvae, external bruising or emergence holes that are distinctly characteristic of weevil damage. Our response variable

was the percent of tubers infested in the edge and center of fields, weight-averaged by their respective areas. Tuber infestation was scored within 10 days of the field's intended harvest date.

Explanatory variables

We generated a comprehensive list of explanatory variables believed to influence Andean potato weevil infestations (Table 1). Most variables were selected based on (1) a review of the "gray" literature on Andean potato weevils; (2) several unstructured interviews with local farmers; and (3) a workshop of experts convened by our research team that brought together agricultural scientists and NGO practitioners working with Andean agroecosystems for more than two decades.

Data on explanatory variables were collected combining farmer record-keeping and direct field measurements. We designed and distributed standard record-keeping forms for key management activities (e.g., pesticide applications, fertilization, weedings) to each participating farmer. To facilitate compliance with our record-keeping requirements, knowledge workers visited each farmer at least twice during the growing season, assisting them as needed to accurately fill out the forms (Fig. 3). The remaining data were gathered directly by our knowledge workers, often following their visits to farmers. Landscape and geographic data were gathered with GPS receivers and GIS software. We mapped three features within 100 m of each focal field: current potato fields, previous-season potato fields and storage units (Fig. 2, Bottom). Potato fields were expressed as the percentage of the surveyed area they occupied, whereas storage units were expressed as counts. We considered the 100-m scale adequate, because weevils only disperse by walking, they have no effective natural enemies that disperse long distances, and the area considered often contained one or more natural streams preventing their dispersal. Accordingly, landscape features separated from focal fields by streams were omitted (see [15] for details of landscape analysis). Soil data were obtained through laboratory analyses. For each field, roughly 100 grams of soil was collected from the base of each plant evaluated for weevil infestations, the samples mixed, and a consolidated sub-sample submitted to the laboratory (Laboratorios Analíticos del Sur; Arequipa, Peru).

Statistical modeling

Analyses were conducted using JMP 7 statistical software [23]. Percent tubers infested was square-root transformed to meet assumptions of parametric statistics. We conducted two separate analyses. First, we conducted a multiple regression analysis to assess the efficacy of pesticide treatments over time. For this analysis, we regressed infestations against the number of treatments applied each month from December until March.

Second, following the information theoretic approach [24], we set out to identify the subset of explanatory variables that produced the best multiple regression model of weevil infestations. The information theoretic approach is thought to be less vulnerable to finding spurious effects due to over-fitting, and more informative than methods based on null hypothesis testing [25]. Multiple models, representing alternative explanations for the patterns observed, are compared to evaluate which one is best supported by the data. When many models are similarly-supported by the data, inferences can be made using all models considered important. In such cases, parameter estimates are averaged across models, weighted by the probability that their originating model is the best in the set. Hence, the relative importance of a parameter is not based on p-values, but rather on its occurrence in the model(s) that are best supported by the data.

Table 1. Local and landscape features of hypothesized to exert important influences on Andean potato weevil infestations (*Premnotrypes* spp.).

Variable category	Variable description	Mean ± SD or mode
Chemical control	Carbofuran at planting (yes/no)	no = 106/138
	Insecticide treatments December	0.51±0.61
	Insecticide treatments January	1.12±0.74
	Insecticide treatments February	0.34±0.52
	Insecticide treatments March	0.02±0.15
Cultural	Ash application at planting[a] (yes/no)	no = 134/138
	Chemical fertilization at planting[b] (g/plant)	8.13±7.03
	Day of first potato hilling[c] (days after Oct 1st)	104.10±8.57
	Harvest day[d] (days after Oct 1st)	201.34±9.50
	Height of first potato hilling[c] (cm)	18.30±3.79
	Height of row at harvest[c] (cm)	29.02±5.17
	Manure fertilization at planting[b] (g/plant)	75.87±21.80
	Number of hillings[c]	2 = 131/138
	Perimeter/area ratio[e] (m^{-1})	0.26±0.10
	Planting day[f] (days after Oct 1st)	48.96±8.39
	Plants/5 meters row[g]	14.81±1.74
	Rotation 2006[h] (potato/other)	other = 114/138
	Rotation 2007[h] (potato/other)	other = 136/138
	Row distance[g] (cm)	95.35±11.72
	Weed removal[i] (yes/no)	yes = 75/138
Geographic	Elevation[j] (m)	3747±148.40
	Field slope[k] (cm)	23.04±10.79
Host related	Potato cultivar[l] (Yungay/Larga)	Yungay = 87/138
Soil	Clay[m] (%)	21.08±5.71
	Loam[m] (%)	34.01±8.01
	Sand[m] (%)	44.91±9.62
	K[m] (ppm)	331.58±201.75
	P[m] (ppm)	28.22±21.94
	Organic matter[m] (%)	5.87±2.38
	pH[m]	4.97±0.98
Landscape	Neighboring current potato (%)[n]	8.79±9.71
	Neighboring previous potato (%)[o]	4.08±4.32
	Neighboring storage units[p]	1.01±1.01

[a]Farmers apply a layer of ash directly below the potato seed at the time of planting; this practice is intended to kill potato weevils.
[b]Fertilization can influence crop defenses against herbivores [17].
[c]Many agronomists recommend hilling the plants (piling dirt up around the stem of the plant) higher to lengthen the distance weevil larvae must travel to find tubers.
[d]Early harvest shortens the exposure of tubers to neonate larvae [18].
[e]Larger fields have lower perimeter to area ratios and have been suggested to have lower infestations [19].
[f]Early emerging plants may experience greater infestations [20].
[g]Planting density may influence the abundance of many insect pests [21].
[h]Planting potatoes following a potato planting should lead to very high infestations [19], but implementing a single host free period should eliminate this risk. Rotation 2007 indicates if potatoes were sown in the field the previous season while Rotation 2006 indicates if potatoes were sown there two seasons before the study.
[i]The study hypothesized that weeds may serve as refuges for adult weevils before potato plants emerge.
[j]Weevils are poorly adapted to elevations above 3,700 meters [18].
[k]The study hypothesized that greater soil erosion in steeper slopes may increase tuber exposure to weevils.
[l]Modern cultivars like Yungay may be more susceptible to insect pests [22].
[m]The study was interested in exploring any soil influences on weevil infestations without any strong *a priori* expectations.
[n]A measure of potato fields sown the within 100 m of the focal potato field; these current fields dilute the effect of immigrating weevils [15].
[o]A measure of potato fields harvested the previous season that lie within 100 m of the focal potato field; these previous fields may be sources of overwintering weevils that immigrate into focal fields [14], [15].
[p]Potato storage units are facilities adjacent to farmer houses and are known to concentrate high densities of overwintering weevils that may immigrate into focal fields [14–15].

Figure 3. Community knowledge worker assisting farmers with record-keeping activities associated with their potato harvest. The individual pictured in this manuscript has given written informed consent (as outlined in the PLoS consent form) to appear in the published photo. Photo credit: Soroush Parsa.

To be conservative in our initial selection of variables, we first developed the least parsimonious model that was best supported by the data, as assessed by Akaike's information criterion (AIC) [24]. Variables were added to this initial model one at the time in the order dictated by their influence on the AIC (variables that reduced the AIC the most were entered first). We continued adding variables even when their addition penalized (or increased) the AIC, as long as the resulting model had an AIC value no more than 2.0 larger than the lowest AIC values reached in previous steps. We chose this threshold because models within 2.0 AIC values are thought to be similarly supported by the data [24]. Our analysis included four categorical variables to control statistically for possible observer effects associated with our four knowledge workers. Interaction terms were omitted because including them would have made the analyses too computationally intensive, and we lacked specific hypotheses linking them to weevil infestations.

We then evaluated models resulting from all possible additive combinations of the selected variables. This process identified 51 models that were within 2 AIC values of the best, indicating substantial model selection uncertainty [24]. Under these circumstances, the information theoretic approach advocates model averaging. Accordingly, three steps were followed to average estimates across the 51 models. First, we computed each model's Akaike weight (w_i), which is interpreted as the relative probability that a given model is the best in the set. Second, we computed the weight for each parameter (w_p), which is interpreted as the probability that the parameter is included in the best model in the set, and is obtained by summing Akaike weights across all models where the parameter occurs. Hence, the parameter weight is a measure of the relative importance of a parameter. Finally, we weight-averaged parameter estimates. To do so, we multiplied the estimates of each model where a parameter occurred by the corresponding Akaike weights for the model; the resulting products were summed; and the sum divided by the parameter weight. This computation yields a "natural" average of the parameter estimate, because it considers only those models where the parameter occurs. However, multi-model predictions also need

to take into account the evidence from models where the parameter does not occur. Accordingly, a second average for the parameter estimate was derived for predictions, by multiplying the "natural" average by the corresponding parameter weight.

To assess the importance of jointly modeling local and landscape variables, we followed the procedure above to derive two additional models: (1) the best model with only local predictors and (2) the best model with only landscape predictors. Then, we computed Akaike weights for the three to estimate their relative support.

We tested for spatial autocorrelation in model residuals with the ncf package [26] for R 2.9.1 (www.r-project.org) using the correlog() function to assess autocorrelation via Moran's I index [27]. We tested for autocorrelation with a 1,000 permutation test for fields up to 3 kilometers apart at intervals of 250 meters. No evidence for spatial autocorrelation of residuals was detected. Unless otherwise stated, mean values are presented with their standard deviation.

Results

Observations and descriptive statistics

For each potato field, we evaluated an average of 654 ± 241 tubers. Infestations averaged $25.1 \pm 20.9\%$ on field "edges" and $16.1 \pm 17.4\%$ on field "centers," yielding a weight-averaged infestation of $18.3 \pm 18.1\%$. Fields were small, averaging 424.6 ± 282.4 m^2. Farmers who applied carbofuran at planting did so at roughly 25 kg of active ingredient per hectare. After plant emergence insecticide treatments varied little, generally consisting of applications of methamidophos with a manual backpack sprayer at roughly 534 ml (320 g) of active ingredient per hectare. Summary statistics for the explanatory variables are presented in Table 1.

Insecticide efficacy model

The insecticide treatment model revealed a temporal decay in the efficacy of treatments, with treatments made after January having no significant effect on potato weevil infestation (Fig. 4)

Development of global (least parsimonious) model

The forward stepwise development of the global model, using AIC as the criterion for variable selection, is shown in Figure 5. Because the value of the AIC has no direct interpretation (i.e. it is a

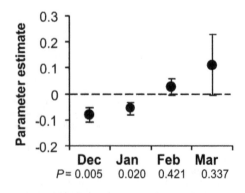

Figure 4. Temporal decay in the efficacy of insecticide treatments against Andean potato weevils (*Premnotrypes* spp.), as applied by farmers. The x-axis shows the parameter estimate ± SEM associated with the effect of a single insecticide application on the proportion of tubers infested with weevils (sqrt-transformed). The y-axis shows the month of the insecticide treatment.

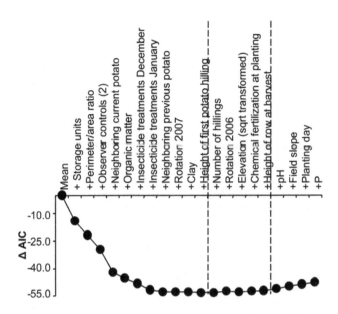

Figure 5. Forward stepwise development of the global (least parsimonious) statistical model explaining Andean potato weevil infestations. The x-axis shows the progressive addition of explanatory variables in order of their contributions to lowering AIC. The y-axis shows cumulative reductions in AIC from the AIC associated with using only the mean to estimate infestations. The first dashed line shows the point where the addition of variables started to penalize the AIC, whereas the second dashed line shows the point where this penalty started to exceed two AIC values. The global model included all variables before the second dashed line.

Table 2. Parameter estimates weight-averaged across the 51 "best" models predicting Andean potato weevil infestations (square root of proportion infested tubers).

Variable	Estimate	SEM	95% CI Lower	95% CI Upper	ω_p
Intercept	1.774	0.936	−0.061	3.608	1
Observer effect 1	−0.043	0.019	−0.081	−0.005	1
Observer effect 2	−0.070	0.016	−0.102	−0.038	1
Perimeter/area ratio	0.519	0.150	0.225	0.814	1
Neighboring storage units	0.044	0.015	0.015	0.073	1
Neighboring current potato	−0.004	0.002	−0.007	−0.001	1
Insecticide treatments January	−0.053	0.021	−0.094	−0.013	1
Insecticide treatments December	−0.054	0.025	−0.103	−0.004	1
Neighboring previous potato	0.007	0.004	0.000	0.014	0.91
Soil clay	0.005	0.003	0.000	0.010	0.91
Elevation (sqrt transformed)	−0.031	0.013	−0.057	−0.005	0.83
Number of hillings	0.127	0.076	−0.022	0.275	0.71
Rotation 2007 [Not potato]	−0.096	0.062	−0.216	0.025	0.71
Rotation 2006 [Not potato]	0.030	0.020	−0.008	0.069	0.49
Height of first potato hilling	−0.006	0.004	−0.014	0.002	0.43
Height of row at harvest	−0.005	0.003	−0.012	0.001	0.42
Chemical fertilization at planting	−0.002	0.002	−0.006	0.002	0.23
Soil organic matter	−0.012	0.008	−0.026	0.003	0.21

Notes: Estimates are followed by their standard errors, their 95% confidence intervals and their Akaike parameter weights (ω_p). Given a set of similarly-adequate predictive models, parameter weights estimate the probability that the parameter is included in the best model in the set.

comparative measure of model adequacy), our "starting point" was the AIC associated with using the mean to predict infestations, and we evaluated changes in this AIC as we added each explanatory variable. The best model derived from this method included 13 variables (2 of them control variables) and reduced the AIC by 53.3. From this point we added four more explanatory variables, which collectively penalized the best AIC by less than 2.0. Hence, the resulting global model included 15 possibly-important explanatory variables and 2 control variables (17 total variables) that collectively reduced the AIC by 52.0. We considered the remaining variables "unimportant" given the dataset.

Model averaging

The combinations of 17 variables generated 131,071 total models. The top 51 models (i.e. within 2.0 AIC from the model with the lowest AIC) explained 40.2–46.4% of the variance in our dataset. Within this subset of well-supported models, none had a high probability of being the single "best" ($0.036 > w_i > 0.013$); thus, all 51 models were similarly well-supported by the data. Results from averaging estimates across all 51 models are presented in table 2. All 51 models included three landscape variables (i.e., perimeter to area ratio, current potato fields, storage units), two local variables (i.e., insecticides December and January), and the controls for observer effects, as evident by their parameter weights of 1. The least commonly-included variables were chemical fertilizer and soil organic matter %, whose parameter weights were 0.23 and 0.21 respectively. The relative impact of each explanatory variable on weevil infestations is projected in Figure 6.

Relative support for local, landscape, and joint models

The best models considering the influence of local and landscape factors in isolation had no probability of outperforming the model considering them jointly (Table 3).

Discussion

Our objective was to identify a key set of variables explaining Andean potato weevil infestations in farmers' fields. We started by generating a comprehensive list of variables, including both local and landscape factors deemed important by farmers and researchers. The explanatory importance of these variables was screened statistically using an information theoretic approach, affording a simultaneous evaluation and contrast of local and landscape factors explaining weevil infestations.

Our results support the importance of studying local and landscape processes jointly. The best models considering either landscape or local factors in isolation had no probability of outperforming the best joint model. Although still rare, the number of pest management studies considering local and landscape factors jointly is increasing (e.g., [6–8]). To our knowledge, however, our study is the first to use crop management records as a source of local-level data.

Our findings confirm the suspected influence of some factors, fail to support the suspected influence of others, and also reveal altogether unsuspected patterns that deserve further investigation. At the local level, our results support (1) the efficacy of foliar

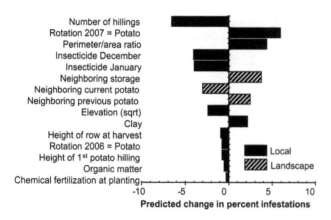

Figure 6. Standardized predicted impacts of explanatory variables on Andean potato infestations. The model is initially set to predict infestations for a field with no pesticide applications and with mean (for continuous variables) or most common (for ordinal and categorical variables) values for all other explanatory variables. For continuous explanatory variables, bars reflect predicted changes in infestations in response to a one standard deviation increase in the explanatory variable. For ordinal explanatory variables, the bars reflect predicted changes in infestations in response to a single unit increase in the explanatory variable; except for the number of hillings, for which only a decrease could maintain predictions within observed bounds. To obtain multi-model predictions, parameter estimates were multiplied by their corresponding parameter weights. Hence, predicted effects are "attenuated" for explanatory variables with parameter weights smaller than 1.

insecticide treatments before - but not after - February, (2) the efficacy of implementing a single potato-free rotation period, and the trends of decreasing infestations with increasing (3) elevation, (4) field perimeter-to-area ratio and (5) hilling height. At the landscape level, our results support the suspected influence of (6)

storage units and (7) previous-season potato fields as sources of weevil infestation. Given what is already known about Andean potato weevils, confirming these patterns is unsurprising, but it provides confidence in our results. More importantly, however, estimating these factors simultaneously for the first time affords a contrast of their predicted influences on weevil infestations. For example, multi-model estimates predict that, on average, our farmers may offset the risk of infestation from a neighboring storage with a single pesticide treatment before February (Fig. 6).

Unsupported factors included manure fertilization, planting day, planting density, row distance, weeding, slope, potato cultivar, and many soil factors. Most of these variables had been included in our study without strong *a priori* expectations. Lack of support for factors of reported importance, including several tactics farmers target at weevils, is harder to interpret conclusively. For example, our study failed to support the efficacy of carbofuran and ash treatments at planting, as well as the efficacy of foliar insecticide treatments after January. We suspect these practices are in fact ineffective. Previous observational evidence suggested that the systemic effect of carbofuran treatments at planting is lost too early in the season to be effective against the progressively-immigrating Andean potato weevils [28]. By contrast, because most weevils colonize fields before February [29], treatments in March or April should not be expected to be efficacious. A field experiment one of us conducted to test the validity of ash treatments also failed to demonstrate impacts on weevil infestations (R. Ccanto, unpublished data). These observations point to the need to address local gaps in Andean potato weevil knowledge.

Despite being one of the most widely-used preventive tactics against Andean potato weevils, we found no link between early harvest and infestations. Early harvest is thought to prevent infestations by late-hatching weevils and to intercept the development of those already within tubers before significant damage is incurred. In a post-hoc analysis, we considered the possibility that farmers with greater expected infestations were harvesting earlier.

Table 3. Parameter estimates ± SEM for the best models predicting Andean weevil infestations based on local factors only (i.e. Best local), landscape factors only (i.e. Best landscape) or both local and landscape factors together (i.e. Best combined).

	Best combined	Best local	Best landscape
ΔAIC relative to best combined	0	+16.03	+18.18
w_i	1.00	0.00	0.00
Perimeter/area ratio	0.501±0.149**	0.548±0.158**	
Neighboring storage units	0.040±0.014**		0.048±0.015**
Neighboring current potato	−0.004±0.002**		−0.006±0.002**
Insecticide treatments January	−0.054±0.020**	−0.066±0.021**	
Insecticide treatments December	−0.051±0.025*	−0.066±0.026**	
Neighboring previous potato	0.008±0.004*		
Clay	0.006±0.003*	0.006±0.003*	
Elevation (sqrt transformed)	−0.028±0.013*		
Number of hillings	0.107±0.066	0.126±0.068	
Rotation 2007 [Not potato]	−0.083±0.060	−0.134±0.063*	
Rotation 2006 [Not potato]	0.032±0.019		
Height of first potato hilling	−0.007±0.004	−0.007±0.004	
Organic matter		−0.012±0.006	

Notes: Lower AIC values suggest better model performance. The Akaike weight, w_i, is interpreted as the relative probability that a given model is the best in the set. The models included control variables for observer effects (not presented in the table).
*P≤0.05.
**P≤0.01.

We thought this was possible, because farmers were observed to sample some of their potatoes several days before harvest to assess their maturity, and presumably also infestation levels. This could allow farmers to adjust their harvest date based on the level of infestation that they observed in their early samples of tubers. Indeed, we found that farmers with greater expected infestations, as estimated by our model, harvested earlier ([harvest date] = 207.4+−15.97*[expected infestations]; d.f. = 136, $P=0.01$).

The previous analysis highlights the special care that must be taken when interpreting observational studies in pest management. Many management tactics are best used adaptively, i.e., in response to pest population densities observed during routine sampling. For example, if pest populations are adequately monitored and pesticides applied when needed, one might expect to find a positive association between the pest densities and the number of pesticide treatments. This adaptive behavior can eliminate what otherwise might be a negative association between the use of an effective pesticide and pest densities. We would not want to conclude from such a correlation that pesticide applications were ineffective. This particular problem did not apply to insecticide treatments in our study, because farmers were not applying pesticides adaptively (adult weevils are not monitored). However, we failed to see a correlation between infestations and harvest date presumably because farmers were adjusting their harvest date based on monitoring tuber infestations. Accordingly, observational studies in pest management demand cautious interpretation, and they are most powerful when coupled with manipulative experiments conclusively-testing variables of interest [5].

This study revealed two unsuspected results we believe deserve empirical attention. At the landscape level, we demonstrated that infestations are negatively correlated with the abundance of neighboring potato fields. The implications of this finding have been thoroughly discussed in a previous article [15]. At the local level, our results suggest that manipulating the hilling of potato plants may contribute to improving weevil management. There was a 0.84 probability that at least one variable relevant to hilling was included in the best model predicting weevil infestations (Table 2), and the effect size of the number of hillings was one of the largest (Table 2; Fig. 6). Our observations cannot elucidate exact mechanisms, but some possibilities may be explored with manipulative experiments. First, higher hilling could lengthen the distance neonate larvae need to migrate to find tubers, potentially decreasing infestations by physical isolation or by increasing exposure to mortality factors such as soil-dwelling natural enemies. Second, the height of the hilling also determines the depth of the furrow, which could inhibit the immigration of the (flightless) adult weevils, especially when heavy rains fill furrows with water. Later in the season, hilling could have a negative effect, because it could protect weevil eggs from mortality factors such as desiccation or

consumption by generalist predators. This could explain why farmers who only hilled once (always early in the season) experienced lower infestations than farmers who hilled a second time (Fig. 6). If this is the case, a management tactic that targets eggs or neonate larvae before the second hilling, for example an application of entomopathogenic nematodes [30], could produce good results.

Ecoinformatics approaches may be particularly helpful during the exploratory phase of pest management research, when a key goal is to screen a large number of potentially important variables. We describe our approach as "facilitated ecoinformatics," because farmers were included in the generation of hypotheses and accompanied in the field monitoring and record-keeping activities needed to test them. This methodological novelty adds to the literature advocating the use of local knowledge in ecological research [31]. A similar approach to research and extension, based on structured field monitoring, record-keeping, and benchmarking, has been implemented successfully to enhance productivity in facilitated learning collaboratives (e.g., [32], [33]). Based on our experience, we suspect ecoinformatics approaches can synergize these programs to streamline learning and development in agriculture.

A greater reliance on ecoinformatics in agriculture is well justified by its unique ability to generate large datasets that capture the true spatial and temporal scale of commercial agriculture [5]. As illustrated above, however, ecoinformatics datasets are observational, and therefore are poorly suited to elicit definitive causal inferences. In addition, ecoinformatics datasets may be particularly subject to observer error, potentially introduced by farmer participation in data collection. Important advances in statistical science are improving our ability to deal with these challenges [5]. We believe, however, that ecoinformatics approaches will prove most useful when used as a tool that complements traditional experimentation, rather than in competition with experimentation, to facilitate advances in agricultural science.

Acknowledgments

We thank Jackely Bejarano and the Yanapai research team for their technical field support. Mark Grote and Neil Willits kindly helped with statistical analyses. Richard Karban, Frank Zalom, Yao Hua Law, Moran Segoli, Kris Wyckhuys and four anonymous reviewers provided critical comments that substantially improved the manuscript.

Author Contributions

Conceived and designed the experiments: SP JAR RC JA MS. Performed the experiments: SP RC EO. Analyzed the data: SP. Contributed reagents/materials/analysis tools: SP JAR. Wrote the paper: SP JAR.

References

1. Rochester W, Zalucki M, Ward A, Miles M, Murray D (2002) Testing insect movement theory: empirical analysis of pest data routinely collected from agricultural crops. Comput Electron Agr 35: 139–149.

2. Zwald N, Weigel K, Chang Y, Welper R, Clay J (2006) Genetic analysis of clinical mastitis data from on-farm management software using threshold models. J Dairy Sci 89: 330–336.

3. Lees F, Baillie M, Gettinby F, Revie CW (2008) The efficacy of emamectin benzoate against infestations of Lepeophtheirus salmonis on farmed Atlantic salmon (Salmo salar L) in Scotland, 2002–2006. PLoS ONE 3(2): e1549.

4. Allen K, Luttrell R (2009) Spatial and temporal distribution of heliothines and tarnished plant bugs across the landscape of an Arkansas farm. Crop Prot 28: 722–727.

5. Rosenheim JA, Parsa S, Forbes AA, Krimmel WA, Law YH, et al. (2011) Ecoinformatics for Integrated Pest Management: Expanding the Applied Insect Ecologist's Tool-Kit. J Econ Entomol 104: 331–342.

6. Östman Ö, Ekbom B, Bengtsson J (2001) Landscape heterogeneity and farming practice influence biological control. Basic Appl Ecol 2: 365–371.

7. Prasifka JR, Heinz KM, Minzenmayer RR (2004) Relationships of landscape, prey and agronomic variables to the abundance of generalist predators in cotton (Gossypium hirsutum) fields. Landscape Ecol 19: 709–717.

8. Zaller JG, Moser D, Drapela T, Schmöger C, Frank T (2008) Effect of within-field and landscape factors on insect damage in winter oilseed rape. Agric Ecosyst Environ 123: 233–238.

9. Bianchi F, Booij C, Tscharntke T (2006) Sustainable pest regulation in agricultural landscapes: a review on landscape composition, biodiversity and natural pest control. Proc R Soc Lond B 273: 1715–1727.

10. Alcazar J, Cisneros F (1999) Taxonomy and bionomics of the Andean potato weevil complex: Premnotrypes spp. and related genera. In: Arthur C, Ferguson P, Smith B, eds. Impact on a Changing World, Program Report 1997–98. Lima, Peru: International Potato Center. pp 141–151.

11. Parsa S (2010) Native Herbivore Becomes Key Pest After Dismantlement of a Traditional Farming System. Am Entomol 56: 242–251.

12. Alcazar J, Cisneros F (1997) Integrated Management for Andean Potato Weevils in Pilot Units. In: Hardy B, Smith B, eds. Program Report 1995–96. Lima, Peru: International Potato Center, Lima, Peru. pp 169–176.

13. Kaya HK, Alcazar J, Parsa S, Kroschel J (2009) Microbial control of the Andean potato weevil complex. Fruit Veg Cereal Sci Biotech 3(1): 39–45.

14. Ortiz O, Alcázar J, Catalán W, Villano W, Cerna V, et al. (1996) Economic impact of IPM practices on the Andean potato weevil in Peru. In: T. Walker, C. Crissman, eds. Case studies of the economic impact of CIP-related technology. Lima, Peru: International Potato Center. pp 95–110.

15. Parsa S, Ccanto R, Rosenheim JA (2011) Resource concentration dilutes a key pest in indigenous potato agriculture. Ecol Appl 21: 539–546.

16. Tosi J (1960) Zonas de vida natural en el Peru. Peru. Lima, Peru: Instituto Interamericano de Ciencias Agricolas. 271 p.

17. Throop HL, Lerdau MT (2004) Effects of nitrogen deposition on insect herbivory: implications for community and ecosystem processes. Ecosystems 7: 109–133.

18. Kühne M (2007) The Andean potato weevil *Premnotrypes suturicallus*: ecology and interactions with the entomopathogenic fungus *Beauveria bassiana* [Ph.D.]. Göttingen, Germany: Georg-August- Universitätsbibliothek Göttingen. 178 p.

19. Rios A, Kroschel J () Evaluation and implications of Andean potato weevil infestation sources for its management in the Andean region. J Appl Entomol, In press.

20. Calderon R, Franco J, Barea O, Crespo L, Estrella R (2004) Desarrollo de componentes del manejo integrado del gorgojo de los Andes en el cultivo de la papa en Bolivia. Cochabamba, Bolivia: Fundacion PROINPA. 144 p.

21. Speight M (1983) The potential of ecosystem management for pest control. Agric Ecosyst Environ 10: 183–199.

22. Harris MK (1980) Arthropod-plant interactions related to agriculture, emphasizing host plant resistance. In: Harris MK, ed. Biology and breeding for resistance to arthropods and pathogens in agricultural plants. Texas: Texas A&M University, College Station. pp 23–51.

23. SAS Institute (2007) JMP user guide, release 7. SAS Institute, Cary, North Carolina.

24. Burnham K, Anderson D (2002) Model selection and multimodel inference: a practical information-theoretic approach. New York: Springer. 514 p.

25. Anderson D, Burnham K, Thompson W (2000) Null hypothesis testing: problems, prevalence, and an alternative. J Wildl Manage 64: 912–923.

26. Bjornstad O (2004) NCF: a package for analyzing spatial (cross-) covariance. R package version 1.1-3. Available: http://cran.r-project.org/web/packages/ncf/index.html. Accessed 2009 Sep 18.

27. Legendre P, Fortin M (1989) Spatial pattern and ecological analysis. Vegetation 80: 107–138.

28. Ewell PT, Fano H, Raman K, Alcázar J, Palacios M, et al. (1990) Farmer management of potato insect pests in Peru: Report of an interdisciplinary research project in selected regions of the highlands and coast. Lima, Peru: International Potato Center (CIP). 77 p.

29. Kroschel J, Alcazar J, Poma P (2009) Potential of plastic barriers to control Andean potato weevil *Premnotrypes suturicallus* Kuschel. Crop Prot 28: 466–476.

30. Parsa S, Alcázar J, Salazar J, Kaya HK (2006) An indigenous Peruvian entomopathogenic nematode for suppression of the Andean potato weevil. Biol Control 39: 171–178.

31. Huntington HP (2000) Using traditional ecological knowledge in science: methods and applications. Ecol Appl 10: 1270–1274.

32. Lacy J (2011) Cropcheck: Farmer benchmarking participatory model to improve productivity. Agric Syst 104: 562–571.

33. Jiménez D, Cock J, Satizábal HF, Barreto S, Miguel A, et al. (2009) Analysis of Andean blackberry (*Rubus glaucus*) production models obtained by means of artificial neural networks exploiting information collected by small scale growers in Colombia and publicly available meteorological data. Comput Electron Agr 69: 198–208.

5

Improving the Degree-Day Model for Forecasting *Locusta migratoria manilensis* (Meyen) (Orthoptera: Acridoidea)

Xiongbing Tu[1,2], Zhihong Li[2], Jie Wang[1], Xunbing Huang[1], Jiwen Yang[3], Chunbin Fan[3], Huihui Wu[1], Qinglei Wang[4], Zehua Zhang[1]*

1 State Key Laboratory for Biology of Plant Diseases and Insect Pests, Institute of Plant Protection, Chinese Academy of Agricultural Sciences, Beijing, P. R. China, **2** Department of Entomology, College of Agronomy and Biotechnology, China Agricultural University, Beijing, P.R. China, **3** Tianjin Binhai New Area of Dagang Agricultural Service Center, Tianjin, P.R. China, **4** Cangzhou Academy of Agricultural and Forestry Sciences of Hebei, Cangzhou, P.R. China

Abstract

The degree-day (DD) model is an important tool for forecasting pest phenology and voltinism. Unfortunately, the DD model is inaccurate, as is the case for the Oriental migratory locust. To improve the existing DD model for this pest, we first studied locust development in seven growth chambers, each of which simulated the complete growing-season climate of a specific region in China (Baiquan, Chengde, Tumotezuoqi, Wenan, Rongan, Qiongzhong, or Qiongshan). In these seven treatments, locusts completed 0.95, 1, 1.1, 2.2, 2.95, 3.95, and 4.95 generations, respectively. Hence, in the Baiquan (700), Rongan (2400), Qiongzhong (3200), and Qiongshan (2400) treatments, the final generation were unable to lay eggs. In a second experiment, we reared locusts for a full generation in growth chambers, at different constant temperatures. This experiment provided two important findings. First, temperatures between 32 and 42°C did not influence locust development rate. Hence, the additional heat provided by temperatures above 32°C did not add to the total heat units acquired by the insects, according to the traditional DD model. Instead, temperatures above 32°C represent overflow heat, and can not be included when calculating total heat acquired during development. We also noted that females raised at constant 21°C failed to oviposit. Hence, temperatures lower than 21°C should be deducted when calculating total heat acquired during adult development. Using our experimental findings, we next micmiked 24-h temperature curve and constructed a new DD model based on a 24-h temperature integral calculation. We then compared our new model with the traditional DD model, results showed the DD deviation was 166 heat units in Langfang during 2011. At last we recalculated the heat by our new DD model, which better predicted the results from our first growth chamber experiment.

Editor: Daniel Doucet, Natural Resources Canada, Canada

Funding: This article was supported by the earmarked fund for China Agriculture Research System (CARS-35-07) and the Special Fund for Agro-scientific Research in the Public Interest, 201003079. The funders had no role in study design, data collection and analysis, decision to publish, or preparation of the manuscript.

Competing Interests: The authors have declared that no competing interests exist.

* E-mail: lgbcc@263.net

Introduction

Many mathematical models utilizing heat-summation are widely used in Integrated Pest Management (IPM) to forecast and predict pest insect phenology and voltinism [1]. Over the years, numerous authors have worked to improve the accuracy of such models. Simpson [2] proposed the accumulated temperature constant relationship and the inverse symmetry curve. Ludwig [3] illustrated that varying and constant temperatures influenced *Popillia japonica* development differently. Davidson [4] used a logistic curve to illustrate the relationship between growth rate and temperature, and Pradhan [5] proposed a formula index. Yang *et al.* [6] developed a weighted calculation method for variable temperatures and the natural accumulated temperature. Arnold [7] proposed a sine-curve model based on maximum and minimum temperature to estimate heat units. Schoolfield and coworkers [8] modified the Sharpe-Michele model. Wagner *et al.* [9] presented easy instructions for the use of the Sharpe-Michele model and designed a computer program to determine the correct

number of parameters to be used in the model for a given data set. Recently, de Jong and van der Have [10] used the Sharpe-Michele model to assess the temperature dependence of development rate, growth rate, and size from biophysics to adaptation. All of these authors made important contributions to degree-days calculation.

Degree-day (DD) model is widely used in theoretical and basic science [11–14]. For example, to understand development, life history, ecology, species adaptations and biogeography, phenotypic plasticity, and physiological evolution, their widest use is applied; i.e., forecasting pest and crop phenology [10,15]. The ability of the DD model to accurately predict pest occurrence ranges from very good to poor, depending on the specific pest and model used [16–18]. Hence, there is continuous effort to improve the model, which fails for any number of reasons. For example, how one calculates DD can strongly influence the accuracy of results [19–21], e.g., using the highest and lowest temperatures rather than the average temperature when calculating the DD. Likewise, there are numerous other factors that can alter or mitigate the influence of ambient temperature on development, including photoperiod, population density, pathogens, predators, competition, nutrition, moisture, thermoregulation, acclimation, etc [22–29]. In addition, the traditional DD model fails when the relationship between temperature and development rate is not linear over the viable range. For example, in *Chrysopasinica*, development rate increases non-linearly at temperatures between 30 and 32°C [30]. Thus, the upper temperature limit during insect development is important, while temperature trends during the growth season can be simulated by using the Monte Carlo method and then used as input for generating degree-day model [9,31–32].

There are lots of models including Briere model, Lactin model, Logan model, Taylor model, etc., developed to study temperature dependence of development rate. Among the various models, each has advantages and disadvantages [32–33]. For example, they can describe development rate vary trends at different temperatures. However, in the Briere model and Lactin model, the initial values of the parameter are not set based on a reasonable explanation, while in the Logan model and Taylor model, they were unable to estimate the lower development threshold temperature [34]. Thus, with a nonlinear solution, this method obtains an approximate, rather than an exact solution [17–18,35]. Usually the predictions from non-linear models are compared and validated with experimental data, as in the case of the experimental derived development upper cut off. We consider that these models will generate different outcomes when they are used to calculate the upper development threshold temperature and DD for an insect, mainly because the independent variables are set inaccurately [17–18,36–37].

In this paper, we develop a modified DD model for the Oriental Migratory Locust, *Locusta migratoria manilensis* Meyen. This pest is widespread throughout Asia, Africa, Europe, Australia, and New Zealand, where it causes severe damage to cereal crops [38–43]. Occurrence and distribution records of *L. m. manilensis* can be traced back for 3000 years in China, where locust plagues are the three main natural disasters, along with floods and droughts [44–45]. IPM is the main strategy used to control the locust population [46], and DD-forecasting is an important component of this strategy [14].Unfortunately, current locust DD model is inaccurate, with typical discrepancies of 10 d or more between predicted vs. actual field phenology [47].For this reason, we investigated development in the Oriental migratory locust, with the goal to improve the DD model for this insect. We first reared locusts in environmental chambers which simulated the climate in various regions in China. We also reared locusts at various constant temperatures and carefully monitored there development. Based on these laboratory results, we calculated the DD for migratory locust development using the integral calculation method, and developed a new model for predicting locust development. When tested against locust voltinism, the new model was more accurate than the old model.

Materials and Methods

Study organism

We studied the oriental migratory locust, *Locusta migrotoria manilensis* Meyen. Eggs were collected in November (Autumn locust) 2008 from fields in near Cangzhou City (N38°30′33.46″, E117°25′32.85″), Hebei Province, China, a known breeding area for *L. m. manilensis*. Collected eggs were transferred to the Institute of Plant Protection, Chinese Academy of Agricultural Sciences, Beijing. Then, during the next year, we reared several successive generations in the laboratory as per Tu *et al.* [48]. In late 2009, we collected eggs from this laboratory colony and kept them at 4°C for three months to use in this study.

The location (N38°30′33.46″, E117°25′32.85″) which is covered with saline-alkali soil is nearby the Bohai Sea. We have got the permission for us to conduct the field studies by Cangzhou Academy of Agriculture and Forestry Sciences of Hebei province, who is the authority department responsible for pest control in agriculture and forestry land, also with the protection of wildlife in Cangzhou. With the help of Dr. Qinglei Wang (Cangzhou Academy of Agriculture and Forestry Sciences), we collected eggs in Autumn for our laboratory experiment. This location is a natural ecosystem, it is not involving endangered or protected species during the field studies.

Using DD to predict locust voltinism

To investigate the relationship between the degree-days (DD) and voltinism (the number of generations that can be produced in a population in one year), we raised the oriental migratory locust in growth cabinets (PRX-350B-30). We used seven different cabinets, each set to a daily temperature and photoperiod cycle that mimicked the natural daily temperature and photoperiod cycles of a specific location in China: Baiquan (BQ), Chengde (CD), Tumotezuoqi (TM), Wenan (WA), Rongan (RA), Qiongzhong (QZ), and Qiongshan (QS) (Table 1). We chose these seven locations, because they encompassed most of the latitude available in China, and corresponded to seasonal DD of 700, 800, 900, 1600, 2400, 3200, and 4000 heat units, respectively, based on the previously calculated lower thermal development threshold of 14.2°C for *L. m. manilensis*. The seven locations (DD) chosen include four (BQ, RA, QZ, QS) where the oriental migratory locust is unable to breed, and three (CD, TM, WA) where the locust is able to breed. The chosen locations also included latitudes with climates that could support one (CD & TM) or more than one (WA) locust generations per year [47,49]. Hence, we modeled latitudes that lacked natural breeding populations of oriental locust, as well as univoltine and multivoltine sites. This allowed us to test the validity of our DD model for predicting locust biogeography and voltinity. Each chamber tested a different DD, and represented a different treatment.

In the field, temperature and photoperiod vary throughout the season, and temperature cycles on a 24-h basis. Using recorded weather data (http://cdc.cma.gov.cn/home.do), we estimated mean daily temperatures and photoperiods throughout the growing season for each location (Fig. 1). We then set each of our growth cabinets to mimic the hourly and seasonal conditions that occur in the field during the locust growing season. Hence, growth chamber temperatures changed hourly and photoperiod changed every 10 d throughout the experiment (Table 2). The relative humidity (RH) was kept at ~60% for eggs and ~80% for nymph and adults. Note that each chamber (each treatment) ran for a different number of days (Table 2), which matched the local locust growing season (i.e., number of days in the field at that location where the mean daily temperature exceeded 14.2°C). Hence, the growing season for *L. m.manilensis* at high latitude BQ is only about 110 d, whereas low latitude QS provides a 360-d growing season. Therefore, the growing chamber that simulated the BQ climate ran for only 110 d, whereas the QS treatment ran for a year (Table 2).

To start the experiment, we transferred about 300 locust eggs into each of the seven growth cabinets. The resulting nymphs and adults were kept in 50×50×60-cm tall cages and fed twice each day with freshly cut wheat leaves (*Triticum sativa* L.) and once each day with artificial diet (100 g wheat bran+5 ml corn oil+vitamins B and C) [48]. The bottom of each cage contained a 20-cm

Table 1. Seven locations in China modeled in this study, and the average heat units (degree-days) available per growing season at each site, based on an estimated lower thermal threshold for development of 14.2°C for *Locusta migratoria manilensis* Meyen.

Location	BQ	CD	TM	WA	RA	QZ	QS
Longitude	E126°04′	E117°58′	E111°08′	E116°27′	E109°22′	E109°50′	E110°20′
Latitude	N47°37′	N40°57′	N40°43′	N38°52′	N24°14′	N19°03′	N19°59′
Available degree-days	698	820	902	1627	2392	3248	3898

Baiquan (BQ), Chengde (CD), Tumotezuoqi (TM), Wenan (WA), Rongan (RA), Qiongzhong (QZ), and Qiongshan (QS).

diameter round hole, under which was placed a 20-cm diameter ×9-cm tall container with compacted sand (3.5 kg sterile sand+0.6 L sterile water) which allowed females to oviposit ad lib. The container was replaced twice a day to ensure space for eggs-laying. We carefully recorded the duration, mortality of each stage, and observerd voltinisms of each treatment during the experiment. Generation times were calculateded by weighted average method [14].

Effects of temperature on locust development

To examine the effects of temperature (especially higher temperature) on locust development, locust eggs (30 eggs/ duplicate treatment, five duplicates/treatment) were transferred to growth cabinets, which were maintained at constant temperatures of (18, 21, 24, 27, 29, 30, 31, 32, 34, 36, 38, 40, 41, or 42)±0.5°C, with 60±5% RH. We recorded the development durations and the survival rates of eggs, nymphs and adults in each treatment. The nymphs and adults were reared as described at (18, 21, 24, 27, 29, 30, 31, 32, 34, 36, 38, 40, 41, or 42)±0.5°C, with 80±5% RH and a 12:12 L:D photoperiod. The lower threshold temperature (LTT) for oocytes development was estimated.

Effects of temperature on adult egg-laying

We tested the effects of temperature on adult egg-laying. Locust eggs were transferred to growth cabinets which were maintained at seven constant temperatures (18, 21, 22, 23, 24, 27, and 30°C). Ten freshly molted adults (5 ♂, 5 ♀) were obtained from nymphs reared at each of the seven temperatures and maintained in incubators at the same constant temperature and conditions as

before. The adults were confined in pairs (1 ♂+1 ♀) in clear 500 ml plastic containers (five pairs per temperature). For each female adult, we recorded the intervals from adult molt to the first oviposition, percentage of females that laid eggs, number of egg-pods laid and adult longevity. The containers lacked sand, so the females laid their egg pods on the floor or the sides of the containers.

DD integral calculation and DD model improving

We recorded 24-h temperature data on each day using a HOBO Pro v2 logger, where the stability was <0.1°C per year. This instrument also had sufficient memory to record over 42,000 12-bit measurements. Low threshold temperatures for different development stage (i) were defined as 'C_i', so only temperature higher than 'C_i' was considered for analysis and heat accumulation. First, we modified the 24-h temperature change function (f_t) using Matlab R2011b. The program script was as follows:

t = 1:24; %('t' as daily 24 h, 1≤t≤24)

d = [data]; %('d' as daily 24 h temperature data)

p_n = polyfit(t,d,n); % ('n' as the power of the function, generally 1≤ n ≤6)

poly2str(p_n,'t') %(obtains the function 'f_t')

Second, we calculated the area of the temperature higher than 'C_i' in the figure, as follows:

solve ('$f_t = C_i$') %obtains the intersect of 'a' and 'b' between y = C_i and f_t

syms t;

int ('f_t', t, a, b) %get the area S_1 from 'a' to 'b', $S_1 = \int T(a,b)$, 'T' is the temperature at time 't'

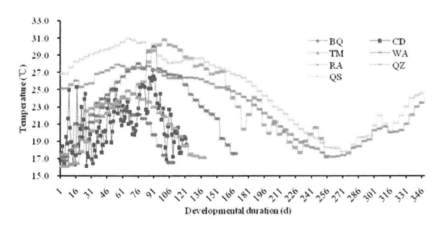

Figure 1. The mimicked temperatures in different growth chambers. Seven locations in China modeled in this study, Baiquan (BQ), Chengde (CD), Tumotezuoqi (TM), Wenan (WA), Rongan (RA), Qiongzhong (QZ), and Qiongshan (QS). Mean temperatures were obtained from (http://cdc.cma. gov.cn/home.do) throughout the growing season for each location based on an estimated lower thermal threshold for development of 14.2°C for *Locusta migratoria manilensis* Meyen. Growth chamber temperatures in Chengde changed daily while in other locations changed every 10 d.To simulate variable environmental temperatures, we designed a variable range of '±5°C' for the daily 24-h temperature change.

Table 2. Photoperiod regimes used in the experiment.

Day (d)	BQ	CD	TM	WA	RA	QZ	QS
1–10	14:10	13.5:10.5	13.5:10.5	13.5:10.5	13:11	12.5:11.5	12.5:11.5
11–20	14:10	13.5:10.5	13.5:10.5	13.5:10.5	13:11	12.5:11.5	12.5:11.5
21–30	14:10	14:10	13.5:10.5	13.5:10.5	13:11	12.5:11.5	12.5:11.5
31–40	14:10	14:10	13.5:10.5	13.5:10.5	13:11	12.5:11.5	12.5:11.5
41–50	14:10	14:10	13.5:10.5	13.5:10.5	13:11	12.5:11.5	12.5:11.5
51–60	15:9	14.5:9.5	14.5:9.5	14.5:9.5	13.5:10.5	13:11	13:11
61–70	15:9	14.5:9.5	14.5:9.5	14.5:9.5	13.5:10.5	13:11	13:11
71–80	15:9	15:9	14.5:9.5	14.5:9.5	13.5:10.5	13:11	13:11
81–90	16:8	15:9	14.5:9.5	14.5:9.5	13.5:10.5	13:11	13:11
90–100	16:8	15:9	14.5:9.5	14.5:9.5	13.5:10.5	13:11	13:11
101–110	16:8	15:9	14.5:9.5	14.5:9.5	13.5:10.5	13:11	13:11
111–120		15:9	14.5:9.5	14.5:9.5	13.5:10.5	13:11	13:11
121–130		15:9	14.5:9.5	14.5:9.5	13.5:10.5	13:11	13:11
131–140			14.5:9.5	14:10	13.5:10.5	13:11	13:11
141–150				14:10	13.5:10.5	13:11	13:11
151–160				14:10	13.5:10.5	13:11	13:11
161–170				14:10	13.5:10.5	13:11	13:11
171–180					13:11	12.5:11.5	12.5:11.5
181–190					13:11	12.5:11.5	12.5:11.5
191–200					13:11	12.5:11.5	12.5:11.5
201–210					12:12	12:12	12:12
211–220					12:12	12:12	12:12
221–230					12:12	12:12	12:12
231–240					11:13	12:12	12:12
241–250					11:13	12:12	12:12
251–260					11:13	12:12	12:12
261–270						11:13	11:13
271–280						11:13	11:13
281–290						11:13	11:13
291–300						11:13	11:13
301–310						11:13	11:13
311–320						11:13	11:13
321–330						10:14	10:14
331–340						10:14	10:14
341–350						10:14	10:14
351–360						13:11	13:11

Each growth chamber was set to simulate the natural seasonal photoperiod pattern of a different location in China: Baiquan (BQ), Chengde (CD), Tumotezuoqi (TM), Wenan (WA), Rongan (RA), Qiongzhong (QZ), or Qiongshan (QS). The first column (Day) divides the experimental period into successive 10-d segments. The other columns show the photoperiod settings for each growth chamber during each 10-d period. Note that each location gives photoperiods for only those weeks in the field when air temperature exceeded 14.2°C. Photoperiod data derived from (http://cdc.cma.gov.cn/home.do).

$S = [S_1 - C_i *(b-a)]/24$ 'S' is the required DD

In this method, we used degree-hours instead of the traditional degree-days to calculate heat units, as follows: $S = \sum (T - C_i)/24$.

Using our experimental findings, we could get some sensitive characteristic parameters of migratory locust (i.e., the upper threshold temperature and prevent egg-laying temperature). These parameters and integral calculation were conducted and used to improve the DD model.

Different DD calculating methods comparision

We compared the accuracy of four different DD calculating methods based on either: (1) the daily mean temperature, (2) max-min temperature [50], (3) 24-hours mean temperature, or (4) integral calculation of 24-hours temperature, based on the previously calculated lower thermal development threshold of 14.2°C for *L. m. manilensis* [51]. Then we analyzed which method could simulate actual temperature variation trend.

Standard heat units calculation and validation based on the Improved DD model

We tested the validity of our calculated DD in the field by recording environmental temperatures and life history of *L. m. manilensis* in Langfang, China in 2011. Because locust eggs survive below ground and locust nymphs and adults live above ground, we used ground temperatures for eggs and air temperatures for nymphs and adults. DD were calculated based on our improved DD model.

We further evaluated the validity of the improved DD model, by recalculating the DD for the oriental migratory locust development at each of our seven focus sites. In this analysis, we used only air temperature, because this was the basis of our growth cabinet studies.

Results and Discussion

Voltinism under different DD

When the oriental migratory locusts were reared in seven different growth chambers providing either 700, 800, 900, 1600, 2400, 3200, or 4000 DD, the number of complete generations that were produced varied dramatically among the different treatments (Table 3). For example, locusts were unable to complete a full generation when reared in the chamber that provided only 700 DD, and which simulated the climate of Baiquan, China (BQ). In contrast, locust completed 1.1 generations in the 900 DD treatment, which mimicked daily temperatures from cool, high latitude Tumotezuoqi (TM). Likewise, locusts reared in the 4000 DD growth chamber, which mimicked daily temperatures from warm, low latitude Qiongshan (QS), completed 4.95 generations (Table 3). Thus, in this laboratory experiment, the number of heat units (DD) available strongly influenced the number of locust generations produced. But there was substantial individual variation in generations completed within treatments. For example, only low proportion eggs laid by the generation I females could keep on developing and reach to the 1st instar nymphs in the cool 900 DD treatment, whereas those reared in the warm 3200 DD growth chamber all locusts of the generation IV reached adult stage and lived as adults for a long time (~156 days), but without laying a single egg-pod. So "0.95" indicates the locusts were not able to complete full generation under those temperature and photoperiod conditions, "0.1" and "0.2" indicate only little of eggs hatched while most of them could stay in egg stage and overwintered. In such a case, the proportion of development reached is given. Hence, for the BQ treatment, locusts developed only 95% (= 0.95) of the way to a full generation. In contrast, the TM growth chamber produced 1.1 generations (Table 3).

Developmental asynchrony within a single generation produced substantial overlap between generations. For example, for locust reared in the 1600-DD Wenan simulation, the overlap period between the 1st and 2nd generation was about 17d. As such, some

Table 3. Voltinism of L.m.manilensis reared in the laboratory under seven different simulated "climates", each representing a different location in China: Baiquan (BQ), Chengde (CD), Tumotezuoqi (TM), Wenan (WA), Rongan (RA), Qiongzhong (QZ), or Qiongshan (QS).

Location and DD modeled	Theoretical generations by traditional DD model	Generations completed in laboratory	Generations Completed in field
BQ (700)	0	0.95 (egg-to-adult)	No locust distribution
CD (800)	1	1 (one egg-to-egg generation)	1
TM(900)	1.1	1.1 (one egg-to-egg generation + egg-to-1st instar nymph of the II generation)	1
WA (1600)	2	2.2 (two egg-to-egg generations + egg-to-2nd instar nymph of the III generation)	2
RA (2400)	3	2.95 (two egg-to-egg generations + egg-to-adult of the III generation)	No locust distribution
QZ (3200)	4	3.95 (three egg-to-egg generations + egg-to-adult of the IV generation)	No locust distribution
QS (4000)	5	4.95 (four egg-to-egg generations + egg-to-adult of the V generation)	No locust distribution

I, II, III, IV, and V refers to the 1st, 2nd, 3rd, 4th, and 5th generation, respectively.

2nd generation hatchings were already 17-d old by the time that the slowest 1st generation female laid her 1st egg pod.

Mismatch between predicted voltinism and realized voltinism

Our laboratory experiment testing seven different temperature treatments found substantial differences between the predicted number of generations and the realized number of generations (Table 3). Previous work estimated that 800 DD above a critical low-temperature threshold of 14.2°C were necessary for the Oriental locust to undergo one complete generation [51]. Hence, the 800, 900, 1600, 2400, 3200, and 4000-DD treatments (Table 3) should have produced 1, 1.1, 2, 3, 4, and 5 generations, respectively. However they did not (Table 3). For example, the QZ treatment provided 3200 DD, which should have produced at least four complete generations, but instead produced only three egg-to-egg generations and the 4th unaccomplished generation. Thus there would be: (1) Mismatch between predicted and realized voltinism according to the traditional DD model. (2) Some characteristics of migratory locust response to temperature change were unknown. We therefore attempted to identify and correct these sources of errors, and improve the accuracy of our predictions.

There was a substantial similarity in generations completed between laboratory simulation and field studies (Table 3), for example, four locations (BQ, RA, QZ, QS) where oriental locust is unable to breed, and three locations (CD, TM, WA) where the locust is able to breed, including the climates could support one in (CD & TM) or two in (WA) locust generations per year [47,49]. However, overwinter eggs will enter diapause without hatching in the same year undergo the climates of (CD & TM) in field, hatchlings emergence mainly because of we have mimicked air temperature in 900 DD (TM) treatment, while the ground temperature would be suitable for them to stay in egg stage. In contrast, climate in Tianjin where could support two locust generations produced parts of hatchling emergence in October in recent years. Thus, temperature increasing in Autumn would enhance overwinter egg-hatching before entering diapause, and decrease population in the next Spring.

Overflow temperature for locust development

To obtain a better understanding of the relationship between temperature and development, we studied locust development rates in growth chambers under 14 different constant temperatures ranging from 18 to 42°C (Fig. 2). Results showed that temperature strongly influenced development rates for both eggs and nymphs (Fig. 2). For example, development was significantly shorter at 32 to 42°C than at 18 to 31°C (F = 65.38, P<0.0001). From 18 to 32°C, development rate was a linear function of temperature (Fig. 2). In contrast, temperatures above 32°C had little effect on development rate (Fig. 2).Thus, 32°C appears to be an important inflection point for oriental locusts: above this value, higher temperatures do not produce faster development which should be defined as overflow temperature for migratory locust development. This is critical for calculating DD and estimating total effective temperatures.

Eggs and nymphs developed faster when reared at higher temperatures till temperature at 32°C (Fig. 2), and this was true for eggs, nymphs and temperatures tested (Fig. 3). Development rate was a linear function of temperature from 18 to 31°C (Fig. 3), the relationship between development rate of egg (V) and temperature (T) was: V = 0.005T - 0.070, r^2 = 0.931, while the function for nymphs (including 1st to 5th instar nymphs) was: V = 0.003T - 0.042, r^2 = 0.963. Extending these development regression lines to the x-axis gives us the theoretical low temperature threshold (LTT) for development. Note that these values converge to ~14°C for the tested eggs and nymphs (Fig. 3).

Many studies have explored the temperature relationships of various subspecies of the migratory locust, Locusta migratoria [40–41,47,52–53]. In general, the biochemical reactions are sensitive to temperature, increasing in rate of locusts with increaseing temperature [48,54]. In the present study, the developmenta rate of locusts increased with the temperature arising from 18 to 31°C (Fig. 3). This conclusion agreed with the results calculated using the traditional DD model [14]. But the overflow temperature was never reported in previous studies [47] and it could induce predicting deviation based on the traditional DD model [14].

Lower threshold temperature (LTT) may differ considerable between immature and mature stages [28,55], so when we predict locust development progress in field studies, the LTT of eggs, nymphs, and adults would be necessary. In the present study, we get the LTT of eggs and nymphs (Fig. 3) seems similar to the previous work as per Tu [51] which was ~14.2°C, to ensure contextual consistence and avoid confusion, we used the LTT of 14.2°C for eggs, nymphs in this paper.

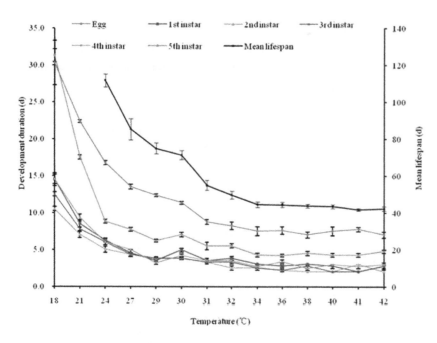

Figure 2. Developmental duration of locust eggs and nymphs at different constant temperatures. Each line represents the mean developmental duration of the same developmental stage at different constant temperatures.

Minimum temperature for adult egg-laying

To further investigate why the traditional DD model failed to accurately predict locust voltinism in our growth chamber experiments (Table 3), we examined the relationship between egg laying and temperature. The results (Table 4) showed that adults failed to oviposit at 18 and 21°C, there are two hypotheses: (1) The low temperature threshold (LTT) for oocyte development is above 21°C. (2) The females developed mature oocytes, but were unable to mate or oviposite, because mating or oviposition

requires neural and muscle action, and 21°C is too low—i.e, the LTT for mating or oviposition (pushing eggs out of the body) muscles and nerves signals is above 21°C [28]. To investigate which hypothesis will be plausible, we examined the relationship between oocyte development and temperature (Fig. 3), result showed the LTT for oocyte development was ~18°C (Fig. 3). As locusts can reach to adult and stay in this stage for a long time, the DD would be enough for oocyte development at 18 and 21°C, so the first hypothesis seems implausible. Thus, females require

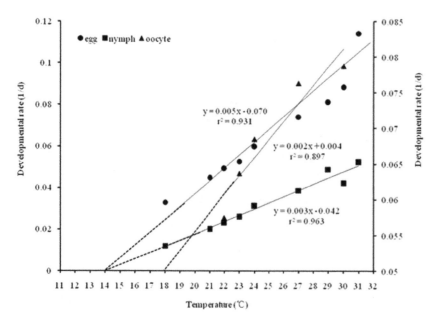

Figure 3. Relationship between temperature and locust (eggs, nymphs, and oocytes) developmental rate. Oocytes developmental rate based on calculating the pre-ovipositing period at 22, 23, 24, 27, and 30°C, while dashed lines show theoretical extension of regression lines to x-axis. The points where the lines intersect the x-axis represent the theoretical low temperature threshold (LTT) for development egg, and nymph was ~14°C, while for oocyte was ~18°C.

Table 4. Oviposition behaviors of migratory locusts at different constant, life time temperatures, given as means ± S.E.

Temperature(°C)	Pre-ovipositing period(d)	Percentage of females that laid eggs (%)	Number of egg pods per female	Adult longevity (d)
18	——	0	0[cB]	75.0±7.2
21	——	0	0[cB]	93.0±2.9
22	17.4±0.3	80	0.8±0.2[cB]	69.8±5.5
23	15.7±0.4	100	2.0±0.3[bcB]	65.8±3.3
24	14.6±1.5	100	7.0±1.2[abAB]	62.8±2.1
27	13.1±0.4	100	10.5±2.3[aA]	45.8±5.4
30	12.7±0.3	100	12.3±2.7[aA]	36.2±2.8

See Methods section for further explanation.
Within each column, the lowercase letters indicate significant differences at $P<0.05$, and capital letters indicate significant differences at $P<0.01$. Note that at 30°C, females laid their 1st egg-pod about 13 d after the adult molt, whereas the 21°C females lived an average of 93 d as adults without laying a single pod.

temperature higher than 21°C for reproduction behavior (mating or ovipositing) at constant temperatures.

For the treatments that produced oviposition (22, 23, 24, 27, and 30°C), the pre-oviposition intervals of female adults were 17.4, 15.7, 14.6, 13.1, and 12.7 d, respectively. While the percentages of females that laid eggs were 0, 0, 80%, 100%, 100%, 100%, and 100%, and numbers of egg-pods per female were 0, 0, 0.8, 2.0, 7.0, 10.5, and 12.3 at 18, 21, 22, 23, 24, 27, and 30°C, respectively. The difference analysis showed the minimum temperature for adult reproduction behavior should be at a point between 21 and 22°C (Table 4), but there was no difference between them ($F = 8.88$, $P = 0.0002$). Thus, 21°C appears to be an important inflection point for the oriental migratory locusts: under this value, lower temperatures do not produce egg-pods which should be defined as ineffective temperature for adult reproduction behavior. This is critical for calculating DD and estimating the distribution locations for the oriental migratory locusts [14].

DD based on 24-h temperature integral calculation

On April 12, 2011, the 24-h temperature at 5 cm underground in Langfang was as shown in Fig. 4. Using Matlab to simulate the temperature change function: $f_t = -3.1195e-007* t^6+7.3636e-006*t^5+0.00047291*t^4-0.020038*t^3+0.23657*t^2-0.60834*t+12.4926$, $R^2 = 0.996$. The results showed that $a = 7.0745$, $b = 18.3312$, and the heat unit was 0.3 DD. The function used by the integral method was: $K=\sum N[\int T(a,b)-14.2*(b-a)]/24$, where K is the total effective temperatures, N is the development duration, and $[\int T(a,b)-14.2*(b-a)]$ is the area of the shaded part in Fig. 4.

Temperatures higher than 32°C did not accelerate the development rate of the migratory locust, should not be included in the DD calculation (Fig. 5). On July 9, 2011, the 24-h temperature change is shown in Fig. 5 and the overflow temperature was 1.9 DD: $a = 4.2758$, $b = 14.0974$. The function used to calculate the overflow DD was: $Ki' = \sum_{n=1}^{Ni}(\int_a^b Tt-32)dt$, where i is the development stage, Ki' is the overflow DD at 'i' stage, Tt is the temperature at time 't' ($1\leq t\leq 24$), 'a' and 'b' are time when the temperature is higher than 32°C during the 24-h period, and Ni is the development duration of 'i' stage. We have calculated the overflow DD was 38 heat units during 2011 in Langfang using this model, whereas there was no overflow DD based on the daily mean temperature. The overflow DD is also part of the invalid heat unit that causes the DD to increase nonlinearly with

generations, so it has an important effect on locust development in the southern population.

Different DD calculating methods comparision

To further investigate the accurancy of 24-hours temperature integral calculation, we compared DD calculation by four different methods. Results showed it was −0.2, 0.2, 0.3 and 0.3 DD based on daily mean temperature (Fig. 6), max-min temperature (Fig. 6), 24-hours temperature (Fig. 6), and 24-hours temperature integral calculation (Fig. 4), respectively. Two criterias were used to assess which method would be more suitable (Table 5). Results showed that 24-hours integral calculation would get the accurate time interval about temperature higher than the lower development threshold temperature and simulate the actual temperature variation trends (Fig. 4, Fig. 6). By taking 24-hours integral calculation as a contrast, the relative error of daily mean, max-min, 24-hours temperature was 167%, 33%, 0% respectively (Table 5).

The integral calculation based on 24-hours temperature was used to calculate the insect development rate considered the effects of low temperatures in the spring and high temperatures in the summer (Fig. 4, 5), which also could describe the temperature vary trend compared to other data (Fig. 4, Fig. 6; Table 5) and get the more accurately results [20–21,32,50]. This method has solved the problem of how to set the initial values as we have 24 temperature data each day, it has eliminated the error by the Briere model, Lactin model, and several other popular models [6,32] and got an exact solution [17–18,35]. We have conducted the concept of degree-hours to calculate DD, which is more precise and available in field conditions. In theory, using degree-half an hours or degree-minutes to calcualte DD would be more accurancy. Unfortunately, they need cumbersome sampling processes. Then it is better for us to take degree-hours in field studies, because it has simplified the overall calculation process and enhanced the precision of 0.1 DD when compared with the method of Zan et al. [32]. Thus, the integral method could have broad applications in insect forecasting and predicting.

Improvement of the DD model

The overflow temperature did not accelerate the development rate of locusts (Fig. 3) and temperature lower than 21°C was invalid for females reproduction behavior (Table 4), so we modified the DD model for migratory locusts in different stage as: $Vi = \sum_{n=1}^{Ni}[\int_1^t (Tt-Ci)dt]/\{Ki-\sum_{n=1}^{Ni}[\int_a^b (Tt-32)dt]\}$, where V_i

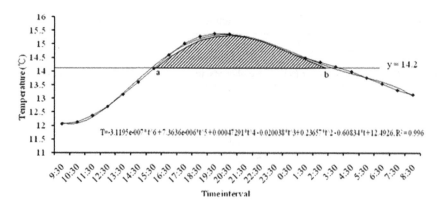

Figure 4. DD for locusts on April 12, 2011, in Langfang based on the integral calculation method. 24-hours ground temperature was recorded by HOBO Pro v2 logger which were used to model temperature change curve by Matlab R2011b. 'a' & 'b' represent the intersect points between y = 14.2 and the curve. The shade area was the DD for locust development in this day.

is the development rate for one day at stage 'i', C_i is the development threshold temperature for migratory locusts at stage 'i' (for eggs and nymphs, $C_i = 14.2°C$; for adults, $C_i = 21°C$), and K_i is the total effective temperatures at stage 'i' or thermal constant.

Empirical DD values are not absolutely constant, even if the feeding environment and all other environmental influences are the same. Environmental influences other than temperature, such as food, population density, and all specifics, influence the number of DD, and, in addition, development rates sometimes deviate slightly from linearity. However, the number of DD to reach maturity is constant to be of biological interest. It implies that the linearity of development rate as a function of temperature is more than a statistical first approximation: it seems a biological property. Therefore, it is a biological question how linearity of development rate is caused [10]. So we proposed a standard DD calculating method based on studied the biological property of the oriental migratory locust. In the present study, invalid DD (i.e., overflow DD ineffective DD for egg-laying) should not be considered when applying the DD model to the oriental migratory locust.

Standard heat units calculation based on the improved DD model

The improved DD model showed that invalid DD should be deducted, so standard DD calculation for locust development should deduct two parts, one was to take off the overflow DD during the whole generation, while the other one was the ineffective DD when mean daily temperature was lower than 21°C for egg-laying, as depicted in Fig. 7. We analyzed the life history of *L. m. manilensis* and temperature changes in Langfang, 2011, which can support two locust generations in field [47]. Results showed that the overflow DD and ineffective DD for egg-laying was 38 and 128 heat units, respectively (Fig. 7). In other words, the DD error in this location was 166 (= 38 +128) heat units between the improved and traditional DD model.

When ground temperature and air temperature were seperated to calculate different development stage of migratory locust, the total effective temperatures for migratory locust was 1437 DD based on 24-hours integral calculation, so the standard DD was 1271 heat unit when the overflow DD and ineffective DD were taken off, as follow: standard DD = (1437-38-128), and it would take ~635 DD at least to finish a life cycle for *L. m. manilensis* in this location (Fig. 7).

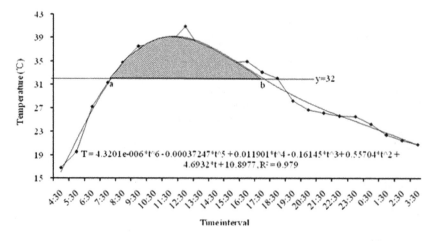

Figure 5. Overflow DD for locusts on July 9, 2011, in Langfang based on the integral calculation method. 24-hours air temperature was recorded by HOBO Pro v2 logger which were used to model temperature change curve by Matlab R2011b. 'a' & 'b' represent the intersect points between y = 32 and the curve. The shade area was the overflow DD for locust development in this day.

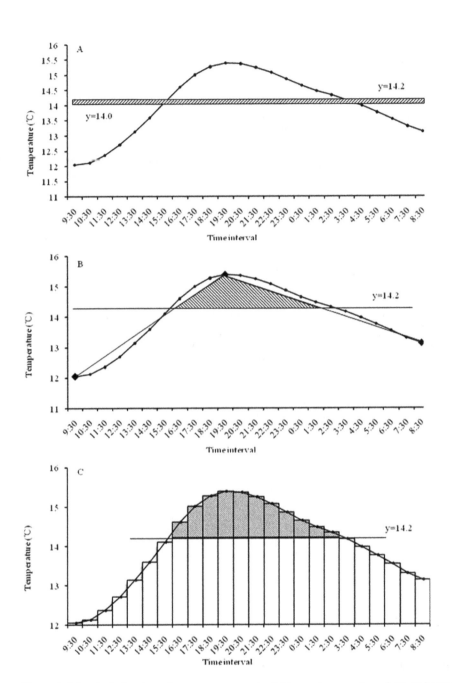

Figure 6. DD calculation based on mean, max-min, 24-hours temperature data. 24-hours temperature data was shown in Fig. 4. Minimum, mean and maximum temperature of this day was about 12.1, 14.0 and 15.4°C. *(A)* DD calculating based on daily mean temperature. *(B)* DD calculating based on maximum-minimum temperature. *(C)* DD calculating based on 24-hours temperature, it should be calculated as: $\sum(T-14.2)/24$, not the same as 24-hours integral calculation in Fig. 4.

In addition, we predicted hatchlings emergence mostly by air temperature as we lacked ground temperature data in field, so the DD was 1566 heat units when only based on air temperature (Fig. 7). Thus, it would take ~700 DD to finish a life cycle for *L. m. manilensis*, as follow: standard DD = (1566-38-128)/2.

Validation the improved DD model

To further investigate locust voltinism in our growth chamber experiments (Table 3), we recalculated the actual DD in each chamber by the imporved DD model. The results (Table 6) showed that the DD for migratory locust development at Baiquan (BQ), Chengde (CD), Tumotezuoqi (TM), Wenan (WA), Rongan

(RA), Qiongzhong (QZ), and Qiongshan (QS) was only 583, 706, 741, 1512, 2040, 2674, and 3351 heat units during locust growing season. They could produce 0.83, 1.01, 1.06, 2.16, 2.91, 3.82, and 4.79 generations, respectively, which were compared to the standard DD (~700DD). This conclusion matched with the realized voltinism (Table 3), and revealed females require temperature higher than 21°C for reproduction behavior (Table 4).

For the voltinism of BQ, it was ~0.83 as we have deleted invalid DD (Table 6). While it was ~0.95 by locusts were developing to adult and staying in this stage in a long time without laid eggs (Table 3). Thus, we considered there was no difference between them, because the oriental migratory locusts were unable to breed

Table 5. Comparision the different DD calculating methods based on the daily mean, max-min, 24-hours temperature, and 24-hours temperature integral calculation, whether it could describe the accurate time interval about temperature higher than the lower development threshold temperature (C) and simulate actual temperature variation trends accurately.

Calculating methods	Simulating actual temperature variation trends accurately	Describe time interval about temperature higher than C accurately	Relative error (%)
Daily mean	−	−	167
Max-min	−	−	33
24-hours temperature	−	+	0
24-hours integral calculation	+	+	0

Note '−' represents 'not' while '+' represents 'yes'. Relative error represents the results (X) by different methods compared to 24-hours integral calculation (Y), as follow: Relative error (%) = |X-Y|/Y*100%.

in BQ, the same as RA, QZ, and QS. For the overflow DD in table 6, they were not revealing the actual values in each location. We have set the 24-hours temperature changes as Fig. 1 of each day as (mean value ±5°C), which were according with normal distribution, however, they were not same as temperature vary trends in China. It is known that temperature different with latitude moves, i.e.for the vary range of 24-hours temperature, high temperature inteval in southern China is more than northern China [56]. Thus, we speculated the overflow DD would be higher than the calculated values in Table 6, especially in southern China. For applying our model, we need to monitor the 24-hours temperature changes of migratory locust at different locations in future.

Conclusions

In this report, we study various characteristics of the temperature response of the migratory locust including overflow temperature and the effective temperature requirements for egg-laying. Results show that development rate increases with the temperature from 18 to 32°C, temperatures>32°C is overflow temperature for migratory locust development and if not considered as a cut of for heat accumulations may be a cause of overflow in the DD model. Temperature <21°C is not suitable for adults eggs laying and should be deducted when calculating the DD for adults ovipositing. They are defined as invalid DD which are the key factors that affect the prediction of migratory locust occurrence and result in DD nonlinear increasing with generations arising in field.

Moreover, we propose an integral calculation method to calculate the DD of migratory locust, which also can be used to calculate overflow DD. This metod records 24-h temperature as the basic data to simulate the daily temperature changes and the accuracy is higher than using the daily mean, maximum and minimum temperatures in the data simulation. We also introduce the concept of degree-hours, which improved the accuracy of the DD calculation in areas with temperature variations.

Thirdly, to eliminate the calculating error by traditional DD model, we improve it by studying temperature response of the migratory locust and proposing integral calculation method. Then we constructed a new DD model as: $Vi = \sum_{n=1}^{Ni} [\int_{1}^{t} (Tt - Ci)dt] / \{Ki - \sum_{n=1}^{Ni} [\int_{a}^{b} (Tt - 32)dt]\}$. The new model is according with the

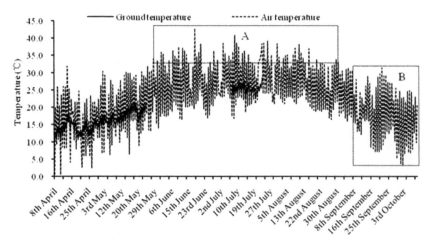

Figure 7. Standard DD calculating for locust accomplishing full generations based on life history of *L. m. manilensis* and temperature changes in Langfang, 2011. In field, we predicted hatchlings emergence mostly by air temperature as we lacked ground temperature data, so the DD was 1566 DD when only used air temperature (the dashed line). Instead, when we used ground temperature (the solid line) to calcualte DD for eggs, the total DD for migratory locust was 1437 heat units based on 24-hours integral calculation. (A) Overflow DD was 38 heat units from 28th May to 5th September. (B) Useless DD for females egg-laying was 128 heat units from 8th September to 10th October (these days were ovipositing periods for females, but temperature lower than 21°C).

Table 6. DD for migratory locust development at Baiquan (BQ), Chengde (CD), Tumotezuoqi (TM), Wenan (WA), Rongan (RA), Qiongzhong (QZ), and Qiongshan (QS) were recalculated by the improved DD model.

	n	Total DD	Overflow DD	Ineffective DD for egg-laying	Actual DD	Generations completed
BQ	1	698	0	115	583	0.83
CD	1	820	0	114	706	1.01
TM	1	902	0	161	741	1.06
WA	2	1627	7	108	1512	2.16
RA	3	2392	20	332	2040	2.91
QZ	4	3248	5	569	2674	3.82
QS	5	3898	47	500	3351	4.79

The lowercase letter 'n' represents the theory generations of *L. m. manilensis* at each location. Actual DD = Total DD-(Overflow DD + Ineffective DD for egg-laying). Generations completed as follow: (n-1)+[actual DD -700*(n-1)]/700DD.

principle of the traditional DD model, meanwhile it would be more precisely when forecasting the occurrence period of migratory locust. This method can also be used to predict the occurrence period for other pests.

Acknowledgments

We thank Douglas Whitman (Department of Biological Sciences, Illinois State University, Normal, Illinois, USA) and Yonglin Chen (Insect Ecology Laboratory, Institute of Zoology, Chinese Academy of Science, Beijing, P.R. China) for their helpful suggestions. We have got the permission for us to conduct the field studies by Cangzhou Academy of Agriculture and Forestry Sciences of Hebei province, who is the authority department responsible for pest control in agriculture and forestry land, also with the protection of wildlife in Cangzhou.

Author Contributions

Conceived and designed the experiments: ZZ XT. Performed the experiments: XT JW XH. Analyzed the data: XT ZL. Contributed reagents/materials/analysis tools: JY CF HW QW. Wrote the paper: ZZ XT ZL.

References

1. Kontodimas DC, Eliopoulos PA, Stathas GJ, Economou LP (2004) Comparative temperature-dependent development of *Nephus includens* (Kirsch) and *Nephus bisignatus* (Boheman) (Coleoptera: Coccinellidae) preying on *Planococcus citri* (Risso) (Homoptera : Pseudococcidae): Evaluation of a linear and various nonlinear models using specific criteria. Environmental Entomology 33: 1–11.
2. Simpson CB (1903) The codling moth. US Department of Agriculture, Division of Entomology 41:165.
3. Ludwig D (1928) The effects of temperature on the development of an insect (*Popillia japonica* Newman). Physiological Zoology 1(3): 358–389.
4. Davidson J (1942) On the speed of development of insect eggs at constant temperatures. Australian Journal of Experimental Biology & Medical Science 20:233–239.
5. Pradhan S (1945) Insect population studies II. Rate of insect development under variable temperature of the field. Proceedings of the Natural Institute of Sciences of India 11(2):74–80.
6. Yang HL, Yao GF, Zhang TW, Wan Z (1959) Studies on the *Parnara guttata* Bremen *et* Grey I, testing on effective accumulated temperature. Acta Entomologica Sinica 9(2):137–148.
7. Arnold CY (1960) Maximum and minimum temperatures as a basic for computing heat units. American Society for Horticultural Science 76: 682–692.
8. Schoolfield RM, Sharpe PJH, Magnuson CE (1981) Non-linear regression of biological temperature-dependent rate models based on absolute reaction-rate theory. Journal of Theoretical Biology 88 (4):719–731.
9. Wagner TL, Wu HI, Sharpe PJH, Schoolfield RM, Coulson RN (1984) Modeling Insect Development Rates: A Literature Review and Application of a Biophysical Model. Annals of the Entomological Society of America 77(2): 208–225.
10. de Jong G, van der Have TM (2009) Temperature dependence of development rate, growth rate and size: from biophysics to adaptation, pp.523–588. In: Whitman D & Ananthakrishnan TN (Eds) Phenotypic Plasticity of Insects: Mechanisms and Consequences. Science Publishers, Enfield, NH, USA.
11. de Réaumur R (1735) Observation du thermomètre, faites à Paris pendant l'année 1735, comparée avec celles qui ont été faites sous la ligne, à l'Isle de France, à Alger et en quelques-unes de nos isles de l'Amérique. Mémoire de l'Académie des Sciences de Paris.
12. Wang J (1960) A critique of the heat unit approach to plant science response studies. Ecology 41: 785–790.
13. Zhang XX (1979) Pest forecast principle and method. Beijing, China Agriculture Press. 1–320.
14. Zhang XX (2002) Insect Ecology and Forecast. Beijing, China Agriculture Press.1–323.
15. Eizenberg H, Colquhoun J, Mallory-Smith C (2005) A predictive degree-days model for small broomrape (*Orobanche minor*) parasitism in red clover in Oregon. Weed Science 53 (1): 37–40.
16. Lin Y, Zhu ZQ, Hu JJ, Bo SW, Ye ZC (1959) Studies on effective accumulated temperature of Yellow rice borer *Tryporyza incertulas* Walker I, developmental threshold and effective accumulated temperature of different stages. Acta Entomologica Sinica 9(5):423–435.
17. Lischke H, Loffler TJ, Fischlin A (1997a) Calculating temperature dependence over long time periods: a comparison and study of methods. Agricultural and Forest Meteorology 86:169–181.
18. Lischke H, Loffler TJ, Fischlin A (1997b) Calculating temperature dependence over long time periods: derivation of methods. Ecological Modelling 98:105–122.
19. Watanabe N, Xiao YY (1981) A simple method for calculating effective accumulated temperature by the daily highest and lowest temperature. China Plant Protection, Disease and Insect Pest Observation Reference Material 2:39–44.
20. Frouz J, Ali A, Lobinske RJ (2002) Influence of Temperature on Developmental Rate, Wing Length, and Larval Head Capsule Size of Pestiferous Midge *Chironomuscrassicaudatus* (Diptera: Chironomidae). Journal of Economic Entomology 95(4):699–705.
21. Son Y, Lewis EE (2005) Modelling temperature-dependent development and survival of *Otiorhynchus sulcatus* (Coleoptera: Curculionidae). Agricultural and Forest Entomology 7: 201–209.
22. Popov GB (1959) Ecological studies on oviposition by *Locusta migratoria migratorloides* (Reiche and Fairmaire) in its outbreak area in the French Sudan. Ecology 66: 1084–1085.
23. Dudley BAC (1964) The effects of temperature and humidity upon certain morphometric and colour characters of the Desert Locust (*Schistocerca: gregaria* Forskal) reared under controlled conditions. Transactions of the Royal Entomological Society of London 116: 115–129.
24. Begon M (1983) Grasshopper populations and weather: the effects of insolation on *Chorthippus brunneus*. Ecological Entomology 8: 361–370.
25. Ackonor JB (1988) Effects of Soil Moisture and Temperature on Hatchling Weight and Survival in *Locusta migratoria migratorioides* (Reiche and Fairmaire). International Journal of Tropical Insect Science 9(5):625–628.
26. Chappell MA, Whitman DW (1990) Grasshopper thermoregulation, pp. 143–172. In: Chapman RF, Joern A. (Eds). Biology of Grasshoppers. Wiley, New York.
27. Heinrich B (1993) The Hot-Blooded Insects. Cambridge, MA, Harvard University Press.

28. Stauffer TW, Whitman DW (1997) Grasshopper Oviposition, pp.231–267. In: Gangwere SK, Muralirangan MC & Muralirangan M (Eds) The Bionomics of Grasshoppers, Katydids and Their Kin. CAB International, Wallingford, UK.

29. Chown SL, Nicolson SW (2004) Insect Physiological Ecology Mechanisms and Patterns. Oxford University Press, Oxford.

30. Mu JY, Liu YJ (1987) Testing biological constants of generation of the green lacewing *Chrysopa sinica* Tjeder, by the method of "thermal sums". Acta phytophylacica sinica 14(3):169–172.

31. Nicholas M, Ulam S (1949) The Monte Carlo Method. Journal of the American Statistical Association 44(247): 335–341.

32. Zan QA, Chen B, Sun YX, Chen HR, Li ZY (2010) Evaluation the cumulative temperature of insect using the sine wave model. Journal of Yunnan Agricultural University 25(4):476–482.

33. Briere JF, Pracros P (1998) Comparison of Temperature-Dependent Growth Models with the Development of *Lobesia botrana* (Lepidoptera: Tortricidae). Environmental Entomology 27(1): 94–101.

34. Shi PJ, Ikemoto T, Ge F (2011) Development and application of models for describing the effects of temperature on insects' growth and development. Chinese Journal of Applied Entomology 48(5): 1149–1160.

35. Pruess KP (1983) Day-degree Methods for Pest Management. Environmental Entomology 12(3):613–619.

36. Allen JC (1976) A Modified Sine Wave Method for Calculating Degree days. Environmental Entomology 5(3): 388–396.

37. Higley LG, Pedigo LP, Ostlie KR (1986) Degree-day: A Program for Calculating Degree-days, and Assumptions behind the Degree-day Approach. Environmental Entomology 15(5): 999–1016.

38. Uvarov BP (1931) Insects and Climate. Transactions of the Royal Entomological Society of London 79(1): 1–232.

39. Uvarov BP (1936) The Oriental migratory locust (*Locusta migratoria manilensis* Meyen, 1835). Bulletin of Entomological Research 27: 91–104.

40. Uvarov BP (1966) Grasshoppers and Locusts Vol. 1. London, Cambridge University Press.

41. Uvarov BP (1977) Grasshoppers and Locusts Vol. 2. London, Cambridge University Press.

42. Ma SJ (1958) The population dynamics of the oriental migratory locust (*Locusta migratoria manilensis* Meyen) in China.Acta Entomologica Sinica 8: 1–40.

43. Ma SJ (1965) Study on the distribution of migratory locust in China. Beijing, Science Press. 1–335.

44. Zhou Y (1980) A History of Chinese Entomology. Shanxi, Entomotaxonomia Press. 55–70.

45. Chen YL (2007) The main locust and ecological management of locust plague in China. Beijing, Science Press. 82–88.

46. Chen YL (2000) The main managed achievements and studies of locust in China. Entomological Knowledge 37: 50–59.

47. Guo F, Chen YL, Lu BL (1991) The biology of the migratory locusts in China. Shandong, Shandong Science and Technology Press.

48. Tu XB, Zhang ZH, Johnson DL, Cao GC, Li ZH, et al. (2012) Growth, development and daily change in body weight of *Locusta migratoria manilensis* (Orthoptera: Acrididae) nymphs at different temperatures. Journal of Orthoptera Research 21(2): 133–140.

49. Zhu EL (1999) The occurrence and management of *Locust migratoriamanilensis* in China. Beijing, China Agriculture Press. 5–6.

50. Baskerville GL, Emin P (1969) Rapid Estimation of Heat Accumulation from Maximum and Minimum Temperatures. Ecology 50(3): 514–517.

51. Tu XB (2010) Chinese Academy of Agricultural Sciences Master Dissertation.

52. Hamilton AG (1936) The relation of humidity and temperature to the development of three species of African locusts—*Locusta migratoria migratorioidaes* (R. & F.) and *Schistocerca gregaria* (Forskal), *Nomadacris septemfasciata* (Serv.). Transactions of the Royal Entomological Society of London 85: 1–60.

53. Hamilton AG (1950) Further studies on the relation of humidity and temperature to the development of two species of African locusts—*Locusta migratoria migratorioidaes* (R. & F.) and *Schistocerca gregaria* (Forsk.). Transactions of the Royal Entomological Society of London 101: 1–58.

54. Mohsen M, Bijan H (2007) Effect of temperature on some biological parameters of an Iranian population of the Rose Aphid, *Macrosiphum rosae* (Hemiptera: Aphididae). Europe Journal of Entomology 104: 631–634.

55. Feng CH, Ma L, Wang S, Liao HM, Luo LM (2011) The calculation and application of effective accumulated temperature on plant diseases and pests monitoring. In: Luo LM (2013) Plant protection theory and practice, proceedings of plant protection of Sichuan Agriculture department (2001–2011). Sichuan science press, Chengdu, P.R. China. pp.68–75.

56. Xu XK, Wang XT, Jin XQ (2009) Vegetation response to active accumulated temperature patterns from 1960–2000 in China. Acta ecologica sinica 29 (11): 6042–6050.

6

Habitats as Complex Odour Environments: How Does Plant Diversity Affect Herbivore and Parasitoid Orientation?

Nicole Wäschke[1], Kristin Hardge[1], Christine Hancock[2], Monika Hilker[1], Elisabeth Obermaier[2¤a], Torsten Meiners[1*¤b]

1 Freie Universität Berlin, Institute of Biology, Applied Zoology / Animal Ecology, Berlin, Germany, 2 University of Würzburg, Department of Animal Ecology and Tropical Biology, Würzburg, Germany

Abstract

Plant diversity is known to affect success of host location by pest insects, but its effect on olfactory orientation of non-pest insect species has hardly been addressed. First, we tested in laboratory experiments the hypothesis that non-host plants, which increase odour complexity in habitats, affect the host location ability of herbivores and parasitoids. Furthermore, we recorded field data of plant diversity in addition to herbivore and parasitoid abundance at 77 grassland sites in three different regions in Germany in order to elucidate whether our laboratory results reflect the field situation. As a model system we used the herb *Plantago lanceolata*, the herbivorous weevil *Mecinus pascuorum*, and its larval parasitoid *Mesopolobus incultus*. The laboratory bioassays revealed that both the herbivorous weevil and its larval parasitoid can locate their host plant and host *via* olfactory cues even in the presence of non-host odour. In a newly established two-circle olfactometer, the weevils capability to detect host plant odour was not affected by odours from non-host plants. However, addition of non-host plant odours to host plant odour enhanced the weevils foraging activity. The parasitoid was attracted by a combination of host plant and host volatiles in both the absence and presence of non-host plant volatiles in a Y-tube olfactometer. In dual choice tests the parasitoid preferred the blend of host plant and host volatiles over its combination with non-host plant volatiles. In the field, no indication was found that high plant diversity disturbs host (plant) location by the weevil and its parasitoid. In contrast, plant diversity was positively correlated with weevil abundance, whereas parasitoid abundance was independent of plant diversity. Therefore, we conclude that weevils and parasitoids showed the sensory capacity to successfully cope with complex vegetation odours when searching for hosts.

Editor: Cesar Rodriguez-Saona, Rutgers University, United States of America

Funding: This work was supported by Deutsche Forschungsgemeinschaft within the Priority Programme 1374 "Infrastructure-Biodiversity-Exploratories" (ME 1810/5-1, OB 185/2-1). The funders had no role in study design, data collection and analysis, decision to publish, or preparation of the manuscript.

Competing Interests: The authors have declared that no competing interests exist.

* E-mail: meito@zedat.fu-berlin.de

¤a Current address: University of Bayreuth, Ecological-Botanical Gardens, Bayreuth, Germany
¤b Current address: Helmholtz Centre for Infection Research, Braunschweig, Germany

Introduction

Host location is a crucial event in an insect's life. It is a prerequisite for accessing food or oviposition sites (reviewed by [1,2]). Herbivores as well as parasitoids use volatile cues of the host plant, the host, or the microhabitat for locating their hosts at greater distances (reviewed by [3,4]). However, multitrophic interactions take place in heterogeneous and complex environments, formed primarily by both host and non-host plants [5].

Plant diversity is known to affect host location behaviour of herbivores [6–8] and carnivores [9–11]. The plant species composition of a community may determine the detectability of food plants for herbivores as well as the detectability of host insects for parasitoids [12–14]. The plethora of physical structures in complex and diverse vegetation may affect insect host foraging behaviour [15]. Furthermore, vegetation odour can significantly impact upon olfactory orientation of insects. Non-host plants and

high plant diversity may form a complex odour bouquet which insects have to cope with when foraging for their hosts [12,16–18].

Results of laboratory studies on insect olfactory orientation to host volatiles often differ from insect olfactory behaviour observed in field studies [12]. Many laboratory studies neglect the impact of the complex odour bouquets present in the natural habitat [19]. Thus, combined field and laboratory studies are necessary in order to elucidate the impact of non-host plants on the orientation of herbivores and their natural enemies.

Laboratory studies have revealed that the effects of diverse odorous surroundings of a host plant or host may be manifold, e.g. positive for herbivores and their parasitoids, negative for both or for just one trophic level (reviewed by [8,20]). Non-host plant odours can mask the target odour [12,21,22] or may have a repellent effect [23,24]. However, some insects are not disturbed by the diversity of odours released from other environmental sources present in the habitat where they are searching for a host

[25,26]. Background (habitat) odour may indicate the presence of a host and even lead to the increased attraction of insects [27,28].

Thus far, research on the effects of plant diversity on insects has focused primarily on crop plants and the orientation behaviour of insect pest species [12,16]. However, agricultural systems do not function like natural ecosystems where members of a food web may adapt to each other in the course of evolution. Insects living in natural habitats might respond differently to environmental factors than those in agricultural ecosystems [3,29]. So far, only a few studies have focused on odour-mediated interactions between non-crop plant species and members of higher trophic levels ([30]; and see e.g. [31–34]).

In the present study we combine a laboratory and a field approach to examine the impact of plant (odour) diversity on host location in a tritrophic system by using the perennial herb *Plantago lanceolata* L. (Plantaginaceae), the herbivorous weevil *Mecinus pascuorum* (Gyllenhal) (Coleoptera: Curculionidae), and its larval parasitoid *Mesopolobus incultus* (Walker) (Hymenoptera: Pteromalidae) as a model system. In order to mimic natural odorous conditions in the lab we established a new olfactometer assay and tested (1) whether weevil adults are attracted by odour of their host plant, and if so, whether this attraction is affected by plant diversity (presence of non-host plants). Furthermore, we investigated (2) whether the parasitoid is attracted to odour of the "host complex" consisting of the host plant and the host insect, and how plant diversity affects the parasitoid olfactory orientation to the host complex. In the field we studied (3) the impact of plant diversity on the abundance of the herbivorous weevil and its parasitoid; we surveyed vegetation data (number of plant species and their abundances) and abundances of the weevils and their parasitoids at grassland plots differing in plant diversity within a large scale project in Germany [35]. Abundances of herbivores and parasitoids were determined since they provide information on how successfully a habitat may be colonised, i.e. they are indicators of host resource availability and foraging efficiency for a host (plant).

Materials and Methods

Ethics statement

Field work permits were issued by the responsible state environmental offices of Baden-Württemberg, Thuringia, and Brandenburg (according to §72 BbgNatSchG).

Study system

The ubiquitous herb *P. lanceolata* is native to Europe [36]. It occurs on meadows and pastures and is widespread in habitats with different plant diversities. The specialised weevil *M. pascuorum* oviposits in the seeds of *P. lanceolata* in June and July. Weevil larvae develop within the seeds of *P. lanceolata* inflorescences and emerge from August to September. These larvae are hosts of the generalist parasitic wasp *M. incultus* which attacks larvae feeding inside the seeds of *P. lanceolata* inflorescences [37] as well as other coleopteran insect larvae feeding on *Plantago* and *Trifolium* plants [38].

Laboratory assays

Effect of plant diversity on olfactory orientation of the herbivore. *Insects and plants.* Weevils used for the laboratory olfactometer bioassays were collected from June to July 2009 when female weevils were searching for oviposition sites [37,39]. To reduce a possible impact of sampling in the studied regions, the weevils were not collected at the Biodiversity Exploratory plots (for details see below "Field study"), but at a site called Wuhletal (Berlin Marzahn-Hellersdorf, Germany). The weevils were reared

at 23°C ± 1°C, 48% rh and 14:10 LD. Since weevils were collected in the field, intrinsic factors that may affect oviposition, i.e. age, egg load, mating status or oviposition experience, were unknown. In the laboratory, male and female weevils were kept together for at least one week to ensure mating before separating the sexes. Weevils were fed daily with fresh host plant material.

All plants that were used for the laboratory assays with weevils were grown in a greenhouse at 24°C to 30°C, 20% to 34% rh, and 14:10 LD. They were grown from seeds that were obtained from Botanical Garden Berlin and sown in soil (Einheitserde Typ T Topferde, Einheitserde- und Humuswerke Gebr. Patzer GmbH & Co. KG, Sinntal - Jossa, Germany). Plants were grown individually in pots (6 cm×6 cm×8 cm) after four to five weeks. At the same time, pots were filled with soil and were later used as control. Seven- to nine-week-old plants were used in the bioassays; when plants of this age were used, they still fit into the two-circle olfactometer set-up described below; furthermore, the *P. lanceolata* plants were flowering at this age and thus, have reached a stage at which they display their inflorescences, i.e. the oviposition sites for the weevils.

The two herbs *Achillea millefolium* L. (Asteraceae) and *Agrimonia eupatoria* L. (Rosaceae) (tested when they had developed 11 and 12 fully expanded leaves, respectively) and the grasses *Festuca rubra* L. and *Poa pratensis* L. (both Poaceae) (tested when they had developed 24 and 16 leaves, respectively) occur in the natural habitat of the weevils and co-occur with the host plant *P. lanceolata*. These non-host plant species were used here in order to generate a complex odour blend that mimics the natural habitat odour. The non-host plant species occurred in both the flowering and non-flowering stage in the field during the weevils oviposition period and flowering time of *P. lanceolata*; the flowering time of the non-host plants growing in the different field plots varied due to the different environmental factors the plots were exposed to. In order to ensure consistent conditions for the laboratory bioassays, we decided to use all non-host plants in a non-flowering stage.

General olfactometer setup

In order to study the weevils olfactory orientation behaviour, we built a new type of olfactometer which mimicked an odorous background around the host plant comparable to the field situation (Fig. 1). This static two-circle olfactometer consisted of circular polyamide gauze (mesh width 0.12 mm, ø 180 mm) that served as a walking arena for the weevils and was divided into a central (ø 60 mm) and an ambient circle. The walking arena was stabilised by metal stands of 40 cm height. Test plants were placed below the walking arena either into the central chamber or into the ambient chamber. The wall of the chambers consisted of flexible polyethylene bags (Toppits ®, Cofresco Frischhalteprodukte GmbH & Co. KG, Minden, Germany) clipped to a glass plate (30 cm×30 cm) at the base of the entire set-up. The ambient chamber provided space for four pots and up to three pots could be placed in the central chamber. A light source (60 W, photosynthetic active radiation 6 µmol m^{-2} s^{-1}) was located above the olfactometer in a distance of 50 cm from the walking arena. Prior to each test, the females were allowed to acclimate for one hour in the test room without food. The plants were acclimated in the olfactometer setup for one hour. Bioassays commenced by the release of a single female weevil in the centre of the ambient field at 3 cm distance from the border of the setup. Bioassays were conducted from 10 to 18h under laboratory conditions (22°C±1°C and 43%–65% rh). To avoid diurnal biases, experiments with the same plant arrangement (treatment) were conducted on different days and at different times of day. After every tested female the walking arena was cleaned with

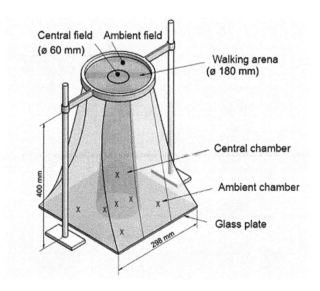

Figure 1. Two-circle olfactometer. The diameter of the whole arena is 180 mm with a central field (ø 60 mm) and an ambient field. Plants and dummies are placed below the walking arena consisting of gauze (possible positions of pots are indicated by (x)). The chamber walls are provided by polyethylene foil (here: cooking bag).

ethanol. Odour from one plant arrangement was offered consecutively to five females. The polyethylene foil was changed for every treatment.

In total, N = 20 females were tested separately for one treatment (plants: N = 4). Each female was observed for 300s. In order to evaluate the weevils host plant finding success and search activity, we recorded behavioural parameters by using the software "The Observer 3.0" (Noldus, Wageningen, The Netherlands) (see below for details on behavioural parameters).

Herbivore olfactory orientation to the host plant in an odorous environment

Table 1 summarizes the different odorous environments in which the weevils orientation to the host plant was tested. Dummy plants (here referred to as dummies) were built from a pot with soil and a green sheet of paper (210 mm×297 mm) that was rolled up and plugged into the soil. In experiments with vacant zones in the ambient chamber, dummies were placed in the olfactometer to provide consistent visual (colour) stimulation [16]. For control, we tested a dummy in the central and four dummies in the ambient chamber. Attraction to the host plant was tested by placing a flowering *P. lanceolata* in the central chamber and four dummies in the ambient chamber (Table 1, experiment (a)). We used only flowering host plants because these are the target hosts of gravid female weevils. We compared the time the weevil stayed in the central circle (duration of stay) in these experiments.

Odours from different plants that are placed in the central chamber might be perceived as a single blend by an insect since the odour sources are placed very closely together. In contrast, odour provided by plants in the central chamber and odour released from plants in the ambient chamber might be perceived as separate blends because the odour sources are further apart than those placed altogether in the central chamber. It is well known that successful host location may depend on whether the odour source of a host is detected separately from other odour sources [40]. Therefore, we tested the effect of plant diversity on olfactory host location first by placing non-host plants (two

herbaceous species) and the host plant in the central chamber, while four dummies were placed in the ambient chamber (Table 1, experiment (b)). Furthermore, the effect of plant diversity was tested by placing the host plant in the central chamber and two non-host plant species (two herbs) plus two dummies in the ambient chamber (Table 1, experiment (c)). In a further bioassay, we tested how the orientation of the weevil to the odour of the host plant in the central chamber is affected by odour from four different non-host plant species (two herb species and two grass species) placed in the ambient chamber (i.e. higher plant complexity than in the abovementioned bioassay with respect to the number of plant species and amount of biomass) (Table 1, experiment (d)). In all three set-ups a flowering *P. lanceolata* plant was positioned in the central field.

In order to evaluate the weevil's host plant finding success and search activity, we recorded the following behavioural parameters: time the weevil spent in the central field (duration of stay), time the weevil needed to enter the central field (target odour) for the first time (latency), number of switches between central and ambient field (frequency), total time walking of a weevil during the observation time (activity). The "latency" was defined as the time the weevil needed to enter the central field for the first time during the observation period of 300s; it was set 300s when the weevil had not reached the central field at all. "Frequency" describes how often a field was visited and serves as an indicator of switches between odour fields. Walking activity was measured as the time during which the weevils were actively walking around rather than resting or cleaning themselves.

Effect of plant diversity on olfactory orientation of the parasitoid

Insects and plants. Parasitoids were obtained from *P. lanceolata* inflorescences collected in the Biodiversity Exploratories in July and August 2010. The inflorescences were kept under the same conditions as the ones obtained for the fieldwork data in 2008 (described below). Emerging unparasitised weevils as well as parasitoids emerging from inflorescences infested with weevil larvae were taken out of the boxes every two days. Thereafter the parasitoids were kept at 10°C ± 1°C, 65% rh, 18:6 LD and fed with aqueous honey solution. They were kept at a long-day-period to mimic European summer time and to retard hibernation. Parasitoids were four to six weeks old when tested. In total, 446 male and female parasitoids emerged. Mating opportunities were given after emergence in inflorescence boxes as well as by keeping both sexes together for at least two weeks. Female parasitoids were tested. Because of the high number of replications and a shortage of parasitoids, we had to test each parasitoid about 1.3 times. Parasitoids were pooled after testing. Female parasitoids were chosen randomly from this pool for the next test. Parasitoids could rest at least two days between two consecutive tests. Bioassays were performed in August and September 2010.

All plants that were used for the laboratory assays with parasitoids were grown under the same conditions and used at the same age and phenotypic stage as described above for the plants used for bioassays with weevils. However, seeds of plants used for the bioassay with parasitoids were obtained from Appels Wilde Samen GmbH (Darmstadt, Germany).

General olfactometer setup

In order to study the parasitoids olfactory orientation behaviour, we used a dynamic Y-tube olfactometer. We tested (i) whether the parasitoid is attracted by odours from the host complex and (ii) how olfactory orientation of the parasitoid is affected by the presence of odour from non-host plants. The two-circle olfactom-

Table 1. Olfactory response of the weevil *Mecinus pascuorum* to host plant odour in the presence of different non-host plant odours in the surroundings.

Treatment		Duration of stay in central field [s]	Time to reach the central field [s]
Central field	**Ambient field**		
(a) HO	DU	61.6	74.1
		(7.1–124.8)	(30.2–293.0)
(b) HO + HE	DU	87.9	122.9
		(0.0–155.6)	(80.2–300.0)
(c) HO	HE + DU	78.0	43.9
		(12.8–139.5)	(17.9–279.6)
(d) HO	HE + GR	71.2	65.2
		(51.1–112.3)	(27.0–125.5)
	Statistics	n.s.	n.s.

Setup: two-circle olfactometer. Duration of stay in the central field and time to reach the central field (latency) [in seconds] are shown for 20 females tested per treatment observed for 300s. Only dummies (DU) or herbaceous species (HE; *Achillea millefolium* and *Agrimonia eupatoria*) and grass species (GR; *Festuca rubra* and *Poa pratensis*) were presented in different combinations in the ambient, or additional to the host plant, (HO: *Plantago lanceolata*) in the central chamber. Dummies consisted of a pot filled with soil and a green sheet of paper. Medians and interquartile ranges (parentheses) are given. n.s. indicates no significant difference (P>0.05) when comparing the different treatments for one behavioural parameter evaluated by Kruskal-Wallis ANOVA followed by Mann-Whitney-*U* test and Bonferroni correction.

eter used for testing the weevils orientation ability could not be adopted for the parasitoid, since the parasitoid individuals did not adapt to the walking arena and showed only frantic activity with erratic movements.

The Y-tube olfactometer used for testing the parasitoids olfactory behaviour consisted of a Y-shaped glass tube (one 20 cm arm and two 14 cm branched arms, ø: 1.2 cm). The open ends of the branched arms were connected by Teflon tubing to glass jars (2100 ml) containing the odour sources (host complex, plants). Air that entered the glass jars was charcoal-filtered and humidified. Air was pumped with a flow of 138 ml/min through the setup. The flow was controlled by flowmeters (Supelco, Bellefonte, PA, USA). Both the tested odour sources and the parasitoids were acclimatised in the test room one hour before testing. One parasitoid was placed into the opening of the long arm of the Y-tube and was observed for a maximum of 300s. We recorded the number of parasitoids which entered one arm and crossed an imaginary border of 5 cm within this arm. Ten parasitoids were tested for each odour source. After testing ten parasitoids, tubes and glass jars were cleaned with 96% ethanol, heated at 100°C for one hour, and odour source sides were exchanged. Prior to the bioassays with plant odours, a blank test was conducted, and the parasitoids showed no side preference.

Parasitoid olfactory orientation to the host and host plants in an odorous environment

The host complex consisted of flowering *P. lanceolata* and five female and five male weevil adults. Although the parasitoid attacks only larval host stages, we did not conduct experiments with host plants infested by weevil larvae as weevils kept in the laboratory did not lay eggs; hence, no host plants with larvae-infested seeds were available in the laboratory. We did not collect host plants infested by weevil larvae from the field since it would have been impossible to determine precisely for how long the plants had been infested and thus, no consistent conditions would have been provided when using these plants. However, we took an alternative approach since in preliminary tests the odour of the host complex consisting of adult weevils plus the host plant was attractive for the parasitoid, whereas odour of the host plant and odour of host

adults tested separately were not attractive (data not shown). It is well known that parasitoids parasitising inconspicuous hosts or host stages (here: larvae hidden within seeds) may also use cues from non-appropriate host stages (here: host adults) (infochemical detour; [34,41]). In our study system, weevil females lay their eggs in the seeds of *P. lanceolata* inflorescences in June and July and stay in the habitat where they have oviposited (pers. observation: N. Wäschke); hence, the host plant plus adult host weevils provide a suitable odour source to test the olfactory orientation abilities of the parasitoid.

In order to test the effect of non-host plants on host location, the parasitoids olfactory response to the following combinations was tested: (a) odour of the host complex: flowering host plant *P. lanceolata* with five female and five male weevil adults *versus* a control (N = 50); (b) odour of two non-host plants (*A. millefolium* and *A. eupatoria*) plus host complex *versus* a control (N = 50); and (c) odour of the host complex (see (a)) *versus* odour of the two non-host plants plus host complex (see (b)) (N = 50). As a control we used a pot filled with soil.

Field study

Effect of plant diversity on herbivore and parasitoid abundance. The field study was conducted as a subproject of a German priority project entitled "Biodiversity Exploratories" (described in detail by [35]). In three geographical regions (exploratories) in Germany (from north to south: Schorfheide-Chorin Biosphere Reserve, Hainich-Dün National Park, and Schwäbische Alb Biosphere Area) 50 grassland plots were assigned to biodiversity research. A plot (50 m×50 m) is almost homogenous with respect to soil type and vegetation properties. The three regions across Germany differ in environmental variables, i.e. in precipitation, altitude, and annual mean temperature [35]. The grassland sites are subjected to different land use and thus show differences in plant diversity. Since land use intensity and plant diversity are negatively correlated with each other [42], we neglected land use intensity and focused on plant diversity effects on insect abundance.

The occurrence of the host plant (*P. lanceolata*) in the three regions determined the number of plots studied per region: N = 21 plots in Schorfheide-Chorin, $N = 22$ in Hainich-Dün, and $N = 34$ in Schwäbische Alb. Within each plot we sampled ten randomly chosen focal *P. lanceolata* plants distributed across the plot. The number of herbaceous plant species and the vertical coverage of each plant species ($r = 15$ cm) as well as host plant density ($r = 100$ cm) around the chosen focal plant were surveyed once in June 2008. To determine the abundance of weevils and parasitoids we collected *P. lanceolata* inflorescences from July to August 2008. Since just a small number of insects might hatch from the ten focal plant inflorescences collected at each plot, we collected additionally 100 randomly chosen *P. lanceolata* inflorescences across the plot by following a random step pattern. Inflorescences of *P. lanceolata* were kept in plastic boxes (17.0 cm×12.5 cm×5.6 cm) with a top cover made of fine-meshed gauze (0.12 mm) under constant conditions (22°C ± 1°C, 50% rh, 11:13 LD). Adult weevils and parasitoids that emerged from the inflorescences in August and September 2008 were identified and counted. In addition to the weevil *M. pascuorum*, adults of a further weevil species emerged, *Mecinus labilis*. We also recorded the number of individuals of this species since it also serves as host for the larval parasitoid. Data are stored at the BExIS database of the Biodiversity Exploratories [43], http://www.biodiversity-exploratories.de/intranet/].

Statistical analysis

Laboratory assays. *Effect of plant diversity on olfactory orientation of the herbivore:* All calculations were performed using R [44]. The Wilcoxon one sample-test was used to evaluate the data obtained by recording the weevils response to the host plant odour (Table 1, a) in the two-circle olfactometer. We tested whether the time spent by the weevils in the central field differed significantly from the null hypothesis (33.3s, assuming equally long residence times in all areas of the olfactometer during an observation period of 300s). If the weevils duration of stay in the central field differed significantly from 33.3s, the weevils discriminated between odour in the central and the ambient field.

Furthermore, we compared the weevils olfactory responses to the odours provided by the different odour source combinations by a Kruskal-Wallis ANOVA followed by Mann-Whitney-*U*-tests with Bonferroni correction to account for non-normality of the data [45]. Variance homogeneity was checked by the Levene-test, and if necessary, logarithmic transformation was conducted.

Effect of plant diversity on olfactory orientation of the parasitoid. Data obtained from the parasitoid bioassay in the Y-tube olfactometer were analysed by the sign test [46]. Only parasitoids that made a decision were included in the analysis.

Field study

Effect of plant diversity on herbivore and parasitoid abundance. Field data were analysed using a generalised linear mixed model. Plant diversity was calculated according to the Shannon-Index $H = - \sum p_i \times \ln p_i$ where p_i is the ratio of the ith species compared to the entire pool. We calculated mean values per plot for host plant density and plant diversity. The region was used as a random effect. Explanatory variables with non-normal distribution were ln transformed for stabilising variance [47]. A term was added (+1) before transformation if necessary. Models were calculated by the lmer function with Laplace approximation in R (package lme4 Version 0.999375-37) with a Poisson error distribution (link = log) for the abundance data as response variables. To account for overdispersion we added an individual based random effect [48]. We started with the full model and discarded terms that were not significantly different from zero. Models were compared by Akaike Information Criterion (AIC, [49]) until we ended up with a minimal adequate model with the

AIC not decreasing anymore or all terms included in the model being significantly different from zero. As fixed effects we added plant diversity and host plant density since the availability of the host plant also impacts upon insect performance and affects the occurrence and abundance of herbivores [6] and their parasitoids [50]. To analyse the effect of vegetation parameters on the parasitoids abundance we corrected for the host abundance by including this variable in the model as a covariate. For calculating host abundance, adults of both weevil species, *M. pascuorum* and *M. labilis*, were counted.

Results

Laboratory assays

Effect of plant diversity on olfactory orientation of the herbivore. The two-circle olfactometer proved to be a suitable laboratory device for testing the orientation of the weevils in complex odorous environments. The weevils did not stay significantly longer in the central field compared to the value expected by the null hypothesis (33.3s; assuming equal duration of stay in the entire arena) when dummies were offered in the central and the ambient chamber (median duration of stay in central field: 0s; interquartile ranges: 0–47.2 s; $N = 20$, $P > 0.05$). Weevils stayed significantly longer in the central field with the host plant odour than expected (median duration of stay: 61.6s; interquartile ranges: 7.1–124.8 s; $N = 20$, $P < 0.05$; Table 1, a) and thus were attracted and/or arrested by odours from the flowering host plant *P. lanceolata*.

When comparing the weevils response to odour of the different plant combinations, the duration of stay in the central field and latency of females to reach the central field did not differ between the treatments with the various plant arrangements tested here ($\chi^2 = 0.432$, $\chi^2 = 2.638$ respectively, both: $df = 3$, $P > 0.05$; Table 1). Neither odour from non-host plants placed additionally to *P. lanceolata* beneath the central field (Table 1, b) nor that of non-host plants offered in the ambient field (Table 1, c) hampered the host finding process. Even though four non-host plants were present in the ambient chamber (Table 1, d), the herbivores still preferred the host plant odour. This indicates that non-host plant odours did not disturb olfactory orientation of female weevils towards flowering host plants.

However, when comparing the different treatments we observed significant differences in the overall walking activity ($\chi^2 = 9.805$, $df = 3$, $P < 0.05$) and the frequency by which the weevils crossed field borders ($\chi^2 = 10.366$, $df = 3$, $P < 0.05$). These behavioural parameters were enhanced when the weevils experienced odours from the four non-host plants in the ambient chamber additional to the host plant in the central chamber compared to the setup with only *P. lanceolata* placed in the central chamber and no other plants in the ambient chamber (Fig. 2).

Effect of plant diversity on olfactory orientation of the parasitoid. Dynamic Y-tube olfactometer studies were conducted to investigate the influence of non-host odours on the orientation of the parasitoid. The parasitoid was attracted to odour from the host complex consisting of flowering *P. lanceolata* and female and male weevils when tested against a control (a pot with soil) (Fig. 3, a). When odour from non-host plants was added to the host complex odour (i.e. increased plant diversity) and tested against a control, this additional non-host plant odour did not affect the attractiveness of the host complex (Fig. 3, b). However, when offering the parasitoids a choice between odour from the host complex only (without non-host plant odour) and odour from the host complex with non-host plant odour added, the parasitoid preferred the simpler odour bouquet (Fig. 3, c). The results

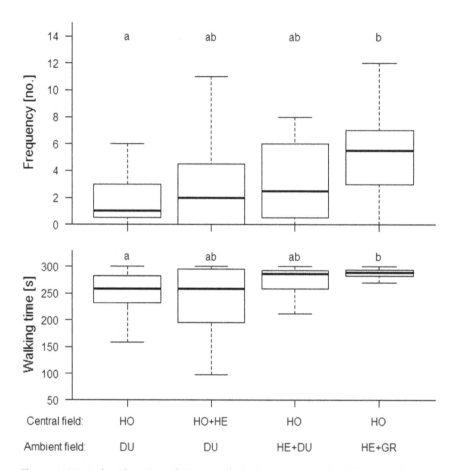

Figure 2. Host plant location of the weevil *Mecinus pascuorum* in different odorous surroundings. The olfactory orientation of female weevils to their host plant *Plantago lanceolata* was tested in the two-circle olfactometer during 300s (N = 20 females per treatment). Number of switches between the fields (frequency) and walking time (in seconds) are shown as medians and quartiles for combinations of host plant (HO) and non-host plant species (HE: herbs; GR: grasses) as well as dummies (DU). Different letters indicate significant ($P \leq 0.05$) differences (Kruskal-Wallis ANOVA followed by Mann-Whitney-*U* test and Bonferroni correction).

indicate that the parasitoid can distinguish between non-host plant and host plant odours.

Field study

Effect of plant diversity on herbivore and parasitoid abundance. To investigate whether abundance of the herbivore

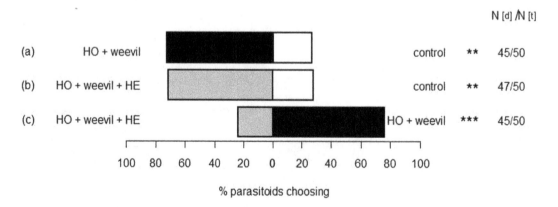

Figure 3. Response of the parasitoid *Mesopolobus incultus* to "host complex" odours in the presence of non-host plants. The % of total number of female parasitoids is shown that responded to odours offered in a Y-tube olfactometer. A pot with soil served as control (white bars). Two different odour sources were used: the host complex (HO: host plant *Plantago lanceolata*, weevils: five male and five female *Mecinus pascuorum*) (black bars) and the host complex plus herbs (HE: *Achillea millefolium* + *Agrimonia eupatoria*) (grey bars). Only parasitoids making a decision were included in the analysis. Numbers of parasitoids making a decision (N_d) and numbers of tested parasitoids (N_t) are given. Data were analysed by the sign test according to MacKinnon: **, $P \leq 0.01$; ***, $P \leq 0.001$.

and its parasitoid is linked with plant species diversity, we recorded vegetation data in three different regions in Germany (Table 2). The field data revealed a positive correlation of plant diversity with the abundance of the weevils; host plant density had no significant effect on weevil abundance (Table 3). The abundance of the parasitoid was independent of plant diversity, and best explained by the abundance of its hosts (both weevil species: *M. pascuorum* and *M. labilis*). Abundances of the parasitoids and the weevils correlated positively (Table 3).

Discussion

We examined the effect of plant diversity on host location behaviour and olfactory orientation of the weevil *M. pascuorum*, a specialist on *P. lanceolata*, and of the pteromalid wasp *M. incultus*, a larval parasitoid of this herbivore. We could demonstrate that (1) female weevils were attracted by odour of flowering *P. lanceolata* plants even when non-host plant odour was present in the surroundings. (2) Similarly, the parasitoid was attracted by odour of its host complex (volatiles from host plant and host) even when this was combined with non-host odour in a no-choice situation. However, the parasitic wasps preferred the "pure" host complex odour when having the choice between the "pure" odour and its combination with non-host plant odour. (3) In the field, increased plant diversity was positively correlated with the herbivores abundance, but not with the parasitoids abundance; the latter was positively linked with the herbivores abundance.

Laboratory assays

In the laboratory we tested the effect of non-host plant odour on the olfactory orientation of the weevil and its parasitoid. A new olfactometer setup was designed, suitable for testing behavioural responses of the weevils to spatially separated odours. It mimics the natural situation where an insect is approaching an odour source surrounded by background odours.

Effect of plant diversity on olfactory orientation of the herbivore

The herbivorous weevil was attracted by volatiles of flowering *P. lanceolata*. Our study shows that odours of various non-host plants

in the presence of host plants did not reduce the weevils success in finding the host plant by olfactory cues. Even when placing non-host and host plants very closely together in the central chamber, the weevils were able to olfactorily detect their host plant within the mixed blend and stayed longer in the central field.

Host location in insects may not only be affected by odours of non-host plants, but also by visual interference with non-host plant neighbours [16,51]. The walking arena of the olfactometer was built of fine-meshed gauze which made it difficult for the weevils to recognize structures of the plants placed several centimetres below the arena. To minimize the effect of colour, dummies made of green paper were used to simulate the green colour of the plants in the plant-free control field. Hence, structural plant cues were almost undetectable and colour cues were almost the same in all olfactometer fields. Thus, we concluded that orientation of the weevils towards a field was due to olfactory orientation rather than to visual orientation.

Although *M. pascuorum* was attracted to the host plant odour independently of non-host plants in the surroundings, high plant diversity with four non-host plants in the ambient chamber induced increased activity in the females. It has been suggested that intensive motion activity may help insects to separate different odour sources [52]; hence, the high locomotion activity of the weevil *M. pascuorum* in the presence of high plant diversity might support host plant location. In line with these thoughts, one might consider a suitable foraging habitat as an environment that elicits intensified host searching behaviour [53], i.e. increased locomotion activity. Since *P. lanceolata* emits only few volatiles [54], background odour released from co-occurring non-host plants might indicate the presence of a suitable habitat.

The effects of non-host plant odour on host foraging in insects vary with the plant – insect system considered. Non-host odour has been shown to impede host foraging in many insect species (e.g. [55]). In contrast, non-host odours or ubiquitous green leaf volatiles were found to have positive effects on host location of other insects (e.g. [28,56,57]). So far, it is not possible to detect common patterns that would allow predicting the impact of non-host odour on insect host foraging success. In our study, a rich odorous environment may stimulate weevils to search more intensively and thus, may improve the likelihood of locating a host

Table 2. Field data: effect of plant diversity on insect abundance.

Region		Host plant density (average per plot) [a]	Plant diversity (H) (average per plot) [a]	Abundance of *Mecinus pascuorum* (per plot) [b]	Abundance of *Mesopolobus incultus* (per plot) [b]
Schorfheide-Chorin	Central value	14.5±3.1	0.9±0.1	8 (0 – 18)	8 (0 – 29)
(north)	N	21	21	21	21
	Range	1.0 – 47.0	0.0 – 1.4	0 – 32	0 – 161
Hainich-Dün	Central value	17.4±3.3	1.5±0.1	1 (0 – 19)	8 (0 – 25)
(central)	N	22	22	22	22
	Range	1.0 – 52.0	0.9 – 2.0	0 – 137	0 – 618
Schwäbische Alb	Central value	22.9±4.5	1.7±0.1	0 (0 – 0)	8 (0 – 5)
(south)	N	34	34	34	34
	Range	2.0 – 111.0	0.9 – 2.4	0 – 4	0 – 31

Plant abundance and diversity (H = Shannon index), herbivore (*Mecinus pascuorum*) and parasitoid (*Mesopolobus incultus*) abundance are shown recorded in the three regions in Germany. Central tendencies ([a] mean (SE), [b] median (interquartile ranges)), number of plots (N), and ranges for the explanatory and the response variables are given for each region.

Table 3. Statistics: effect of plant diversity on insect abundance.

Explanatory variables	Abundance of *Mecinus pascuorum*				Abundance of *Mesopolobus incultus*			
	B	SE	z value	P	β	SE	z value	P
Intercept	−5.6026	2.1219	−2.640	**<0.01**	−0.5121	0.2782	−1.841	<0.1
Plant diversity (H)	2.5085	1.0397	2.413	**<0.05 (+)**	NA	NA	NA	NA
Host plant density [a]	0.5961	0.3311	1.800	<0.1	NA	NA	NA	NA
Host (weevil) abundance [a]	–	–	–	–	1.1133	0.1216	9.159	**<0.001 (+)**
AIC full model	214.7				250.7			
AIC minimal model	214.7				247.5			

Results of a generalised linear mixed model describing the abundances of the herbivorous weevil *Mecinus pascuorum* and the parasitoid *Mesopolobus incultus* in the field. Estimates (β) with standard errors (SE) are given for the minimal adequate model (evaluated by Akaike information criterion (AIC)). P values are marked bold if significant. Direction of relationship is given in parenthesis. Seventy-seven plots were involved in the analysis. [a] In transformed; NA: excluded from the model; –: not included in the full model.

plant. Although we cannot distinguish whether the higher weevil activity in the olfactometer was indeed caused by a higher number of plant species or by higher amount of plant biomass and thus a higher amount of volatiles, our laboratory and field data suggest that odour of vegetation with greater plant diversity presents a "patch of interest" for the weevil. Patches with host plants and high non-host plant diversity in the surroundings may provide enhanced host plant quality or offer more refuge areas allowing weevils to escape from natural enemies or competitors [58].

Effect of plant diversity on olfactory orientation of the parasitoid

The larval parasitoid was attracted by the host complex odour (i.e. odour from the host plant and the host insect, i.e. the weevils) and was capable of discrimination between host complex odour offered with and without non-host plant odour. The parasitoid responded more strongly to the simpler odour bouquet lacking non-host odour, but was also attracted to the odour of the host complex offered together with non-host plants. Parasitoids do not only use volatile cues from the host or the host plant for host location, but also cues emitted by the host habitat [59]. In the tritrophic system studied here the weevils lay their eggs in June and July and remain in the habitat where they have oviposited. The emission of volatiles from *P. lanceolata* was found to significantly increase after herbivory by generalist larvae that severely damaged the plant [54]. However, adult *Mecinus* weevils chew only upon small parts of the flower stem. Therefore, the adult weevils are not expected to induce *P. lanceolata* in such a way that it leads to a significantly higher volatile emission. The quantity of emitted and perceived plant volatiles is important for parasitoids when searching for herbivorous hosts [25]. Thus, it might be beneficial for the parasitoid to positively respond to general cues from the habitat where the host with its host plant occurs. In conclusion, the parasitoid may use habitat odour for long-range orientation and might be attracted to the host complex in combination with non-host plants as habitat odour. As shown for different parasitoid species, the presence of non-host plants does not always hinder the close-range foraging activities [60]. However, the parasitoid may respond specifically to the pure host complex at a short-range scale when a choice between odours of the host complex and the surroundings is possible. This was shown here where the parasitoid distinguishes between the host complex offered alone and in combination with non-host plants.

Field study

Effect of plant diversity on herbivore and parasitoid abundance. In agricultural habitats the abundance of specialised herbivorous insects was shown to decrease with increasing plant species diversity, thus indicating that specialist herbivores are negatively affected by plant diversity (reviewed by [16,61]). In contrast, when considering plant - herbivore interactions of non-crop species in a natural or semi-natural context, beside negative effects of plant diversity [14], several studies found a positive effect of high plant diversity on plant damage by herbivory [62], on the probability of herbivore occurrence [12], and on herbivore abundances [7]. The results of our study corroborate these latter findings.

In our field study the abundance of the parasitoid of weevil larvae was strongly associated with host abundance. The positive correlation found between abundances of weevils and parasitoids might be explained by improved oviposition possibilities for the parasitoids in patches with high host density, thus leading to an aggregation of parasitoids in said patches [63]. The abundance of the herbivorous host considered here is positively correlated with plant diversity which often correlates positively with herbivore diversity [64]. Since the studied parasitoid species can parasitize also other insect hosts in the same habitat, not only higher host density but also higher host diversity could be an explanation for the positive correlation between herbivore and parasitoid abundance in this study [65]. However, parasitoid abundances are often not affected by just a single environmental factor [11,64].

Conclusions

In our laboratory bioassays we have shown that both an herbivorous and parasitic insect are not prevented from successful host location when plant diversity increases. Furthermore, the laboratory study revealed that odour from a species-rich vegetation enhanced the weevils searching activity for host plants. In the field, the abundance of herbivorous weevils was positively correlated with plant diversity. In grasslands, diverse habitats may constitute high quality patches where numerous multitrophic interactions characterise complex food webs. To gain a deeper understanding of the mechanisms which shape the positive relationship between communities of high plant species diversity and the organisms involved in multitrophic interactions, further laboratory and field studies are necessary. These should experimentally alter the chemical diversity of habitat odour and disentangle odour-mediated plant diversity effects on insect abundances from other parameters that vary with changing plant species diversity.

Acknowledgments

We thank the managers of the three Biodiversity Exploratories, Swen Renner, Sonja Gockel, Andreas Hemp, Martin Gorke, and Simone Pfeiffer for their work in maintaining the plot and project infrastructure, and Markus Fischer, the late Elisabeth K. V. Kalko, K. Eduard Linsenmair, Dominik Hessenmöller, Jens Nieschulze, Daniel Prati, Ingo Schöning, François Buscot, Ernst-Detlef Schulze, and Wolfgang W. Weisser for their role in setting up the Biodiversity Exploratories project. We thank Stefan Vidal and Peter Sprick for identifying parasitoids and weevils. Furthermore we thank Franziska Birr, Ivonne Halboth, Sabina Reschke and Christoph Rothenwöhrer for their fieldwork or laboratory assistance. We thank Thomas Dürbye from the seed bank of the Botanical Garden Berlin for seeds. Jeff Powell and Steffen Wagner provided statistical advice. We thank two anonymous reviewers for their comments on the manuscript and Ian Dublon for his linguistic revision.

Author Contributions

Conceived and designed the experiments: TM EO NW MH CH KH. Performed the experiments: NW KH CH. Analyzed the data: NW. Contributed reagents/materials/analysis tools: TM MH. Wrote the paper: NW TM MH. Revised the manusript: EO CH KH.

References

1. Schoonhoven LM, Van Loon JJA, Dicke M (2005) Insect-Plant Biology. 2nd ed. Oxford University Press. 421 p.
2. Vinson SB (1976) Host selection by insect parasitoids. Annu Rev Entomol 21: 109–133. doi: 10.1146/annurev.en.21.010176.000545.
3. Visser JH (1986) Host odor perception in phytophagous insects. Annu Rev Entomol 31: 121–144. doi: 10.1146/annurev.ento.31.1.121.
4. Godfray HCL (1994) Parasitoids. Princeton University Press. 488 p.
5. Hilker M, McNeil J (2008) Chemical and behavioral ecology in insect parasitoids: how to behave optimally in a complex odorous environment. In: Wajnberg E, Bernstein C, van Alphen J, editors. Behavioral Ecology of Insect Parasitoids. Blackwell Publishing, Oxford, UK. 92–112.
6. Root RB (1973) Organization of a plant-arthropod association in simple and diverse habitats: the fauna of collards (Brassica oleracea). Ecol Monogr 43: 95–120. doi: 10.2307/1942161.
7. Unsicker SB, Baer N, Kahmen A, Wagner M, Buchmann N, et al. (2006) Invertebrate herbivory along a gradient of plant species diversity in extensively managed grasslands. Oecologia 150: 233–246. doi: 10.1007/s00442-006-0511-3.
8. Randlkofer B, Obermaier E, Hilker M, Meiners T (2010) Vegetation complexity - the influence of plant species diversity and plant structures on plant chemical complexity and arthropods. Basic Appl Ecol 11: 383–395. doi: 10.1016/j.baae.2010.03.003.
9. Sheehan W (1986) Response by specialist and generalist natural enemies to agroecosystem diversification: a selective review. Environ Entomol 15: 456–461.
10. Bukovinszky T, Gols R, Hemerik L, van Lenteren JC, Vet LEM (2007) Time allocation of a parasitoid foraging in heterogeneous vegetation: implications for host-parasitoid interactions. J Anim Ecol 76: 845–853. doi: 10.1111/j.1365-2656.2007.01259.x.
11. Petermann JS, Müller CB, Weigelt A, Weisser WW, Schmid B (2010) Effect of plant species loss on aphid–parasitoid communities. J Anim Ecol 79: 709–720. doi: 10.1111/j.1365-2656.2010.01674.x.
12. Randlkofer B, Obermaier E, Meiners T (2007) Mother's choice of the oviposition site: balancing risk of egg parasitism and need of food supply for the progeny with an infochemical shelter? Chemoecology 17: 177–186. doi: 10.1007/s00049-007-0377-9.
13. Barbosa P, Hines J, Kaplan I, Martinson H, Szczepaniec A, et al. (2009) Associational resistance and associational susceptibility: having right or wrong neighbors. Annu Rev Ecol Evol Syst 40: 1–20. doi: 10.1146/annurev.ecolsys.110308.120242.
14. Kostenko O, Grootemaat S, van der Putten WH, Bezemer TM (2012) Effects of diversity and identity of the neighbouring plant community on the abundance of arthropods on individual ragwort (Jacobaea vulgaris) plants. Ent Ex Appl 144: 27–36. doi: 10.1111/j.1570-7458.2012.01251.x.
15. Randlkofer B, Obermaier E, Casas J, Meiners T (2010) Connectivity counts – disentangling effects of vegetation structure elements on the searching movement of a parasitoid. Ecol Entomol 35: 446–455. doi: 10.1111/j.1365-2311.2010.01200.x.
16. Finch S, Collier RH (2000) Host-plant selection by insects – a theory based on "appropriate/inappropriate landings" by pest insects of cruciferous plants. Ent Ex Appl 96: 91–102. doi: 10.1046/j.1570-7458.2000.00684.x.
17. Perfecto I, Vet LEM (2003) Effect of a nonhost plant on the location behavior of two parasitoids: the tritrophic system of Cotesia spp. (Hymenoptera: Braconidae), Pieris rapae (Lepidoptera: Pieridae), and Brassica oleraceae. Environ Entomol 32: 163–174. doi: 10.1603/0046-225X-32.1.163.
18. Beyaert I, Hilker M (2013) Plant odour plumes as mediators of plant – insect interactions. Biol Rev, in press, article first published online: 28 May 2013; doi: 10.1111/brv.12043.
19. Knudsen GK, Bengtsson M, Kobro S, Jaastad G, Hofsvang T, et al. (2008) Discrepancy in laboratory and field attraction of apple fruit moth Argyresthia conjugella to host plant volatiles. Physiol Entomol 33: 1–6. doi: 10.1111/j.1365-3032.2007.00592.x.
20. Schröder R, Hilker M (2008) The relevance of background odor in resource location by insects: a behavioral approach. Bioscience 58: 308–316. doi: 10.1641/B580406.
21. Thiery D, Visser JH (1986) Masking of host plant odour in the olfactory orientation of the Colorado potato beetle. Ent Ex Appl 41: 165–172.
22. Gohole LS, Overholt WA, Khan ZR, Vet LEM (2003) Role of volatiles emitted by host and non-host plants in the foraging behaviour of Dentichasmias busseolae, a pupal parasitoid of the spotted stemborer Chilo partellus. Ent Ex Appl 107: 1–9. doi: 10.1046/j.1570-7458.2003.00030.x.
23. Hori M, Komatsu H (1997) Repellency of rosemary oil and its components against the onion aphid, Neotoxoptera formosana (Takahashi) (Homoptera, Aphididae). Appl Entomol Zool 32: 303–310.
24. Sanon A, Dabire C, Huignard J, Monge JP (2006) Influence of Hyptis suaveolens (Lamiaceae) on the host location behavior of the parasitoid Dinarmus basalis (Hymenoptera: Pteromalidae). Environ Entomol 35: 718–724.
25. Dicke M, de Boer JG, Höfte M, Rocha-Granados MC (2003) Mixed blends of herbivore-induced plant volatiles and foraging success of carnivorous arthropods. Oikos 101: 38–48. doi: 10.1034/j.1600-0706.2003.12571.x.
26. Couty A, Van Emden H, Perry JN, Hardie J, Pickett JA, et al. (2006) The roles of olfaction and vision in host-plant finding by the diamond back moth, Plutella xylostella. Physiol Entomol 31: 134–145. doi: 10.1111/j.1365-3032.2006.00499.x.
27. Mozuraitis R, Stranden M, Ramirez MI, Borg-Karlson AK, Mustaparta H (2002) (-)-Germacrene D increases attraction and oviposition by the tobacco budworm moth Heliothis virescens. Chem Senses 27: 505–509. doi: 10.1093/chemse/27.6.505.
28. Mumm R, Hilker M (2005) The significance of background odour for an egg parasitoid to detect plants with host eggs. Chem Senses 30: 337–343. doi: 10.1093/chemse/bji028.
29. van Nouhuys S, Via S (1999) Natural selection and genetic differentiation of behavior between parasitoids from wild and cultivated habitats. Heredity 83: 127–137.
30. Unsicker SB, Kunert G, Gershenzon J (2009) Protective perfumes: the role of vegetative volatiles in plant defense against herbivores. Curr Opin Plant Biol 12: 479–485. doi: 10.1016/j.pbi.2009.04.001.
31. Kessler A, Baldwin IT (2001) Defensive function of herbivore-induced plant volatile emissions in nature. Science 291: 2141–2144. doi: 10.1126/science.291.5511.2141.
32. Pareja M, Moraes MCB, Clark SJ, Birkett MA, Powell W (2007) Response of the aphid parasitoid Aphidius funebris to volatiles from undamaged and aphid-infested Centaurea nigra. J Chem Ecol 33: 695–710. doi: 10.1007/s10886-007-9260-y.
33. Bezemer TM, Harvey JA, Kamp AFD, Wagenaar R, Gols R, et al. (2010) Behaviour of male and female parasitoids in the field: influence of patch size, host density, and habitat complexity. Ecol Entomol 35: 341–351. doi: 10.1111/j.1365-2311.2010.01184.x.
34. Castelo MK, van Nouhuys S, Corley JC (2010) Olfactory attraction of the larval parasitoid, Hyposoter horticola, to plants infested with eggs of the host butterfly, Melitaea cinxia. J Insect Sci 10: 1–16. doi:10.1673/031.010.5301.
35. Fischer M, Bossdorf O, Gockel S, Hänsel F, Hemp A, et al. (2010) Implementing large-scale and long-term functional biodiversity research: The Biodiversity Exploratories. Basic Appl Ecol 11: 473–485. doi: 10.1016/j.baae.2010.07.009.
36. Schubert R, Vent W (1990). Exkursionsflora von Deutschland 4. 8th ed. Volk und Wissen. 812p.
37. Mohd Norowi H, Perry JN, Powell W, Rennolls K (2000) The effect of spatial scale on interactions between two weevils and their parasitoid. Ecol Entomol 25: 188–196. doi: 10.1046/j.1365-2311.2000.00242.x.
38. Universal Chalcidoidea Database website, Natural History Museum, UK. Available: www.nhm.ac.uk/research-curation/research/projects/chalcidoids. Accessed 2013 May 2.
39. Dickason EA (1968) Observations on the biology of Gymnaetron pascuorum (Gyll.) (Coleoptera: Curculionidae). Coleopt Bull 22: 11–15.
40. Bruce RJA, Wadhams LJ, Woodcock CM (2005) Insect host location: A volatile situation. Trends Plant Sci 10: 269–274. doi: 10.1016/j.tplants.2005.04.003.
41. Vet LEM, Dicke M (1992) Ecology of infochemical use by natural enemies in a tritrophic context. Annu Rev Entomol 37: 141–172. doi: 10.1146/annurev.en.37.010192.001041.
42. Blüthgen N, Dormann CF, Prati D, Klaus V, Kleinebecker T, et al. (2012) A quantitative index of land-use intensity in grasslands: Integrating mowing,

grazing and fertilization. Basic Appl Ecol 13: 207–220. doi: 10.1016/j.baae.2012.04.001.

43. Lotz T, Nieschulze J, Bendix J, Dobbermann M, König-Ries B (2012) Diverse or uniform? - Intercomparison of two major German project databases for interdisciplinary collaborative functional biodiversity research. Ecol Inf 8: 10–19. doi: 10.1016/j.ecoinf.2011.11.004.

44. R Development Core Team. (2010) R: A language and environment for statistical computing. R Foundation for Statistical Computing, Vienna, Austria. Available: http://www.R-project.org. Accessed 2013 May 2.

45. Sachs L (1992) Angewandte Statistik. Anwendung statistischer Methoden. 7th ed. Springer Verlag. 846 p.

46. MacKinnon WJ (1964) Table for both the sign test and distribution-free confidence intervals of the median for sample sizes to 1,000. J Am Stat Assoc 59: 935–956.

47. Crawley MJ (2007) The R Book. John Wiley & Sons. 942 p.

48. Elston DA, Moss R, Boulinier T, Arrowsmith C, Lambin X (2001) Analysis of aggregation, a worked example: numbers of ticks on red grouse chicks. Parasitology 122: 563–569.

49. Burnham KP, Anderson DR (2002) Model Selection and Multimodel Inference: A Practical Information-Theoretic Approach. 2nd ed. Springer Verlag. 488 p.

50. Vanbergen AJ, Jones TH, Hails RS, Watt AD, Elston DA (2007) Consequences for a host-parasitoid interaction of host-plant aggregation, isolation, and phenology. Ecol Entomol 32: 419–427. doi: 10.1111/j.1365-2311.2007.00885.x.

51. Hambäck PA, Pettersson J, Ericson L (2003) Are associational refuges species-specific? Funct Ecol 17: 87–93. doi: 10.1046/j.1365-2435.2003.00699.x.

52. Wäschke N, Meiners T, Rostàs M (2013) Foraging strategies of insect parasitoids in complex chemical environments. In: Wajnberg E, Colazza S, editors. Recent Advances in Chemical Ecology of Insect Parasitoids, Wiley-Blackwell, UK. 193–224.

53. Pettersson J, Ninkovic V, Glinwood R, Al Abassi S, Birkett M, et al. (2008) Chemical stimuli supporting foraging behaviour of Coccinella septempunctata L. (Coleoptera: Coccinellidae): volatiles and allelobiosis. Appl Entomol Zool 43: 315–321. doi: 10.1303/aez.2008.315.

54. Fontana A, Reichelt M, Hempel S, Gershenzon J, Unsicker SB (2009) The effects of arbuscular mycorrhizal fungi on direct and indirect defense metabolites

of Plantago lanceolata L.. J Chem Ecol 35: 833–843. doi: 10.1007/s10886-009-9654-0.

55. Zhang Q-H, Schlyter F (2004) Olfactory recognition and behavioural avoidance of angiosperm nonhost volatiles by conifer-inhabiting bark beetles. Agric For Entomol 6: 1–19. doi: 10.1111/j.1461-9555.2004.00202.x.

56. Müller C, Hilker M (2000) The effect of a green leaf volatile on host plant finding by larvae of a herbivorous insect. Naturwissenschaften 87: 216–219. doi: 10.1007/s001140050706.

57. Reinecke A, Ruther J, Tolasch T, Francke W, Hilker M (2002) Alcoholism in cockchafers: orientation of male Melolontha melolontha towards green leaf alcohols. Naturwissenschaften 89: 265–269. doi: 10.1007/s00114-002-0314-2.

58. Gilbert LE, Singer MC (1975) Butterfly ecology. Annu Rev Ecol Syst 6: 365–397. doi: 10.1146/annurev.es.06.110175.002053.

59. Vet LEM, Lewis W, Carde R (1995) Parasitoid foraging and learning. In: Cardé RT, Bell WJ, editors. Chemical Ecology of Insects 2. Chapman & Hall New York. 65–101.

60. Gohole LS, Overholt WA, Khan ZR, Vet LEM (2005) Close-range host searching behavior of the stemborer parasitoids Cotesia sesamiae and Dentichasmias busseolae: influence of a non-host plant Melinis minutiflora. J Insect Behav 18: 149–169. doi: 10.1007/s10905-005-0472-0.

61. Altieri MA (1995) Biodiversity and Biocontrol: Lessons from insect pest management. Advances in Plant Pathology 11. Academic Press Ltd. 191–209.

62. Scherber C, Mwangi PN, Temperton VM, Roscher C, Schumacher J, et al. (2006) Effects of plant diversity on invertebrate herbivory in experimental grassland. Oecologia 147: 489–500. doi: 10.1007/s00442-005-0281-3.

63. Janz N (2002) Evolutionary ecology of oviposition strategies. In: Hilker M, Meiners T, editors. Chemoecology of Insect Eggs and Egg Deposition. Blackwell Publishing Oxford. 349–376.

64. Siemann E, Tilman D, Haarstad J, Ritchie M (1998) Experimental tests of the dependence of arthropod diversity on plant diversity. Am Nat 152: 738–750. doi: 10.1086/286204.

65. Andow DA (1991) Vegetational diversity and arthropod population response. Annu Rev Entomol 36: 561–586. doi: 10.1146/annurev.ento.36.1.561.

7

Potential Use of a Serpin from *Arabidopsis* for Pest Control

Fernando Alvarez-Alfageme[1*¤a], **Jafar Maharramov**[1], **Laura Carrillo**[1¤b], **Steven Vandenabeele**[2,3¤c], **Dominique Vercammen**[2,3¤d], **Frank Van Breusegem**[2,3], **Guy Smagghe**[1*]

1 Laboratory of Agrozoology, Department of Crop Protection, Faculty of Bioscience Engineering, Ghent University, Gent, Belgium, 2 VIB Department of Plant Systems Biology, Ghent University, Gent, Belgium, 3 Department of Plant Biotechnology and Genetics, Ghent University, Gent, Belgium

Abstract

Although genetically modified (GM) plants expressing toxins from *Bacillus thuringiensis* (*Bt*) protect agricultural crops against lepidopteran and coleopteran pests, field-evolved resistance to *Bt* toxins has been reported for populations of several lepidopteran species. Moreover, some important agricultural pests, like phloem-feeding insects, are not susceptible to *Bt* crops. Complementary pest control strategies are therefore necessary to assure that the benefits provided by those insect-resistant transgenic plants are not compromised and to target those pests that are not susceptible. Experimental GM plants producing plant protease inhibitors have been shown to confer resistance against a wide range of agricultural pests. In this study we assessed the potential of AtSerpin1, a serpin from *Arabidopsis thaliana* (L). Heynh., for pest control. *In vitro* assays were conducted with a wide range of pests that rely mainly on either serine or cysteine proteases for digestion and also with three non-target organisms occurring in agricultural crops. AtSerpin1 inhibited proteases from all pest and non-target species assayed. Subsequently, the cotton leafworm *Spodoptera littoralis* Boisduval and the pea aphid *Acyrthosiphon pisum* (Harris) were fed on artificial diets containing AtSerpin1, and *S. littoralis* was also fed on transgenic *Arabidopsis* plants overproducing AtSerpin1. AtSerpin1 supplied in the artificial diet or by transgenic plants reduced the growth of *S. littoralis* larvae by 65% and 38%, respectively, relative to controls. Nymphs of *A. pisum* exposed to diets containing AtSerpin1 suffered high mortality levels ($LC_{50} = 637$ μg ml^{-1}). The results indicate that AtSerpin1 is a good candidate for exploitation in pest control.

Editor: Miguel A. Blazquez, Instituto de Biología Molecular y Celular de Plantas, Spain

Funding: The authors acknowledge the support of Ghent University, particularly the Industrial Research Fund (IOF) and the Multidisciplinary Research Partnership Ghent Bio-economy, of COST Action 862 to LC, and of Fund for Scientific Research-Flanders (FWO-Vlaanderen) to DV. SV was recipient of a BELSPO grant (12TKM106) and a Marie Curie Fellowship (MOIF-CT-2004). The funders had no role in study design, data collection and analysis, decision to publish, or preparation of the manuscript.

Competing Interests: The authors have declared that no competing interests exist.

* E-mail: fernando.alvarez@art.admin.ch (F-AA); guy.smagghe@ugent.be (GS)

¤a Current address: Agroscope Reckenholz-Tänikon Research Station (ART), Zurich, Switzerland
¤b Current address: Centro de Biotecnología y Genómica de Plantas (UPM-INIA), Universidad Politécnica de Madrid, Pozuelo de Alarcón, Madrid, Spain
¤c Current address: Innogenetics N.V., Zwijnaarde, Belgium
¤d Current address: Crop Design N.V., Zwijnaarde, Belgium

Introduction

Herbivorous pests of major crops are estimated to reduce yields by 8–15% worldwide [1]. Engineering crop plants for endogenous resistance to insect pests has been an important success of molecular technology. Currently, genetically modified (GM) plants expressing δ-endotoxins from *Bacillus thuringiensis* (*Bt*) are providing significant control of agricultural insect pests and have reduced pesticide usage and production costs [2], [3]. The area sown with *Bt* crops has increased each year since 1996, when the first *Bt* crops were cultivated; in 2010, *Bt* crops were planted on 58 million hectares [4].

As farmers increasingly plant insect-resistant GM crops, selection pressure for the development of insect pests resistant to *Bt* toxins is also increasing. To date, field-evolved resistance has been documented in populations of five lepidopteran species [5]. Moreover, the efficacy of commercial *Bt* crops for some lepidopteran pests, such as the cotton leafworm *Spodoptera littoralis* Boisduval, is limited [6],[7], and phloem feeding pests including aphids are not susceptible to *Bt* crops [8]. Hence, complementary pest control strategies are necessary both to assure that the benefits provided by insect-resistant transgenic plants are not compromised and to target those pests that are not susceptible to *Bt* toxins. A summary of the strategies currently being investigated can be found in [8–11]. Among these, GM crops producing plant serine or cysteine protease inhibitors have been shown to confer resistance against a wide range of agricultural pests [12]. Protease inhibitors contribute to plant defense by inhibiting invertebrate proteases and, consequently, by reducing the availability of amino acids necessary for invertebrate growth and development. Transgenic plants expressing protease inhibitors, however, rarely achieve the same level of pest control as transgenic plants expressing *Bt* toxins [13] because herbivores are able to use several strategies to adapt to the inhibitors [12]. Still, plant protease inhibitors have the potential to be effective insecticidal proteins if insect adaptation to them can be overcome. For example, the combination of two protease inhibitors can lead to adverse effects on the target species that are not obtained with either inhibitor alone [14].

Serpins (serine protease inhibitors or classified inhibitor family I4) are the largest and most broadly distributed superfamily of protease inhibitors [15]. Serpin-like genes have been identified in animals, plants, bacteria, and some viruses [16]. Most serpins are irreversible inhibitors of serine proteases of the chymotrypsin family, although some have evolved to inhibit other types of serine proteases, and a few are also able to inhibit cysteine proteases [17–21]. Furthermore, some serpins have the ability to form complexes with very divergent proteases [22]. Serpins are involved in a number of fundamental biological processes, and a role in the protection of storage tissue against insects and pathogens has been proposed for plant serpins [23], [24]. Consistent with the idea that serpins protect against plant pests, the survival and fecundity of the green peach aphid *Myzus persicae* (Sulz.) were strongly and negatively correlated with the level of the serpin CmPS-1 in the phloem sap of *Cucurbita maxima* Duchesne [25]. A related serpin from *Cucurbita sativa* L., CsPS-1, is also thought to play a role in defense against herbivores [26].

Here we assessed the potential of AtSerpin1, a serpin from *Arabidopsis thaliana* (L). Heynh., for pest control. *In vitro* assays were conducted to measure the inhibitory activity of AtSerpin1 against a range of pest species that rely mainly on either serine or cysteine proteases for digestion. Because insect-resistant GM plants should ideally control target species without harming non-target arthropods, a decomposer, a pollinator, and a predator were included in these *in vitro* assays. Subsequently, two pest species, *S. littoralis* and the pea aphid *Acyrthosiphon pisum* (Harris), were used in *in vivo* assays on artificial diets containing AtSerpin1. Finally, transgenic *Arabidopsis* plants overproducing AtSerpin1 were tested against *S. littoralis*.

Materials and Methods

Invertebrates

Pest species. A permanent colony of *A. pisum* was reared on broad bean, *Vicia faba* L., plants. A laboratory colony of *S. littoralis* was maintained on an agar-based artificial diet. The two-spotted spider mite, *Tetranychus urticae* Koch, was reared on *Phaseolus vulgaris* L. bean plants in the laboratory, and the Colorado potato beetle, *Leptinotarsa decemlineata* (Say), was reared on fresh leaves of *Solanum tuberosum* L. in the laboratory. Frozen larvae of the Mediterranean corn borer, *Sesamia nonagrioides* Lefèbvre, were provided by Dr. Félix Ortego (Centro de Investigaciones Biológicas, CSIC, Madrid, Spain). All stages of a permanent insect colony of the red flour beetle, *Tribolium castaneum* Herbst, and of the yellow mealworm, *Tenebrio molitor* L., were kept on wheat flour mixed with brewer's yeast (10/1, w/w).

Non-target species. Large earth bumblebees, *Bombus terrestris* (L.), and green lacewings, *Chrysoperla carnea* (Stephens), were purchased from Biobest NV (Westerlo, Belgium) and reared in our laboratory for several generations with commercial sugar water and pollen and eggs of the flour moth *Ephestia kuehniella* Zeller, respectively. Common earthworms, *Lumbricus terrestris* L., were collected in an agricultural field in Ghent (Belgium) and frozen in the laboratory upon arrival.

All laboratory colonies were reared in environmental chambers at 24±2°C, 65±5% RH, and a 16:8 h (L:D) photoperiod.

Preparation of extracts

Adults of *A. pisum*, *T. castaneum*, and *T. molitor*, and a mixture of all stages of *T. urticae* were collected from the rearing colonies, homogenized in 0.15 M NaCl, centrifuged at 10,000 *g* for 5 min and stored frozen at −20°C until needed. Last instar larvae of *S. littoralis*, *S. nonagrioides*, *L. decemlineata*, and *C. carnea*, and adults of *L.*

terrestris, *B. terrestris*, and *C. carnea* were dissected in ice-cold 0.15 M NaCl, and the midguts and contents were removed. Each midgut was subsequently homogenized in 500 µl of 0.15 M NaCl. The suspensions were then centrifuged at 10,000 *g* for 5 min, and the supernatants were stored frozen at −20°C until needed. Before extracts were frozen, total protein content was determined according to the method of Bradford [27] with bovine serum albumin (BSA) as a standard.

Production of recombinant AtSerpin1

Recombinant Atserpin1 was produced and purified as described in Vercammen et al. [33]. The cDNA for the ORF of At1g47710 was obtained by RT-PCR with the following forward and reverse primers, provided with the adequate 5′ extensions by Gateway® cloning (Invitrogen, Merelbeke, Belgium): 5′-ATGGACGTGC-GTGAATC-3′ and 5′-TTAATGCAACGGATCAACAAC-3′. After recombination in pDEST17, the plasmid was introduced into *E. coli* strain BL21(DE3)pLysE, and production of the HIS6-tagged protein was induced by incubation in 0.2 mM isopropyl-β-D-thiogalactopyranoside for 24 h. The protein was purified by metal ion affinity chromatography (TALO™; BD, Franklin Lakes, NJ). Protein concentration and purity were checked by Bradford analysis (Bio-Rad, Nazareth, Belgium) and SDS-polyacrylamide gel electrophoresis (PAGE).

In vitro inhibitory activity of AtSerpin1 against invertebrate digestive proteases

To elucidate the potential of AtSerpin1 to inhibit invertebrate digestive proteases, several species known to rely either on serine or cysteine proteases for protein digestion were selected for *in vitro* experiments. Specifically, the ability of AtSerpin1 to inhibit the trypsin- and chymotrypsin-like serine activities in extracts of *S. littoralis*, *S. nonagrioides*, *T. molitor*, *L. terrestris*, *B. terrestris*, and *C. carnea* was tested using the substrates ZPR-AMC (*N*-carbobenzoxy-Phe-Arg-7-amido-4-methylcoumarin) and SLLVT-AMC (*N*-Suc-Leu-Leu-Val-Tyr-7-amido-4-methylcoumarin), respectively, and 0.1 M Tris-HCl buffer (pH 9.0). Inhibition of the cathepsin B- and L-like cysteine activities by AtSerpin1 in extracts of *A. pisum*, *T. castaneum*, *L. decemlineata*, and *T. urticae* was determined using the substrates ZRR-AMC (*N*-carbobenzoxy-Arg-Arg-7-amido-4-methylcoumarin) and ZPR-AMC, respectively, and 0.1 M phosphate buffer (pH 5.0). The standard assay used 5 µg of protein extract in a volume of 100 µl. AtSerpin1 was added at different final concentrations, ranging from 0.15 to 10 µM, and was incubated with the extracts for 15 min at room temperature. The substrate was then added to a final concentration of 0.2 mM. The reaction was incubated at 30°C for 45 min, and the emitted fluorescence was measured with a 365 nm excitation wavelength filter and a 465 nm emission wavelength filter. Results were expressed as a percentage of protease activity relative to that in the absence of the inhibitor. All assays were carried out in duplicate with pooled extracts.

Generation of *Arabidopsis* plants overproducing AtSerpin1

Transgenic *Arabidopsis* plants overproducing AtSerpin1 were generated to further investigate the potential of the serpin against *S. littoralis* larvae. The cDNA for the ORF of At1g47710 was obtained by reverse transcription-PCR with the following forward and reverse primers, provided with the adequate 5′ extensions for Gateway® cloning: 5′-ATGGACGTGCGTGAATC-3′ and 5′-TTAATGCAACGGATCAACAAC-3′. The ORF was cloned into the binary vector pB7GW2 [28] via Gateway® recombination. In the resulting vector, the ORF was under transcriptional control of the promoter of the cauliflower mosaic virus 35S (CaMV 35S); the glufosinate ammonium resistance gene was present to allow for transgene selection. Binary constructs were transformed into *Agrobacterium tumefaciens* strain C58C1RifR[pMP90], and transgenic *Arabidopsis* Columbia-0 were obtained via floral dip transformation [29] and subsequent selection. Serpin overexpression was assessed by immunoblotting using antisera against AtSerpin1 [19]. Three single-locus homozygous lines with high transgenic protein expression were selected for further analysis by Western blot.

In vivo effect of AtSerpin1

In vivo experiments with *S. littoralis* and *A. pisum* were used to assess the potential of AtSerpin1 for pest control. These two herbivorous species were selected because (i) they are serious pests of several agricultural crops, (ii) they rely on different proteolytical enzymes for protein digestion, and (iii) our *in vitro* studies demonstrated that they are both highly susceptible to AtSerpin1 (see Results).

Effect of purified AtSerpin1 on *S. littoralis*. Third-instar *S. littoralis* larvae (8–10 mg each) from the laboratory colony were starved for 4 h before the start of the bioassay. Subsequently, four larvae were placed in a Petri dish (9 cm diameter) and fed *ad libitum* for 6 days with artificial diet containing 0 (control), 65, or 650 μg g^{-1} AtSerpin1. Larvae were weighed on day 2, 4, and 6. Each treatment was represented by 12 replicate Petri dishes.

At the end of the feeding assay, 24 larvae from the control and the 65 μg g^{-1} AtSerpin1 treatment were selected randomly and the midguts were dissected. Susbsequently, the serine-like proteolytic activites trypsin, chymotrypsin, and elastase were quantified as described by Ortego et al. [30].

Effect of transgenic *Arabidopsis* overproducing AtSerpin1 on *S. littoralis*. Second-instar *S. littoralis* larvae (2.5–3.0 mg each) from the laboratory colony were starved for 4 h and transferred to pots planted with 4-week-old *Arabidopsis*: transgenic lines overproducing AtSerpin1 (lines AtSerpin^{OE1}, AtSerpin^{OE2}, and AtSerpin^{OE3}) or the non-transformed line Col-0. Four larvae were confined per pot and allowed to feed for 4 days. Larvae were weighed on the second and the fourth day. Six pots per line were used, resulting in 24 larvae per treatment.

Both bioassays with *S. littoralis* were carried out in a growth chamber at 24±2°C, 65±5% RH, and a 16:8 h (L:D) photoperiod.

Effect of purified AtSerpin1 on *A. pisum* survival. Reproductive adults from the *A. pisum* laboratory colony were collected and transferred to fresh bean leaves, where they were allowed to produce nymphs for 12 h. Experimental arenas consisted of sachets containing 130 μl of artificial diet as described by Shahnaz et al. [31]. Neonate (<12 h) *A. pisum* nymphs were then brushed carefully onto sachets containing AtSerpin1 at concentrations ranging from 0 to 1000 μg ml^{-1}. Fifteen nymphs were confined in each sachet, and three to six sachets were used per treatment. Nymphal survival was recorded after 3 days, and Abbott's correction for natural mortality was applied [32].

Effect of purified AtSerpin1 on *A. pisum* proteolytic activities. Neonate nymphs (<12 h) from the permanent *A. pisum* culture were placed on sachets containing 0 or 1000 μg ml^{-1} AtSerpin1, and allowed to feed for 24 h. Three sachets containing 15 nymphs were used per treatment. After the feeding period, aphids from every sachet were collected, homogenized in 0.15 M NaCl, and stored frozen at −20°C until required. Finally, digestive enzyme activities were measured as described by Carrillo et al. [33].

Both bioassays with *A. pisum* were performed in a growth chamber at 24±2°C, 65±5% RH, and a 16:8 h (L:D) photoperiod.

Statistical analysis

A one-way analysis of variance (ANOVA) followed by a Student-Newman-Keuls test was used to compare *S. littoralis* larval growth among the different treatments in both bioassays and to compare the proteolytic activities of *A. pisum* fed with artificial diet with or without AtSerpin1. Proteolytic activities of *S. littoralis* larvae fed either with control diet or diet incorporating AtSerpin1 were analyzed using the Mann-Whitney *U* test because data were not normally distributed. Differences between treatments were considered significant at $P<0.05$. The concentration of AtSerpin1 causing 50% mortality (LC$_{50}$) on aphid nymphs was analyzed using nonlinear sigmoid curve fitting using the GraphPad Prism 4.0 software (GraphPad Software Inc., La Jolla, CA).

Results

In vitro inhibitory activity of AtSerpin1 against invertebrate digestive proteases

The inhibitory activity of AtSerpin1 was tested *in vitro* against serine or cysteine proteases from several invertebrate pest and non-target species (Table 1). The inhibition of trypsin- and chymotrypsin-like serine activities was investigated in extracts of the pests *S. littoralis*, *S. nonagrioides*, and *T. molitor*, and in extracts of the non-targets *L. terrestris*, *B. terrestris*, and *C. carnea* (Table 2). For all species, inhibition of trypsin activity by 10 μM AtSerpin1 was higher than 80%. The trypsin activities of the non-targets *C. carnea* larvae and *B. terrestris* were highly susceptible to AtSerpin1, with an inhibition of 70% and 90%, respectively, at the lowest concentration tested (0.15 μM). AtSerpin1 also inhibited chymotrypsin activity in all species tested, except in the case of *S. nonagrioides*.

The inhibition of cathepsin B- and L-like cysteine activities was determined in extracts of the pest species *A. pisum*, *T. castaneum*, *L. decemlineata*, and *T. urticae* (Table 2). AtSerpin1 inhibited the hydrolysis of the substrate ZRR-AMC in all species studied, although it never caused more than 75% inhibition, suggesting that cathepsin B activity is much less susceptible than trypsin activity to AtSerpin1. Inhibition of cathepsin L activity was also detected in these four species.

Effect of AtSerpin1 on S. littoralis

Bioassay with artificial diet. Ingestion of artificial diets containing the protease inhibitor AtSerpin1 markedly reduced the weight gain of *S. littoralis* (Figure 1). A significant difference ($P<0.001$) occurred after only 2 days of exposure when third instars were reared on artificial diet containing 650 μg g^{-1} AtSerpin1. This difference continued throughout the bioassay, and on day 6, the weight increase was 65% ($P<0.001$) lower for *S. littoralis* larvae ingesting the inhibitor than for the control. For larvae exposed to 65 μg g^{-1} AtSerpin1, weight gain was significantly reduced by 20% on day 4 and by 33% on day 6 relative to the control (Figure 2).

To investigate the physiological background, biochemical analysis were carried out on guts of *S. littoralis* larvae dissected at the end of the feeding assay. Trypsin and chymotrypsin activities were significantly reduced in those fed on artificial diet incorporating 65 μg g^{-1} AtSerpin1 compared to those feeding on control diet, whereas no differences were observed for elastase activity (Figure 2).

Bioassay with transgenic *Arabidopsis*. Three transgenic *Arabidopsis* lines overproducing AtSerpin1 were tested against *S.*

Table 1. Ecological function and main digestive proteases of the invertebrate species tested *in vitro* against AtSerpin1.

Species name	Ecological function	Main proteases	Reference
Spodoptera littoralis (Lepidoptera: Noctuidae)	Herbivory	SEP	[34]
Sesamia nonagrioides (Lepidoptera: Noctuidae)	Herbivory	SEP	[30]
Tenebrio molitor (Coleoptera: Tenebrionidae)	Herbivory	SEP, CEP*	[35]
Lumbricus terrestris (Annelida: Lumbricidae)	Decomposition	SEP	[36]
Bombus terrestris (Hymenoptera: Apidae)	Pollination	SEP	[37]
Chrysoperla carnea (Neuroptera: Chrysopidae)	Predation	SEP	[38]
Acyrthosiphon pisum (Homoptera: Aphididae)	Herbivory	CEP	33
Tribolium castaneum (Coleoptera: Tenebrionidae)	Herbivory	CEP	[39]
Leptinotarsa decemlineata (Coleoptera: Chrysomelidae)	Herbivory	CEP	[40]
Tetranychus urticae (Acari: Tetranychidae)	Herbivory	CEP	[41]

Abbreviations: SEP = serine endoproteases; CEP = cysteine endoproteases.
*Only SEP were tested against AtSerpin1.

littoralis Expression of AtSerpin1 in leaves of *Arabidopsis* was confirmed by Western blot using increasing concentrations of purified AtSerpin1 (Figure 3). Differences in the AtSerpin1 expression among the transgenic lines was observed, being higher in AtSerpin^{OE2} and AtSerpin1^{OE3}. In those lines, about 5 ng AtSerpin ug^{-1} of total protein content was measured.

Second-instar larvae were fed for 4 days on transgenic or non-transformed plants, and the increase of weight was measured (Figure 4). No significant differences were observed when the transgenic lines AtSerpin^{OE1} and AtSerpin^{OE2} were compared

with the control plants. However, the increase of weight was 25% lower ($P<0.001$) on day 2 and 38% lower ($P<0.001$) on day 4 for *S. littoralis* larvae reared on the transgenic AtSerpin^{OE3} than for larvae fed on non-transformed plants.

Effect of AtSerpin1 on *A. pisum*

A. pisum nymphs reared for 3 days on diets containing 100 to 1000 µg ml^{-1} AtSerpin1 were highly susceptible to the inhibitor (Figure 5). Mortality reached 77.4% when *A. pisum* were fed 1000 µg ml^{-1} of the serpin. The effective AtSerpin1 concentra-

Table 2. *In vitro* inhibitory activity of the protease inhibitor AtSerpin1 against trypsin- and chymotrypsin-like serine, and cathepsin B- and cathepsin L-like cysteine activities in extracts of several pest and non-target invertebrate species.

	Inhibition (%)					
	Trypsin activity			Chymotrypsin activity		
Species name	0.15 µM	1.25 µM	10 µM	0.15 µM	1.25 µM	10 µM
*Spodoptera littoralis**	11.8±0.8	52.3±0.3	90.4±0.7	18.2±4.0	30.7±15.3	39.7±14.8
*Sesamia nonagrioides**	36.0±1.3	89.8±1.3	97.5±0.3	ni	ni	ni
*Tenebrio molitor**	51.2±3.5	94.9±0.6	92.3±3.6	7.2±0.8	14.1±2.3	45.3±3.8
Lumbricus terrestris†	ni	ni	82.7±1.2	-	-	-
Bombus terrestris†	91.8±0.5	97.8±0.3	98.5±0.1	-	-	-
Chrysoperla carnea†						
Larvae	66.6±3.5	94.0±3.2	98.8±1.3	ni	51.1±7.2	75.5±1.3
Adults	56.2±1.5	99.2±0.4	99.5±1.0	ni	15.8±2.5	66.1±5.7
	Cathepsin B activity			Cathepsin L activity		
	0.15 µM	1.25 µM	10 µM	0.15 µM	1.25 µM	10 µM
*Acyrthosiphon pisum**	38.6±1.3	49.9±0.1	55.5±1.7	38.3±2.2	42.4±2.7	39.2±4.3
*Tribolium castaneum**	ni	ni	75.1±3.6	ni	ni	69.8±3.5
*Leptinotarsa decemlineata**	13.5±13.4	40.6±8.5	45.8±10.6	ni	3.9±9.9	47.8±3.8
*Tetranychus urticae**	ni	ni	46.7±25.0	ni	ni	24.6±3.1

The percentage of inhibition was calculated as [(1 – activity with AtSerpin1/activity without AtSerpin1)×100]. Values represent mean+SE for duplicated independent determinations from a unique pool of extracts.
*pest species;
†non-target species.
"ni" denotes no inhibition.

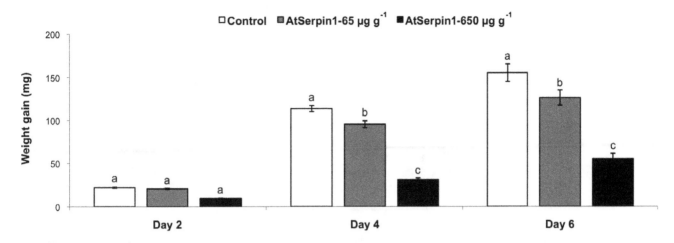

Figure 1. Weight gain of *Spodoptera littoralis* larvae fed on a diet containing 65 or 650 µg g^{-1} AtSerpin1 or control diet without inhibitor. Feeding assays were performed for 6 days with third-instar larvae. Bars represent mean ± SE. Bars with different letters on the same day are significantly different ($P<0.05$; one-way ANOVA followed by Student-Newman-Keuls) ($N=48$).

tion for 50% mortality (LC$_{50}$) at the third day of feeding was 637 µg ml^{-1} (95% confidence limits = 367–1105; $R^2 = 0.91$) (Figure 5). Hence, it appears that AtSerpin1 not only inhibits cysteine proteases in *A. pisum* extracts *in vitro* but also has a strong insecticidal effect on nymphs.

To investigate the response of proteolytical enzymes of *A. pisum* to the ingestion of AtSerpin1, nymphs were fed with a diet containing 1000 µg ml^{-1} AtSerpin1 or a control diet without the inhibitor for 24 h, and proteolytic activities were subsequently quantified (Table 3). The cathepsin B- and L-like cysteine activities were significantly reduced (by 37% and 47%, respectively) in nymphs fed with AtSerpin1. In contrast, leucine aminopeptidase activity was enhanced by 42% when aphids were exposed to the inhibitor. Lastly, no differences were observed in carboxypeptidase A and B activities in *A. pisum* nymphs that were fed a diet with or without AtSerpin1.

Discussion

Although many plant protease inhibitors from the serpin superfamily have been identified and hypothesized to have a role in host defense, to our knowledge only Yoo et al. [25] and the current study have investigated the potential of a serpin for pest control.

In vitro inhibitory activity of AtSerpin1 against invertebrate digestive proteases

In vitro studies revealed that AtSerpin1 has a broad spectrum of activity because it inhibited both serine and cysteine proteases from a wide range of organisms, including the common earthworm (*L. terrestris*), the two-spotted spider mite (*T. urticae*), and eight insect species belonging to five different orders. Two

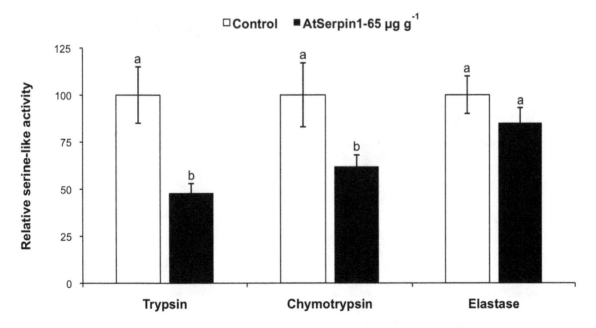

Figure 2. Serine-like proteolytic activities of *Spodoptera littoralis* third-instar larvae fed for 6 days on a diet containin 65 µg g^{-1} AtSerpin1 or control diet without inhibitor. Bars represent mean ± SE. Bars with different letters are significantly different ($P<0.05$; Mann-Whitney U test) ($N=24$).

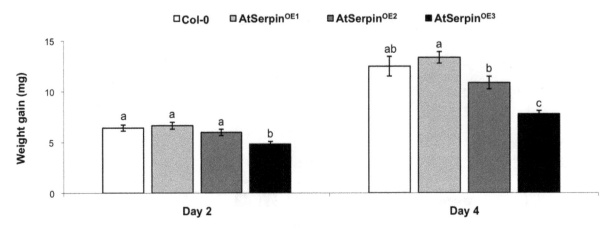

Figure 3. Western blot immunoassay showing the expression of AtSerpin1 in leaves of the transgenic *Arabidopsis* lines AtSerpin1^{OE1}, AtSerpin1^{OE2}, and AtSerpin1^{OE3}, and the non-transformed line Col-0. Lanes: (1) Page ruler plus protein standard; (2) 100 ng AtSerpin1; (3) 50 ng AtSerpin1; (4) 25 ng AtSerpin1; (5) 12.5 ng AtSerpin1; (6) 5 ng AtSerpin1; (7) 0 ng AtSerpin1; (8) overproducing line AtSerpin^{OE3} (6 ng); (9) overproducing line AtSerpin^{OE2} (6 ng); (10) overproducing line AtSerpin^{OE1} (6 ng); (11) non-transformed line Col-0. In lanes 7–9, the upper band is the full-length and active form of AtSerpin1, while the lower band is the cleaved form after interaction with a protease.

recent studies have demonstrated the ability of AtSerpin1 to inhibit cysteine proteases [19], [42]; the potential of AtSerpin1 to target serine proteases, however, has never been reported before. Although most serpins inhibit either serine or cysteine proteases [16], some can inhibit proteases from several families. For example, the mouse serpin SON-5 is a dual inhibitor of both chymotrypsin-like serine and the papain-like cysteine proteases [43]. Brüning et al. [44] showed that a serpin, Spn4, from the fruit fly *Drosophila melanogaster* Meigen inhibits proteases from three different families.

The role of plant serpins in the protection of crops against insects has been proposed [23], [24], but very little is known about the potential of such protease inhibitors to control agricultural

pests. To address this question, we selected two species, *S. littoralis* and *A. pisum*, for further *in vivo* studies (discussed in the next two sections). We selected these species in part because *S. littoralis* relies mainly on serine proteases, *A. pisum* relies mainly on cysteine proteases, and both were susceptible to AtSerpin1 in the *in vitro* experiments.

In vivo effect of AtSerpin1 on *S. littoralis*

Serine proteases provide the major midgut endoproteolytic activities in *S. littoralis* larvae [34], and previous studies have demonstrated that transgenic plants expressing serine protease inhibitors can confer resistance against *S. littoralis* [45], [46]. When the protease inhibitor AtSerpin1 was incorporated into an artificial

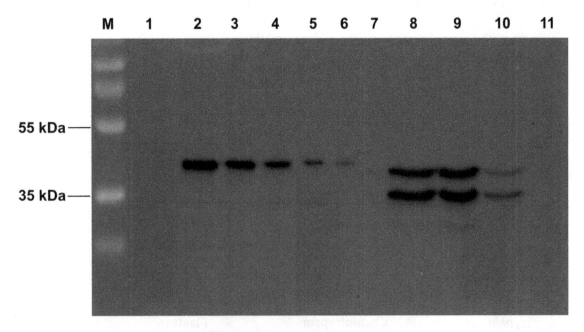

Figure 4. Weight gain of *Spodoptera littoralis* larvae fed on transgenic *Arabidopsis* plants overproducing AtSerpin1 (lines AtSerpin^{OE1}, AtSerpin^{OE2}, and AtSerpin^{OE3}) or on non-transformed plants (line Col-0). Feeding assays were performed for 4 days with second-instar larvae. Bars represent mean ± SE. Bars with different letters on the same day are significantly different (*P*<0.05; one-way ANOVA followed by Student-Newman-Keuls) (*N* = 24).

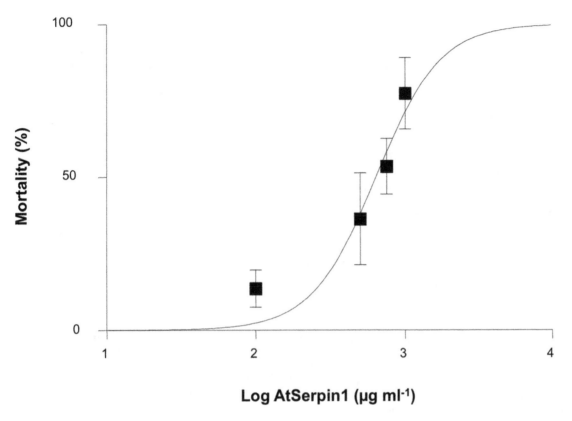

Figure 5. Concentration-response curve for mortality of newborn *Acyrthosiphon pisum* **nymphs fed for 3 days with artificial diet containing increasing concentrations of the protease inhibitor AtSerpin1.** Points represent mean ± SE. Three to six replicates with 15 nymphs each were used per concentration.

diet, the weight gain of *S. littoralis* larvae was substantially reduced relative to the control. The observed effects of AtSerpin1 on larval weight gain were correlated with a significant decreased of midgut trypsin activity. Weight gain also was reduced when *S. littoralis* were fed with transgenic *Arabidopsis* plants overproducing the serpin.

Table 3. Proteolytic activities of *Acyrthosiphon pisum* adults after 1 day of feeding on a control diet (AtSerpin1−) or a diet containing 1000 µg ml^{-1} AtSerpin1 (AtSerpin1+).

		Specific activity[a]	
Protease	pH	AtSerpin1−	AtSerpin1+
Cysteine protease			
Cathepsin B	6.5	4.0±0.36[a]	2.5±0.19[b]
Cathepsin L	3	18.3±0.87[a]	9.6±0.34[b]
Cathepsin L	5.5	13.9±1.43[a]	7.8±0.47[b]
Leucine amino peptidase	7	8.7±0.93[a]	12.4±0.84[b]
Carboxypeptidase A	7	9.2±0.50[a]	8.4±0.40[a]
Carboxypeptidase B	8	12.6±0.44[a]	12.3±0.90[a]

[a]Specific activities as nmoles of substrate hydrolyzed min^{-1} mg protein^{-1}.
Values are mean ± SE of triplicate measurements from three independent replicates.
Means followed by the same letter within a row are not significantly different from each other (P≤0.05; one-way ANOVA followed by Student-Newman-Keuls).

The results obtained in our bioassays are in agreement with many studies that have shown the potential of different plant serine protease inhibitors to interfere with the performance of lepidopteran species, either when the inhibitors are incorporated into an artificial diet or when they are expressed in transgenic plants [8], [12], [47]. However, the level of pest control that is routinely provided by *Bt* toxins is rarely provided by serine protease inhibitors, including AtSerpin1. It is well known that lepidopteran pests possess a remarkable ability to adapt their digestive proteolytic metabolism to the dietary material ingested and, therefore, to counteract the inhibitory activity of protease inhibitors [48], [49]. For this reason, researchers have suggested that a combination of two or more inhibitors may be required to overcome the capacity of such species to adapt to protease inhibitors. For example, Dunse et al. [14] recently demonstrated that growth of larvae of the cotton bollworm *Helicoverpa armigera* (Hübner) was substantially decreased on an artificial diet containing two serine protease inhibitors but not on a diet containing one serine protease inhibitor.

In vivo effect of AtSerpin1 on *A. pisum*

Cysteine proteases have been identified in several aphid species [50], [51], including *A. pisum* [33], [52], [53]. Our *in vitro* assays showed that AtSerpin1 strongly inhibits cathepsin B and L protease activities of whole *A. pisum* extracts, and when administered into an artificial diet, AtSerpin1 was toxic to *A. pisum* nymphs with 50% mortality at 637 µg ml^{-1}. Researchers previously suggested that a serpin from *C. maxima* (CmPS-1) plays a role in plant defence against aphids; feeding assays established a correlation between increase in CmPS-1 within the phloem sap and the reduced ability of *M. persicae* to survive and reproduce on *C. maxima* plants [25]. However, survival of neonate *M. persicae* nymphs fed on a sucrose solution supplemented with 200 μg ml^{-1} of purified CmPS-1 was not reduced. This might be because CmPS-1 requires additional phloem proteins to form an active complex [25].

Some studies have reported deleterious effects of plant cysteine protease inhibitors on aphids fed on artificial diets. The cystatin OC-I induced moderate but significant growth inhibition on three aphid species: *A. pisum*, the cotton aphid *Aphis gossypii* Glover, and *M. persicae* [54]. Likewise, diets supplemented with OC-I (ranging from 20 to 500 μg ml^{-1}) significantly reduced nymphal survival of the potato aphid *Macrosiphum euphorbiae* (Thomas) and prevented aphids from reproducing [55]. Artificial diets containing either a modified version of OC-I or the recombinant chicken egg white cystatin (CEWc) reduced the survival and growth of *M. persicae* nymphs [56]. The barley cystatin HvCPI-6 was toxic to *A. pisum* nymphs (LC$_{50}$ = 150 μg ml^{-1}) [33]. Moreover, the developmental time of *A. pisum* was significantly delayed when newborn nymphs were fed for 1 day on diet containing HvCPI-6 at 400 μg ml^{-1} and were subsequently placed on bean plants until they reached adulthood [33].

In the current study, the effect of AtSerpin1 on nymphal mortality was correlated with a significant decrease of cathepsin B and L protease activities after the nymphs fed on artificial diet containing serpin. In addition, leucine aminopeptidase activity was enhanced, suggesting a compensatory response to the inhibitory effect mediated by AtSerpin1. The overproduction of non-target proteases as a response to plant defense proteins is common in herbivorous arthropods [48], [57]. In a bioassay similar to the one described here, the ingestion of HvCPI-6 by *A. pisum* and *M. persicae* nymphs was correlated with a decrease of cathepsin B and L protease activities, and in the case of *M. persicae*, an increase of leucine aminopeptidase activity [33]. Because the artificial diet used in both studies was protein free, the results suggest that the toxicity of the serpin was not linked to disruption of food protein digestion but to the disruption of non-digestive proteases involved in other physiological processes. The cysteine protease inhibitor from rice, oryzacystatin (OC-I), not only affected the aphid *M. persicae* through digestive tract targets but also inhibited extra-digestive proteolytic activities in the hemolymph and internal organs [54]. Similar to our findings, the effects of OC-I on *M. persicae* were correlated with a reduction of a major cysteine-like protease activity in whole adult extracts [54].

Concluding remarks

Before commercial release, GM crops must undergo an environmental risk assessment to ensure that they do not cause unacceptable detrimental effects to non-target organisms. This is especially relevant in the case of plants producing protease inhibitors, given that these inhibitors may affect many different organisms. Our *in vitro* assays showed that the serine proteases of the three non-target species tested were highly inhibited by AtSerpin1. Therefore, if GM plants producing AtSerpin1 are to be deployed in the future, the impact on non-target organisms should be taken into account and special attention should be given to the routes of exposure.

In vivo assays with *S. littoralis* and *A. pisum* showed very promising results for pest control by AtSerpin1. Artificial diet and plant bioassays have demonstrated that AtSerpin1 reduces the growth of *S. littoralis* larvae but does not cause mortality. For AtSerpin1 to make a meaningful contribution to plant resistance against *S. littoralis*, the efficacy of the serpin must be increased either by protein engineering [12] or by using it in combination with other protease inhibitors (see above) or with other pesticidal proteins. Interestingly, *A. pisum* nymphs incurred high mortality levels when exposed to AtSerpin1 through artificial diet. Some studies have previously shown that transgenic plants producing cysteine protease inhibitors can confer partial resistance against aphid species [33], [54], [58]. Future experiments with *A. pisum* should therefore determine whether the detrimental effect observed with an artificial diet bioassay in the current study is obtained with AtSerpin1-expressing transgenic plants.

Acknowledgments

We are grateful to Félix Ortego and Olivier Sanvido for their valuable comments on earlier versions of this manuscript. We also thank Didier Van de Velde for his technical assistance and Jurgen Haustraete (Protein Service Facility, Department for Molecular Biomedical Research, VIB, Ghent, Belgium) for assistance in protein purification methodology.

Author Contributions

Conceived and designed the experiments: FAA JM LC SV DV FVB GS. Performed the experiments: FAA JM LC SV DV. Analyzed the data: FAA JM LC GS. Contributed reagents/materials/analysis tools: FVB GS. Wrote the paper: FAA LC FVB GS.

References

1. Oerke EC (2006) Crop losses to pests. J Agri Sci 144: 31–43.
2. Toenniessen GH, O'Toole JC, DeVries J (2003) Advances in plant biotechnology and its adoption in developing countries. Curr Opin Plant Biol 6: 191–198.
3. Brookes G, Barfoot P (2005) GM crops: the global economic and environmental impact: the first nine years 1996–2004. Ag Bio Forum 8: 15.
4. James C (2010) Global status of commercialized biotech/GM Crops: 2010. ISAAA Brief No. 42, International Service for the Acquisition of Agri-Biotech Applications, Ithaca, NY, USA.
5. Carrière Y, Crowder DW, Tabashnik BE (2010) Evolutionary ecology of insect adaptation to Bt crops. Evolut Appl 3: 561–573.
6. Pilcher CD, Rice MD, Obrycki JJ, Lewis LC (1997) Field and laboratory evaluation of transgenic *Bacillus thuringiensis* corn on secondary Lepidopteran pests (Lepidoptera: Noctuidae). J Econ Entomol 90: 669–678.
7. Dutton A, Romeis J, Bigler F (2005) Effects of Bt maize expressing Cry1Ab and Bt spray on *Spodoptera littoralis*. Entomol Exp Appl 114: 161–169.
8. Malone LA, Gatehouse AMR, Barratt BIP (2008) Beyond *Bt*: Alternative strategies for insect-resistant genetically modified crops. In: Integration of insect-resistant genetically modified crops within IPM programs J. Romeis, AM. Shelton, GG. Kennedy, eds. Springer Science+Business Media B.V.: pp 357–417.
9. Ferry N, Edwards MG, Gatehouse JA, Gatehouse AMR (2004) Plant-insect interactions: molecular approaches to insect resistance. Curr Opin Biotech 15: 155–161.
10. Christou P, Capell T, Kohli A, Gatehouse JA, Gatehouse AMR (2006) Recent developments and future prospects in insect pest control in transgenic crops. Trends Plant Sci 11: 302–308.
11. Gatehouse JA (2008) Biotechnological prospects for engineering insect-resistant plants. Plant Physiol 146: 881–887.
12. Schlüter U, Benchabane M, Munger A, Kiggundu A, Vorster J, et al. (2010) recombinant protease inhibitors for herbivore pest control: a multitrophic perspective. J Exp Bot 61: 4169–4183.
13. Ferry N, Edwards MG, Gatehouse JA, Capell T, Christou P, et al. (2006) Transgenic plants for insect pest control: a forward looking scientific perspective. Transgenic Res 15: 13–19.
14. Dunse KM, Stevens JA, Lay FT, Gaspar YM, Heath RL, et al. (2010) Coexpression of potato type I and II proteinase inhibitors gives cotton plants protection against insect damage in the field. Proc Natl Acad Sci USA 107: 15011–15015.
15. Rawlings ND, Tolle DP, Barrett AJ (2004) MEROPS: the peptidase database. Nucleic Acids Res 32 Database issue: D160–D164.
16. Gettins PGW (2002) Serpin structure, mechanism, and function. Chem Rev 102: 4751–4803.
17. Schick C, Brömme D, Bartuski A, Uemura Y, Schechter N, et al. (1998) The reactive site loop of the serpin SCCA1 is essential for cysteine proteinase inhibition. Proc Natl Acad Sci USA 95: 13465–70.

18. McGowan S, Buckle A, Irving J, Ong P, Bashtannyk-Puhalovich T, et al. (2006) X-ray crystal structure of MENT: evidence for functional loop-sheet polymers in chromatin condensation. EMBO J 25: 3144–55.
19. Vercammen D, Belenghi B, van de Cotte B, Beunens T, Gavigan JA, et al. (2006) Serpin 1 of *Arabidopsis thaliana* is a suicide inhibitor for metacaspase 9. J Mol Biol 364: 625–636.
20. Ong PC, McGowan S, Pearce MC, Irving JA, Kan WT, et al. (2007) DNA accelerates the inhibition of human cathepsin V by serpins. J Biol Chem 282: 36980–36986.
21. Roberts TH, Hejgaard J (2008) Serpins in plants and green algae. Funct Integr Genomics 8: 1–27.
22. Huntington JA (2006) Shape-shifting serpins- advantages of a mobile mechanism. TiBS 31: 427–435.
23. Dahl SW, Rasmussen SK, Hejgaard J (1996) Heterologous expression of three plant serpins with distinct inhibitory specificities. J Biol Chem 271: 25083–25088.
24. Rasmussen SK, Dahl SW, Norgard A, Hejgaard J (1996) A recombinant wheat serpin with inhibitory activity. Plant Mol Biol 30: 673–677.
25. Yoo B-C, Aoki K, Campbell LR, Hull RJ, Xoconostle-Cazares B, et al. (2000) Characterization of *Cucurbita maxima* Phloem Serpin-1 (CmPS-1). J Biol Chem 275: 35122–35128.
26. Kehr J (2006) Phloem sap proteins: their identities and potential roles in the interaction between plants and phloem-feeding insects. J Exp Bot 57: 767–774.
27. Bradford MM (1976) A rapid and sensitive method for the quantitation of microgram quantities of protein utilizing the principle of protein-dye binding. Anal Biochem 72: 248–254.
28. Karimi M, Inzé D, Depicker A (2002) GATEWAY vectors for *Agrobacterium*-mediated plant transformation. Trends Plant Sci 7: 193–195.
29. Clough SJ, Bent AF (1998) : Floral dip: a simplified method for *Agrobacterium*-mediated transformation of *Arabidopsis thaliana*. Plant J 16: 735–743.
30. Ortego F, Novillo C, Castañera P (1996) Characterization and distribution of digestive proteases of the stalk corn borer, *Sesamia nonagrioides* Lef. (Lepidoptera: Noctuidae). Arch Insect Biochem Physiol 33: 163–180.
31. Shahnaz SN, van EJM, Damme E, Smagghe G (2008) Carbohydrate-binding activity of type-2 ribosome-inactivating protein SNA-I from elderberry (*Sambucus nigra*) is a determine factor for its insecticidal activity. Phytochem 69: 2972–2978.
32. Abbott WS (1925) A method of computing the effectiveness of an insecticide. J Econ Entomol 18: 265–267.
33. Carrillo L, Martinez M, Alvarez-Alfageme F, Castañera P, Smagghe G, et al. (2011a) A barley cysteine proteinase inhibitor reduces the performance of two aphid species in artificial diets and transgenic *Arabidopsis* plants. Transgenic Res 20: 305–319.
34. Lee MJ, Anstee JH (1995) Endoproteases from the midgut of larval *Spodoptera littoralis* include a chymotrypsin like enzyme with an extended binding site. Insect Biochem Molec Biol 25: 49–61.
35. Zwilling R, Medugorac I, Mella K (1972) The evolution of endopeptidases-XIV. Non-tryptic cleavage specificity of a baee-hydrolyzing enzyme (β-protease) from *Tenebrio molitor*. Comp Biochem Physiol B 43: 419–424.
36. Bewley GC, DeVillez EJ (1968) Isolation and characterization of the digestive proteinases in the earthworm *Lumbricus terrestris* Linnaeus. Comp Biochem Physiol B 25: 1061–1066.
37. Malone LA, Burgess EPJ, Stefanovic D, Gatehouse HS (2000) Effects of four protease inhibitors on the survival of worker bumblebees, *Bombus terrestris* L. Apidologie 31: 25–38.
38. Mulligan EA, Ferry N, Jouanin L, Romeis J, Gatehouse AMR (2010) Characterisation of adult green lacewing (*Chrysoperla carnea*) digestive physiology: impact of a cysteine protease inhibitor and a pyrethroid. Pest Manag Sci 66: 325–336.
39. Murdock LL, Brookhart G, Dunn PE, Foard DE, Kelley S, et al. (1987) Cysteine digestive proteinases in Coleoptera. Comp Biochem Physiol B 87: 783–787.
40. Novillo C, Castañera P, Ortego F (1997) Characterization and distribution of chymotrypsin-like and other digestive proteases in Colorado potato beetle larvae. Arch Insect Biochem Physiol 36: 181–201.

41. Carrillo L, Martinez M, Ramessar K, Cambra I, Castañera P, et al. (2011b) Expression of a barley cystatin gene in maize enhances resistance against phytophagous mites by altering their cysteine-proteases. Plant Cell Reports 30: 101–112.
42. Lampl N, Budai-Hadria O, Davydov O, Joss TV, Harrop SJ, et al. (2010) *Arabidopsis* AtSerpin1, crystal structure and in vivo interaction with its target protease RESPONSIVE TO DESSICATION-21 (RD21). J Biol Chem 285: 13550–13560.
43. Al-Khunaizi M, Luke CJ, Cataltepe S, Miller D, Mills DR, et al. (2002) The serpin SON-5 is a dual mechanistic class inhibitor of serine and cysteine proteinases. Biochemistry 41: 3189–3199.
44. Brüning M, Lummer M, Bentele C, Smolenaars MMV, Rodenburg KW (2007) The Spn4 gene from *Drosophila melanogaster* is a multipurpose defence tool directed against proteases from different peptidase families. Biochem J 401: 325–331.
45. De Leo F, Bonadé-Bottino MA, Ceci LR, Gallerani R, Jouanin L (1998) Opposite effects on *Spodoptera littoralis* larvae of high expression level of a trypsin proteinase inhibitor in transgenic plants. Plant Physiol 118: 997–1004.
46. Marchetti S, Delledonne M, Fogher C, Chiaba C, Chiesa F, et al. (2000) Soybean Kunitz, C-II and PI–IV inhibitor genes confer different levels of insect resistance to tobacco and potato transgenic plants. Theor Appl Genet 101: 519–526.
47. Carlini CR, Grossi-de-Sá MF (2002) Plant toxic proteins with insecticidal properties. A review on their potentialities as bioinsecticides. Toxicon 40: 1515–1539.
48. Lara P, Ortego F, Gonzalez-Hidalgo E, Castañera P, Carbonero P, et al. (2000) Adaptation of *Spodoptera exigua* (Lepidoptera: Noctuidae) to barley trypsin inhibitor BTI-CMe expressed in transgenic tobacco. Transgenic Res 9: 169–178.
49. De Leo F, Bonade-Bottino M, Ceci LR, Gallerani R, Jouanin L (2001) Effects of a mustard trypsin inhibitor expressed in different plants on three Lepidopteran pests. Insect Biochem Mol Biol 31: 593–602.
50. Rahbé Y, Ferrasson E, Rabesona H, Quillien L (2003b) Toxicity to the pea aphid *Acyrthosiphon pisum* of anti-chymotrypsin of anti-chymotrypsin isoforms and fragments of Bowman-Birk protease inhibitors from pea seeds. Insect Biochem Mol Biol 33: 299–306.
51. Deraison C, Darboux I, Duportets L, Gorojankina T, Rahbé Y, Jouanin L (2004) Cloning and characterization of a gut-specific cathepsin L from the aphid *Aphis gossypii*. Insect Mol Biol 13: 165–177.
52. Cristofoletti PT, Ribeiro AF, Deraison C, Rahbé Y, Terra WR (2003) Midgut adaptation and digestive enzyme distribution in a phloem feeding insect, the pea aphid *Acyrthosiphon pisum*. J Insect Physiol 49: 11–24.
53. Rispe C, Kutsukake M, Doublet V, Hudaverdian S, Legeai F, et al. (2008) Large gene family expansion and variable selective pressures for cathepsin B in aphids. Mol Biol Evol 25: 5–17.
54. Rahbé Y, Deraison C, Bonade-Bottino M, Girard C, Nardon C, et al. (2003a) Effects of the cysteine protease inhibitor oryzacystatin (OC-I) on different aphids and reduced performance of *Myzus persicae* on OC-I expressing transgenic oilseed rape. Plant Sci 164: 441–450.
55. Azzouz A, Cherqui A, Campan EDM, Rahbé Y, Duport G, et al. (2005) Effects of plant protease inhibitors, oryzacystatin I and soybean Bowman-Birk inhibitor, on the aphid *Macrosiphum euphorbiae* (Homoptera, Aphididae) and its parasitoid *Aphelinus abdominalis* (Hymenoptera, Aphididae). J Insect Physiol 51: 75–86.
56. Cowgill SE, Wright C, Atkinson HJ (2002) Transgenic potatoes with enhanced levels of nematode resistance do not have altered susceptibility to nontarget aphids. Mol Ecol 11: 821–827.
57. Álvarez-Alfageme F, Martinez M, Pascual-Ruiz S, Castañera P, Diaz I, et al. (2007) Effects of potato plants expressing a barley cystatin on the predatory bug *Podisus maculiventris* via herbivorous prey feeding on the plant. Transgenic Res 16: 1–13.
58. Ribeiro APO, Pereira EJC, Galvan TL, Picanco MC, Picoli EAT, et al. (2006) Effect of eggplant transformed with oryzacystatin gene on *Myzus persicae* and *Macrosiphum euphorbiae*. J Appl Entomol 130: 84–90.

8

Impact of Community-Based Larviciding on the Prevalence of Malaria Infection in Dar es Salaam, Tanzania

Mathieu Maheu-Giroux, Marcia C. Castro*

Department of Global Health and Population, Harvard School of Public Health, Boston, Massachusetts, United States of America

Abstract

Background: The use of larval source management is not prioritized by contemporary malaria control programs in sub-Saharan Africa despite historical success. Larviciding, in particular, could be effective in urban areas where transmission is focal and accessibility to *Anopheles* breeding habitats is generally easier than in rural settings. The objective of this study is to assess the effectiveness of a community-based microbial larviciding intervention to reduce the prevalence of malaria infection in Dar es Salaam, United Republic of Tanzania.

Methods and Findings: Larviciding was implemented in 3 out of 15 targeted wards of Dar es Salaam in 2006 after two years of baseline data collection. This intervention was subsequently scaled up to 9 wards a year later, and to all 15 targeted wards in 2008. Continuous randomized cluster sampling of malaria prevalence and socio-demographic characteristics was carried out during 6 survey rounds (2004–2008), which included both cross-sectional and longitudinal data (N = 64,537). Bayesian random effects logistic regression models were used to quantify the effect of the intervention on malaria prevalence at the individual level. Effect size estimates suggest a significant protective effect of the larviciding intervention. After adjustment for confounders, the odds of individuals living in areas treated with larviciding being infected with malaria were 21% lower (Odds Ratio = 0.79; 95% Credible Intervals: 0.66–0.93) than those who lived in areas not treated. The larviciding intervention was most effective during dry seasons and had synergistic effects with other protective measures such as use of insecticide-treated bed nets and house proofing (i.e., complete ceiling or window screens).

Conclusion: A large-scale community-based larviciding intervention significantly reduced the prevalence of malaria infection in urban Dar es Salaam.

Editor: James G. Beeson, Burnet Institute, Australia

Funding: This research was supported by Award Number R03AI094401-01 (PI MCC) from the National Institute of Allergy and Infectious Diseases. The content is solely the responsibility of the authors and does not necessarily represent the official views of the National Institute of Allergy and Infectious Diseases or the National Institutes of Health. MMG was supported by an International Fulbright Science & Technology Award sponsored by the Bureau of Educational and Cultural Affairs of the U.S. Department of State, and a Doctoral Foreign Study Award from the Canadian Institutes of Health Research. The funders had no role in study design, data collection and analysis, decision to publish, or preparation of the manuscript.

Competing Interests: The authors have declared that no competing interests exist.

* E-mail: mcastro@hsph.harvard.edu

Introduction

The Ross-Macdonald model of malaria transmission suggests that control methods that reduce adult mosquitoes' longevity can achieve greater malaria reduction than strategies that target larval stages. Yet, Larval Source Management (LSM), such as the use of larvicides and the draining of breeding habitats, has historically been a very successful tool to reduce mosquito density [1] – examples include the elimination of *Anopheles arabiensis* from Egypt [2] and Brazil [3], malaria control in the Zambian copperbelt (1930–1950) [4], Dr. Gorga's work during the construction of the Panama canal [5], and the vector control program of the Tennessee Valley Authority [6]. With the discovery of DDT, however, such approaches where disfavored as exemplified by the almost exclusive use of this potent insecticide during the Global Malaria Eradication Program (1955–1969) [7]. In addition, LSM programs were often associated with vertical, authoritarian management. Currently, there are few examples of LSM initiatives in post-colonial Africa [8–10]. LSM is often perceived as a secondary malaria control strategy, labor-intensive, requiring strong managerial support and oversight for monitoring and evaluation [11,12], and often beyond the financial and operational capabilities of most malaria endemic areas in sub-Saharan Africa [13].

Such considerations might explain the insufficient evidence-base of LSM in post-colonial Africa, and the contemporary prioritization of malaria control programs that rely on Insecticide-Treated Nets (ITNs) and Insecticide Residual Spraying (IRS) as the main vector control measures. Nevertheless, a renewal of interest in applications of LSM within the sub-Saharan context has been observed recently [14–18]. In fact, in April of 2012, the World Health Organization (WHO) released an interim position statement [19] on the use of larvicides for malaria control in sub-Saharan Africa, recognizing that larviciding should be considered for malaria control but only in areas where breeding sites are 'few, fixed and findable' [19]. Larval control is regarded as being of secondary importance in comparison with IRS and ITNs. Although the WHO acknowledges that larvicides could be effective as one of the leading methods of vector control in urban areas of sub-Saharan Africa, it highlights the lack of recent and sound evidence of its effectiveness. Few contemporary studies have assessed the effectiveness of larvicides on malaria infection. Studies in highland valley communities of Kenya [20] and urban Tanzania [21] demonstrated substantial reduction in malaria prevalence, while no reductions were observed in a study conducted in a rural setting in The Gambia [22]. Strong empirical evidence on the causal effect of larviciding on malaria infection is difficult to obtain since larviciding interventions need to be implemented and scaled-up over large areas, appropriate control groups with similar malaria ecology are difficult to find, and the cost of such trials can be prohibitively expensive [14].

The rationale for adding larvicides to the arsenal of malaria control tools in urban areas is manifold. First, in contrast to rural areas, vector breeding habitats are generally fewer and much easier to reach in highly densely populated areas [10]. Second, the most potent malaria vector in Africa, An. gambiae, has been shown to exhibit exophagic behavior in some urban areas - although the majority of bites still take place indoors [23]. If this behavior intensifies over time, and therefore more biting and resting start to occur outside of homes, the efficacy of both IRS and ITNs would be reduced. Mathematical models have provided evidence that the outdoor biting rate defines what is achievable in terms of malaria reduction with IRS and ITNs [24]. LSM is one of the few strategies that could contribute to further reduce malaria when Anopheles are partially exophagic [14]. Third, insecticide resistance has emerged for the primary malaria vectors in many areas of the African continent [25–28] and combining IRS and ITN with larviciding could become more desirable in such settings. Finally, relying solely on IRS and ITNs may be insufficient to achieve malaria elimination in much of sub-Saharan Africa [29,30]. As such, larviciding may be part of an integrated vector management (IVM) approach [31] that could help hinder malaria transmission [18]. Such informed use of larvicides, based on local malaria ecology, is in line with WHO's current position on IVM [31,32].

Africa is the fastest urbanizing continent in the world and its share of urban population is expected to double between 2000 and 2030 [33]. Malaria intensity is generally much lower in urban areas and transmission is highly focal [34,35]. A corollary of this reduced endemicity is that urban dwellers will develop lower levels of clinical immunity to the disease, which can pose public health challenges. It has been estimated that about 28% of the malaria burden in sub-Saharan Africa is attributable to urban malaria [34]. Malaria control in urban settings offers more options than for rural areas because logistical constraints are alleviated by relatively good transportation, education, communication, and health infrastructures [36].

Following this rationale, the Dar es Salaam Urban Malaria Control Program (UMCP) was launched in 2004, targeting 15 of the city's 73 wards, covering 56 km^2 of the city, and a population of more than 610,000 residents [36]. The goal was to develop a sustainable larval control intervention as one of the main components of a malaria control strategy. Regular application of microbial larvicides was initiated in 2006 through vertically managed community-based delivery systems [36]. Initial results, restricted to children under five years of age and comprising data from the first period of larviciding (2006–2007) in three wards of the city (N = 4,450), demonstrated that this intervention reduced by 72% the odds of malaria infection [21]. In addition, rigorous

monitoring of larval population in the same period showed that larviciding reduced anopheline larval abundance by 96% [36]. The larviciding intervention was scaled-up to 9 wards in 2007 and to all 15 wards in 2008.

In this paper, we will comprehensively investigate the effectiveness of the larviciding intervention on reducing malaria prevalence using 4.6 years of data, including individuals of all ages, and combining both cross-sectional and longitudinal data (N = 64,537). This will provide crucial evidence on the potential contribution of larvicide use for reducing population-level malaria burden in urban areas of sub-Saharan Africa.

Materials and Methods

Study site

Dar es Salaam is the largest city and economic capital of the United Republic of Tanzania with an estimated population of 2.7 million in 2005 [37]. The climate is tropical humid with two rainy seasons – the long rains during the months of April and May and the short rains of October and November. Malaria transmission is year-round [38] with peaks in incidence after the two rainy seasons. Plasmodium falciparum accounts for more than 90% of cases and the principal vectors involved in malaria transmission are An. gambiae s.s. and An. funestus [10]. An. coustani's contribution to malaria transmission is believed to be marginal [21]. Dar es Salaam is composed of three municipalities: Illala, Temeke, and Kinondoni. These municipalities are further divided in 73 wards (Figure 1). Each ward is comprised of administrative sub-units called mtaa (plural mitaa) which are further divided in ten-cell units (TCU) – the smallest administrative unit that contains approximately 10–20 houses, but may also contain as many as 100 [10].

Design of the larviciding intervention

The Dar es Salaam's UMCP was launched in 2004, and targeted 15 wards, five in each of the three municipalities, totaling 67 mitaa. During the first phase of the project (May 2004 to February 2006), systems for extensive mapping [39,40] and surveillance of potential mosquito breeding sites were developed [36]. In 2005, routine surveillance of immature and adult mosquitoes was fully operationalized. Comprehensive larviciding of the identified breeding habitats debuted in March 2006 in three wards (Figure 2). The program was community-based but the UMCP remained responsible for vertical management and supervision. This entailed that responsibility for routine mosquito control and surveillance was delegated to modestly paid community members referred to as Community-Owned Resource Person (CORP) [11,12,41]. After 13 months of larviciding in these three wards, operations were extended to six additional wards: two in each municipality, totaling 9 wards covered by larviciding activities. Finally, about 12 months later, in April of 2008, the intervention was scaled-up to all 15 wards of the UMCP. The order in which wards were chosen to receive the larviciding intervention was not randomly allocated. Rather, the choice was the result of careful consideration of the following two criteria: (i) the availability of comprehensive and detailed maps of the ward, and (ii) the proven ability of the ward supervisor and CORPs to efficiently undertake the required tasks.

The biological agents Bacillus thuringiensis var. israelensis (Bti; VectoBac® Valent BioSciences Corporation, VBC, USA) and Bacillus sphaericus (Bs; VectoLex®, VBC, USA) were used to control the aquatic stages of anopheline mosquitoes. Each mtaa, or portion of a mtaa, was under the responsibility of a designated CORP who was instructed to treat breeding habitats on a weekly basis. The dosage was 0.04 grams per m^2 and 1 gram per m^2 for Bti and Bs,

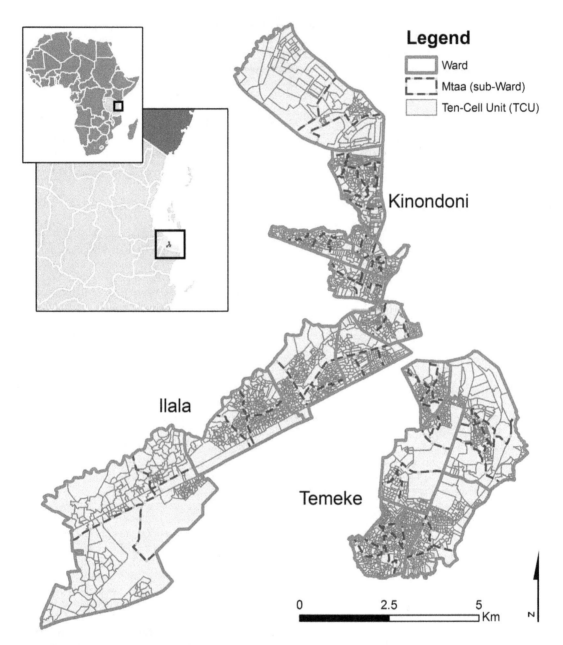

Figure 1. Map of the study area and administrative units. The northern portion belongs to the municipality of Kinondoni, the south-eastern portion to Temeke, and the south-western part to Ilala.

respectively. Closed habitats that mainly breed *Culex quinquefaciatus* were treated with *Bs* every three months by a separate team of CORPs (although *Culex* mosquitoes play no role in malaria transmission, this was a programmatic decision to gain support from the community).

UMCP Data collection

During the study period, a total of six randomized cluster-sampled household surveys were carried out (Figure 3). A list of TCUs was assembled for each ward before March of 2004 and was regularly updated throughout the study duration. During the first round of the survey, ten TCUs were randomly sampled from each of the 15 wards. All households located in the sampled TCUs were invited to participate in the survey. From the second round onwards, the TCUs sampled in the first round were followed-up

longitudinally, and another ten TCUs per ward were selected for cross-section surveys. Since loss to follow-up is non-negligible in urban areas, starting from the 3rd survey round, the list of subjects to be followed-up also included randomly selected subjects interviewed in previous cross-section surveys. This was implemented in order to guarantee that the minimum required sample size would be met. Sample size calculations used a significance level of 5% and 80% power to detect a 5% absolute difference in malaria prevalence from 10% baseline prevalence. This is equivalent to a ±50% relative risk of infection. Calculations were based on mean TCU population size [21].

Upon consenting to the interview, each household was geo-referenced using a hand-held global positioning system (GPS) device. A detailed questionnaire was administered, collecting information grouped in four modules: (i) house characteristics (e.g.,

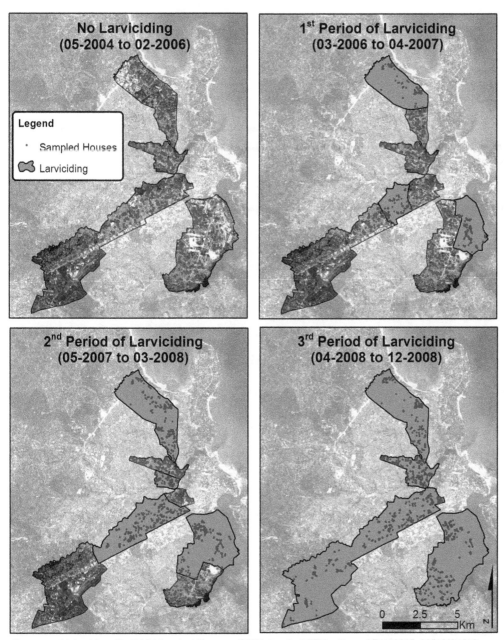

Basemap obtained from the NASA Landsat Program (2003), Landsat ETM+, Scene L71166065_06520070618, United States Geological Survey, Sioux Falls SD, 06-18-2007.

Figure 2. Map control and intervention wards and location of sampled households for each larviciding period.

location, conditions, number of habitants); (ii) head of the household (e.g., occupation, education, knowledge of malaria transmission and disease symptoms, assets, agricultural practices); (iii) use of preventive measures (e.g., bednet, mosquito repellent, coil); and (iv) individual characteristics (e.g., age and sex of all household members, occurrence of fever in the past two weeks, treatment-seeking behavior, use of antimalarial drug, sleeping habits, travel history). A proxy for socio-economic status was constructed using an asset-based index calculated by performing Principal Component Analysis [42] of the households' possession, excluding protective assets such as bednets and window screenings. Table 1 describes the variables selected for this study, their type and, if appropriate, the way they were categorized.

Malaria infection status was ascertained for all household members for whom written informed consent was provided. Finger-pricked blood samples were analyzed using Giemsa-stained thick smear microscopy. Quality check was conducted on a 10% sample of blood slides at the Muhimbili University of Health and Allied Sciences – MUHAS (a center of excellence in laboratory analysis), indicating a 94.5% specificity rate and 95.7% sensitivity rate [43]. Individuals found to be infected with malaria were treated with appropriate front-line regimens (sulphadoxine-pyrimethamine until August 2006, after which it was replaced with artesunate-amodiaquine). In order to minimize selection bias and achieve full coverage for each house and TCU, up to three attempts were made to enroll subjects.

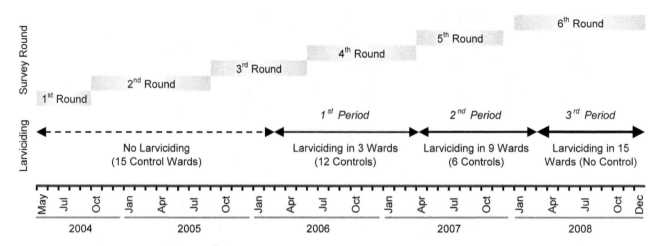

Figure 3. Timeline of data collection activities and larviciding intervention. The first survey round was conducted form 05/2004 to 09/2004, the second from 10/2004 to 08/2005, the third from 09/2005 to 05/2006, the fourth from 06/2006 to 03/2007, the fifth from 04/2007 to 11/2007, and the sixth and last survey round from 01/2008 to 12/2008. The first period of the intervention started on March 1st 2006, the second period of larviciding on May 1st 2007, and the last period of larviciding on April 1st 2008.

Information was collected from a total of 48,525 unique individuals and the great majority of them (39,146) were interviewed once. A total of 5,223 participants were followed up twice, 2,349 three times, 1,236 four times, 472 five times, and 99 subjects participated in every round of the survey. Including follow-up data, our sample is thus composed of 64,537 observations, which were drawn from 913 unique TCU and 6,796 households. The small number of subjects who participated in more than two rounds results from two main factors. First, the high mobility observed among urban dwellers; in the second survey round 25.6% of the subjects had moved or were travelling. Second, 13.9% of those interviewed in round 1 declined to participate in the second survey round. Reasons for refusal included pain inflicted by the finger prick, misconceptions about malaria transmission, and the mistrust of the malaria counts provided in the precedent survey round. Sensitization efforts addressed these issues and refusal decreased in subsequent survey rounds.

Rainfall data

Rainfall estimates were obtained from the National Oceanic and Atmospheric climate prediction center. This data source combines modeling of satellite-based infrared data collected each 30-minute and station rainfall data to estimate the quantity of daily precipitation over the African continent, and has a spatial resolution of 8 kilometers [44]. Given the biology of the *Anopheles* mosquito and of the *Plasmodium* parasite, the effect of rainfall on malaria transmission is expected to be lagged in time. Previous empirical studies suggested that the effect of rainfall on malaria transmission is lagged by approximately 8 weeks [45–47]. For each observation, we therefore calculated total weekly precipitation (cm) and lagged this estimate by 8 weeks.

Statistical analyses

The main outcome for this study is malaria infection status (a binary variable – Table 1) as determined by the Giemsa-stained thick smear. Malaria transmission is most directly related to the density of sporozoites-infected adult anophelines, which are not targeted by the larviciding activities. Therefore, a decline in the prevalence of malaria infection is not expected to be observed until the existing pool of infected mosquitoes dies off, and the overall

density of mosquitoes is reduced. Based on observations of entomological indices and malaria incidence, it has been estimated that peaks in vector density are followed by peaks in malaria incidence after approximately 1–2 months [48]. Also, the implementation of larviciding activities requires fine-tuning before CORPs became fully familiar with the routine procedures, which could further lag any potential impacts. Based on programmatic and biological considerations, a lag of five weeks was deemed most appropriate and is consistent with results from a previous larviciding study in urban Cameroon [49].

The effects of the microbial larviciding activities on malaria occurrence were first examined using univariate statistics. Malaria prevalence was calculated for each survey round, stratifying by larviciding intervention status, if applicable. Confidence intervals for malaria prevalence were constructed using 9,999 bootstrapped replicates. Clustering of standard errors was taken into account by defining the sampling unit as the TCU [50].

Bayesian random effects logistic models where used to take into account clustering of observations at the household and TCU levels in multivariable analyses. We assumed that our binary outcome followed a Bernoulli distribution, $Y_i \sim Bernoulli(p_i)$, where p_i is the probability of an individual harboring malaria parasites, which is itself a function of covariates modeled with a *logit* link. Our model has the following form:

$$\text{logit}(p_{itjk}) = \alpha + \beta(\text{Intervention}_{it}) + \delta X_{it} + f(\text{Rainfall}) +$$

$$f(\text{Time}) + \mu_j + \upsilon_k + \varepsilon_{itjk}$$

$$\mu_j \sim N(0,\sigma_\mu^2), \upsilon_k \sim N(0,\sigma_k^2), \text{and} \varepsilon_{itjk} \sim N(0,\sigma^2)$$

where p_{itjk} is the probability of individual i at time t living in TCU j and household k to be infected with malaria; β is the coefficient of the larviciding intervention; δ is a vector of coefficients for control variables in vector X (age, sex, sleeping outside of ward in previous weeks, taking antimalarial drug in previous two weeks, individuals treated for malaria in a previous

Table 1. Characteristics of study participants stratified by survey round and intervention group (lagged by 5 weeks).

Variables	Survey round #1	Survey round #2	Survey round #3		Survey round #4		Survey round #5		Survey round #6	
	Control	Control	Control	Larvicide	Control	Larvicide	Control	Larvicide	Control	Larvicide
Outcome: Prevalence of malaria infection	20.8%	16.9%	10.2%	13.1%	7.1%	4.6%	5.2%	4.4%	2.3%	1.7%
Individual-Level Variables	n = 5,809	n = 11,149	n = 10,791	n = 697	n = 9,951	n = 2,385	n = 6,461	n = 5,663	n = 744	n = 10,887
Age										
0 to <5 years of age	16.0%	14.9%	15.3%	15.1%	13.5%	12.7%	12.3%	11.5%	18.4%	10.3%
5 to <15 years of age	27.7%	27.7%	27.2%	30.1%	28.3%	28.9%	28.0%	31.0%	26.7%	30.3%
15 to <30 years of age	28.5%	29.2%	28.4%	29.1%	29.0%	29.3%	28.6%	28.6%	30.4%	29.3%
30 to <45 years of age	15.8%	16.3%	16.8%	14.5%	17.0%	16.6%	19.2%	18.7%	14.7%	18.4%
45 to <60 years of age	7.1%	7.2%	7.2%	8.3%	7.3%	7.6%	7.8%	6.6%	5.2%	7.5%
≥ 60 years of age	4.9%	4.7%	5.1%	2.9%	4.8%	4.9%	4.2%	3.7%	4.6%	4.2%
Missing	0.2%	0.2%	0.1%	0%	0.1%	0%	0%	0%	0%	0%
Place slept in previous two weeks										
Outside the ward	2.9%	2.1%	6.2%	12.1%	8.4%	9.3%	4.7%	8.5%	29.2%	5.1%
Missing	0.1%	0.1%	0.1%	0%	0.2%	0.1%	0%	0%	0%	0.1%
Male sex	36.7%	35.0%	34.5%	35.4%	35.3%	37.0%	36.2%	38.4%	35.2%	39.0%
Slept under a bed net the night before	78.7%	88.9%	85.3%	97.6%	87.8%	78.9%	86.0%	82.2%	94.2%	91.5%
Slept under an ITN the night before	20.5%	23.4%	27.8%	23.7%	24.8%	20.5%	20.9%	20.7%	14.2%	29.3%
Use of coil the night before	4.9%	5.8%	6.6%	8.9%	7.4%	5.1%	8.6%	8.0%	2.2%	5.7%
Use of repellent the night before	0.3%	1.3%	1.6%	4.9%	5.0%	3.4%	3.0%	3.0%	0.5%	3.3%
Use of spray the night before	8.4%	10.5%	15.8%	16.8%	21.0%	18.2%	30.8%	30.6%	6.6%	29.2%
Took malaria drug in previous two weeks	7.4%	3.7%	5.4%	3.0%	8.2%	3.9%	4.9%	6.9%	6.3%	2.0%
Interviewed during wet season	10.7%	49.7%	56.2%	100.0%	27.4%	37.5%	64.2%	12.8%	99.5%	46.5%
Follow-up observation	0%	17.0%	26.5%	24.1%	31.2%	32.1%	35.2%	32.8%	17.2%	27.2%
Household-Level Covariates	N = 1,240	N = 2,107	N = 2,038	N = 124	N = 1,824	N = 396	N = 1,046	N = 827	N = 103	N = 1,549
Occupation of household head/designated										
Business / Government / Formal sector	63.1%	58.2%	59.8%	67.7%	67.0%	64.4%	60.7%	68.0%	37.9%	76.7%
Farmer / Fisherman	3.3%	1.6%	2.1%	0.0%	0.9%	2.0%	1.4%	0.7%	0%	0.8%
Informal sector	16.9%	17.8%	21.1%	22.6%	19.7%	16.7%	22.8%	17.9%	53.4%	12.5%
Retired / No job / Domestic	15.2%	20.5%	16.3%	9.7%	11.3%	15.2%	13.3%	12.9%	7.8%	9.0%
Missing	1.5%	1.9%	0.8%	0%	1.0%	0.8%	1.7%	0.5%	1.0%	1.0%
Socio-Economic Status										
Lowest quintile	32.0%	32.3%	29.7%	12.9%	20.4%	24.0%	7.3%	7.3%	3.9%	8.4%
Second quintile	29.4%	28.7%	26.2%	20.2%	23.6%	15.4%	20.9%	16.3%	11.7%	15.0%
Third quintile	13.6%	12.1%	16.0%	20.2%	19.9%	18.9%	14.1%	15.7%	57.3%	18.1%
Fourth quintile	12.1%	11.2%	12.5%	21.8%	19.7%	19.9%	29.2%	30.7%	23.3%	29.1%
Highest quintile	12.9%	15.7%	15.5%	25.0%	16.3%	21.7%	28.5%	30.0%	3.9%	29.4%
Education of Household Head/Designated										
Illiterate	6.0%	4.5%	9.4%	0.8%	6.4%	5.3%	2.6%	2.7%	13.6%	1.6%
Primary	64.4%	60.6%	51.0%	50.0%	46.2%	48.0%	35.9%	30.8%	48.5%	35.6%

Table 1. Cont.

Variables	Survey round #1	Survey round #2	Survey round #3		Survey round #4		Survey round #5		Survey round #6	
	Control	Control	Control	Larvicide	Control	Larvicide	Control	Larvicide	Control	Larvicide
Secondary	26.9%	28.2%	33.3%	37.9%	42.0%	39.6%	57.4%	60.2%	37.9%	59.3%
Tertiary	1.7%	3.6%	4.9%	11.3%	4.5%	5.8%	3.4%	5.4%	0%	2.9%
Other	0.2%	0.2%	0.4%	0%	0.1%	0.5%	0%	0%	0%	0.1%
Missing	1.0%	2.9%	1.0%	0%	0.9%	0.8%	0.7%	0.8%	0%	0.6%
Know how malaria is transmitted	68.7%	62.4%	78.4%	83.9%	82.9%	84.3%	90.2%	90.1%	81.6%	88.6%
House has window screening	22.0%	19.7%	29.5%	37.9%	23.7%	48.0%	21.5%	28.3%	31.1%	39.1%
House has complete ceiling	27.6%	24.8%	24.1%	35.5%	29.4%	36.4%	42.4%	46.8%	14.6%	33.2%
Own house	51.9%	63.1%	72.4%	66.1%	76.4%	80.3%	81.2%	80.2%	85.4%	85.7%
Household cultivates crops	19.4%	11.0%	10.3%	12.1%	8.7%	11.4%	5.8%	6.8%	13.6%	5.6%

survey round, sleeping under an ITN the night before, living in a house with a complete ceiling, and living in a house with window screens) – in the case of longitudinal observations, many of these variables are time variant; μ_j is a TCU-level random effect; v_k is an household random effect; and ε_{ijk} are the residuals. Rainfall was modeled using a smooth function where the spline penalty follows a second-order random walk process (where second-order increments are assumed to be independent with mean of zero and variance σ_t^2). This is appropriate when one wants to model smooth curves with small curvatures [51,52], which is likely to be the case for the relationship between malaria and rainfall. Finally, the time trend was accounted for with $f(.)$ and modeled as a first order autoregressive process [53]. It was chosen over other type of process based on the Deviance Information Criterion (DIC) [54], which provides information on the model's fit while penalizing for model complexity.

Potential effect modification of the intervention by other determinants of malaria infection was also investigated for a number of covariates (e.g., age, use of ITN, house proofing, etc.). Variable selection for the final multivariable models was achieved through the consideration of a number of issues: (i) subject-matter knowledge about confounders, (ii) variable exhibiting sufficient variation, and (iii) extent of potential measurement errors.

In order to investigate the robustness of our results to modeling assumptions, we used three additional model specifications by including: (i) individual random effects, (ii) ward fixed effects, and (iii) spatially-structured random effects. We also performed a number of sensitivity analyses. Specifically, we tested for potential spillover effects of the intervention, used different lags for the larviciding intervention and for the rainfall estimates, further covariate adjustments (socio-economic status, educational level, and occupation), and varied the choice of penalty for the semi-parametric time trend (first and second-order random walk). Technical details and results are presented in the Supplemental online material (Text S1).

Models were fitted using Integrated Nested Laplace Approximations (INLA) [55]. A major advantage of INLA is that it calculates posterior marginal distributions in very short computational time as compared to more traditional Markov Chain Monte Carlo (MCMC) approaches. Further, INLA has been shown to yield very high accuracy that is comparable to MCMC [55,56].

Non-informative priors for the regression parameters and hyperparameters were used (see Text S1 for details). All analyses were performed using the R statistical software [57] and estimation of the marginal posterior distribution of the parameters of interest was performed using the *INLA* library [58]. Observations with missing data for age (n = 44), place slept in previous two weeks (n = 52), occupation of the household head (n = 134), and education level of the household head (n = 136) were retained in the analysis using the missing indicator method [59].

Ethical considerations

Ethics approval was obtained from the Medical Research Coordination Committee of the National Institute for Medical Research, Ministry of Tanzania (Reference number NIMR/HQ/R.8a/Vol. IX/279 &234). Approval from Harvard School of Public Health Institutional Review Board was also obtained (Protocol # 20323-101). Written informed consent was obtained from all study participants after being provided with information regarding the goal, objectives, risk and benefits of the study. Parents or designated guardians provided signed informed consent on behalf of children under 18 years of age. These procedures were approved by the ethics committees.

Results

Throughout the study period, malaria prevalence exhibited a considerable decline. Malaria prevalence was highest during the first round of data collection in 2004, with 20.8% prevalence (95% CI: 16.8–24.9%). It decreased to 16.9% (95% CI: 15.1–18.8%) in the second survey round, 10.4% (95% CI: 9.7–11.0%) in the third, 6.6% (95% CI: 6.0–7.1%) in the fourth, 4.8% (95% CI: 4.3–5.4%) in the fifth, and 1.7% (95% CI: 1.4–2.1%) in the last survey round. Stratifying malaria prevalence by survey round and larviciding intervention status, we observed that prevalence was slightly lower in the intervention wards as compared to the control ones, with the notable exception of the third survey round (Figure 4). Note that the start of the larviciding phases did not precisely coincide with the beginning of the survey rounds due to operational issues (as shown in Figure 3, phase 1 of larviciding was launched in March 2006, while the fourth survey round started in June 2006; phase 2 in May 2007; and phase 3 in April 2008). Hence, median dates of

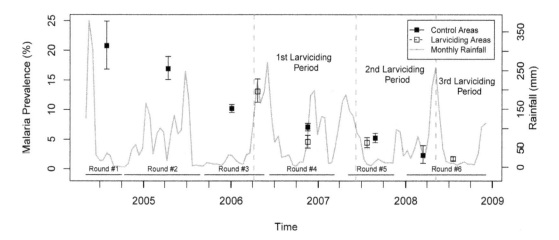

Figure 4. Crude prevalence of malaria infection stratified by survey round and larviciding status. Confidence intervals are based on 9,999 bootstrap replicates and account for clustering at the ten-cell unit level. Monthly rainfall variation is also shown.

interviews in larviciding and control areas do not necessarily coincide, and seasonality in malaria transmission could confound the observed differences in prevalence shown in Figure 4.

For each survey round, the socio-demographic characteristics of study participants and households, stratified by larviciding

Table 2. Univariate and multivariable effect size estimates of the larviciding intervention on malaria prevalence in Dar es Salaam, 2004–2008 (N = 64,537).

	Univariate		Multivariable	
	OR*	95% CrI†	OR*	95% CrI†
LARVICIDING INTERVENTION	**0.79**	**0.66–0.93**	**0.79**	**0.66–0.93**
Age				
Under five years of age	-	-	1.00	-
≥5 and <15 years of age	-	-	**0.82**	**(0.76–0.90)**
≥15 and <30 years of age	-	-	**0.67**	**(0.61–0.73)**
≥30 and <45 years of age	-	-	**0.60**	**(0.54–0.66)**
≥45 and <60 years of age	-	-	**0.55**	**(0.48–0.63)**
≥60 years of age	-	-	**0.47**	**(0.40–0.56)**
Male sex	-	-	**1.08**	**(1.01–1.15)**
Slept outside ward (previous 2 weeks)	-	-	0.90	(0.77–1.04)
Treated for malaria (previous round)	-	-	**0.65**	**(0.56–0.75)**
Took malaria drug (previous 2 weeks)	-	-	1.02	(0.90–1.16)
ITN used the night before	-	-	**0.93**	**(0.86–0.99)**
House has closed ceiling	-	-	0.93	(0.85–1.01)
House has window screens	-	-	**0.90**	**(0.83–0.98)**
Trend for time (AR1§)	Yes		Yes	
Semi-parametric smooth for rainfall	Yes		Yes	
Random effects (TCU & Household)	Yes		Yes	

Statistically significant results are bolded.
*OR = Odds Ratio.
†CrI = Credible Intervals.
§AR1 = First Order Autoregressive Process.

intervention status, are presented in Table 1. Use of bednet was highly variable through time and seems to be correlated with rainfall and, probably, abundance of nuisance insects. The proportion of interviews performed during the wet seasons also differs between larviciding and control groups. Interestingly, the proportion of individuals reporting having taken anti-malarial drug in the previous two weeks remained relatively constant through time despite the overall decline in malaria prevalence. Finally, we note that socio-economic status seems to be increasing with time, as exhibited by the rising proportion of individuals in the upper quintiles. Overall, individuals in control and larviciding areas do not seem to differ dramatically in their socio-demographic characteristics. Most differences are observed in either the third or sixth survey rounds where the sample sizes in the larviciding and control groups, respectively, are notably smaller.

Taking into account the previously stated limitations of our univariate analysis, we present in Table 2 the results from the random effects logistic regression models that account for clustering of observations within household and TCU. These analyses suggest a significant protective effect of larviciding, with a point estimate for the odds ratio of 0.79 (95% Credible Intervals (CrI): 0.66–0.93) in both univariate and multivariable analyses. When considering potential effect modification of the larviciding intervention by season, we see that larviciding activities achieved maximum programmatic impact during the dry season (Table 3) with an odds ratio of 0.60 (95% CrI: 0.47–0.75). The dry season is defined as the months of January, February, and June through September. The effect of the larviciding intervention also had synergistic effects with other malaria protective measures such as houses with window screens (OR = 0.68; 95% CrI: 0.54–0.85), houses with complete ceiling (OR = 0.66; 95% CrI: 0.53–0.83), and using an ITN the night before (OR = 0.63; 95% CrI: 0.48–0.82). Finally, the effect of the intervention was also heterogeneous among age groups with the larviciding intervention exhibiting a greater protective effect for children under five (OR = 0.61; 95% CrI: 0.46–0.80).

Model specifications seem to have little bearing on the estimates of the posterior marginal for the larviciding intervention (see Tables S1 and S2 in Text S1). Importantly, including fixed effects at the ward level, which would control for any time-invariant measured or unmeasured confounders of the larviciding-malaria

Table 3. Effect modification of the larviciding intervention by selected determinants of malaria prevalence in Dar es Salaam, 2004–2008 (N = 64,537).

Effect modification of the larviciding intervention by selected determinants of malaria infection (Odds Ratio and 95% Credible Intervals)*

	Control	Larviciding	Effect of Larviciding Within Strata
Wet Season	1.00	1.06 (0.84–1.33)	1.06 (0.84–1.33)
Dry Season §	0.97 (0.69–1.10)	**0.57 (0.41–0.77)**	**0.60 (0.47–0.75)**
	Control	Larviciding	Effect of Larviciding Within Strata
No Screen	1.00	0.84 (0.70–1.02)	0.84 (0.70–1.02)
Window Screens	0.93 (0.85–1.02)	**0.80 (0.65–0.99)**	**0.68 (0.54–0.85)**
	Control	Larviciding	Effect of Larviciding Within Strata
Open Ceiling	1.00	0.84 (0.70–1.01)	0.84 (0.70–1.01)
Complete Ceiling	0.97 (0.88–1.06)	**0.78 (0.63–0.97)**	**0.66 (0.53–0.83)**
	Control	Larviciding	Effect of Larviciding Within Strata
No ITN	1.00	**0.83 (0.69–0.99)**	**0.83 (0.69–0.99)**
ITN used	0.96 (0.88–1.04)	**0.77 (0.61–0.96)**	**0.63 (0.48–0.82)**
	Control	Larviciding	Effect of Larviciding Within Strata
Aged ≥5 years	1.00	**0.83 (0.69–0.99)**	**0.83 (0.69–0.99)**
<5 years of age	1.35 (1.23–1.47)	**0.73 (0.56–0.94)**	**0.61 (0.46–0.80)**

Statistically significant results are bolded.
All models are adjusted for age, sex, sleeping outside of the ward (previous 2 weeks), being treated for malaria in a previous round, use of malaria drugs (previous 2 weeks), use of ITN, complete ceiling, window screen, precipitation, time trend. Random effects at household and TCU levels are also included.
§Dry season is defined as the months of January, February, and June through September.

relationship, had little impact on the point estimate of the larviciding intervention (adjusted OR = 0.80; 95% CrI:0.66–0.97).

Finally, our sensitivity analyses (see Table S3 and S4 in Text S1) demonstrated that spillover effects were not biasing our effect size estimate towards the null. As expected, effect size estimates were somewhat sensitive to variation in the assumed lag length between initiation of larviciding activities and malaria transmission but the effect remained statistically significant over lag lengths varying between 28 and 60 days. Results were also robust to changes in other model parameters.

Discussion

This study has shown that a community-based larviciding program, centrally managed by the UMCP, provided significant protection to individuals living in areas covered by the larviciding operations. The strength of association was robust to model specifications and consistently approximated a 21% reduction in the odds of malaria infection. Further, the larviciding intervention achieved maximum effectiveness during the dry season and had synergistic effects with other protective measures such as use of ITN, houses with windows screens, and houses with complete ceilings. In addition, we found no evidence of spillover effects between intervention and control areas.

Our estimated effect size for the larviciding intervention is much lower, but not statistically different, than the one previously reported for the first larviciding period of the UMCP, where the odds ratio of living in areas treated with larvicides and being infected with malaria was estimated to be 0.28 (95% CI: 0.10–0.80) as compared to individuals living in control areas [21]. This can be explained in part by the fact that our study considered all age ranges, while Geissbühler et al [21] restricted their analysis to children under five years of age. While there is no reason to believe that larviciding should be more protective for children than for adults, since the intervention acts at the population level by

reducing vector density, children might be more likely to spend evenings and nights at or close to their home, a period of the day when most of malaria transmission occurs. There is thus less potential misclassification of exposure for this age group as compared to adults, who might visit friends or spend time during evenings near high exposure areas not covered by larviciding activities. Indeed, we found that the product term between the larviciding intervention and age was statistically significant. The estimated odds ratio for the larviciding intervention was of 0.61 (95% CrI: 0.46–0.80) for children under five years of age which is closer to the one reported by Geissbühler et al [21] but insufficient to explain this differential. Another reason which could explain this difference in impact is that our analysis covered all three phases of the intervention with a total of 33 months of larviciding activities, while Geissbühler et al [21] analyzed only the first phase, when the intervention was operational in only three wards for 12 months. Analyses over a longer period may be impacted by programmatic fatigue, coupled with the potential impact that other unmeasured and/or unknown interventions could have on the prevalence of malaria infection and overall transmission dynamics (e.g., artemisin-based combination therapy – ACT started to be the first line of treatment in 2007).

Larviciding during the dry season was shown to be more effective at lowering the prevalence of malaria infection than during the rainy season (when the stratified effect was not significant). This result is especially interesting since 49% of malaria cases were sampled during the dry season. Since larval habitats are less numerous and easier to access when rainfall is low, larviciding activities could have been more effective at suppressing larval production due to operational issues. This highlights one of the key aspects of successful larviciding programs: the ability to locate and access all potential breeding habitats in the targeted area. Also, larviciding should not be deployed alone, but in conjunction with other appropriate vector control activities [60]. The fact that we have estimated larviciding to be more effective than ITNs in Dar es Salaam should not be taken at face value, since the effect size estimate for ITNs does not take into account potential community effects that extend to non-users [61,62], and that the use of ITNs and other protective measures is likely a function of perceived risk by household members. The combination of different vector control strategies is also supported by our findings of significant synergistic effects between larviciding and use of ITNs, window screens, and houses with a complete ceiling.

With renewed impetus for the long-term goal of malaria eradication [63], the need for tailored programs is imperative, including vector control [30]. Vector control programs should not be established as stand-alone entities. Rather, intersectoral collaboration, health system strengthening, and community mobilization are instrumental to vector control program success. Integrated Vector Management (IVM), as endorsed by WHO [31,64], emphasizes rational decision making processes to efficiently use resources and attain health-based targets [65]. IVM specifically acknowledges that a 'one size fits all' strategy for malaria control will be ineffective. Larviciding should be considered as part of an IVM approach in other urban areas of sub-Saharan Africa, if the local malaria ecology warrants its use. Our study provides a number of important lessons regarding the implementation of larval control: (i) breeding habitats can, and should, be mapped at high resolution using low-cost technology [36], (ii) locally relevant entomological information should be collected to inform operational activities, (iii) monitoring and evaluation systems should be implemented to ensure effective and appropriate delivery and fine-tuning of interventions, and (iv) community involvement and sensitization can be beneficial to programmatic activities. Other strategies included in an IVM approach could facilitate the use of larviciding. For example, in Dar es Salaam 33% of *Anopheles* breeding habitats are found in clogged drains [66]. In this context, the use of environmental management to restore the functionality of drains would result in fewer breeding habitats [43], and therefore reduce the area to be covered with larviciding.

Strengths of this study include its large sample size, longitudinal design, large temporal and spatial extent of larviciding activities that limited potential spillover effects, and availability of reliable baseline information. This study also has some limitations. First, the wards targeted by the UMCP were not randomly allocated to the larviciding intervention. This entails that our effect size estimates for the larviciding intervention could be biased by residual confounding. This is unlikely to be the case as including fixed effects at the ward level, which would control for such time-invariant non-measured confounders, did not impact our results. Second, ACTs were effectively introduced in Dar es Salaam in January 2007. With its gametocidal proprieties, this drug, if used on a large scale, has the potential to significantly reduce the reservoir of malaria in the general population. Although attempts

were made at collecting information on ACT use from health facility data, we were not able to assemble reliable temporal information for the targeted 15 wards. Thus, some of the secular decline in the prevalence of malaria infection observed in control areas before the introduction of larviciding may be a result of ACT use (and possibly of other unobserved activities that could potentially impact the risk of malaria transmission).

Our results have important implications for malaria control in sub-Saharan Africa. Specifically, we have provided evidence that a community-based application of microbial larvicides was effective in reducing malaria transmission in urban Dar es Salaam. Microbial larvicides have been shown to be environmentally safe, specific in their action, and highly effective in killing *Anopheles* larvae under field conditions [67–69]. With important projected increases in urban population in sub-Saharan Africa, mosquitoes' behavioral adaptation to current control strategies, and the already recorded emergence of resistance to pyrethroid insecticides, larval source management, and larviciding in particular, should be given careful consideration by managers of malaria control programs.

Supporting Information

Text S1   Supplemental online information. Description of results from additional model specifications (Tables S1 and S2), potential spillover effects (Table S3), sensitivity analyses performed (Table S4), and detailed information on the prior distributions used.

Acknowledgments

We acknowledge the team work involved in activities of the UMCP during 2004–2008. Particularly, we thank Gerry F. Killeen, Ulrike Fillinger, Burton Singer, Marcel Tanner, Steve Lindsay, Michael Kiama (*in memoriam*), and Deo Mtasiwa. We recognize the financial support for the UMCP provided by the Swiss Tropical Institute, the Bill & Melinda Gates Foundation, Valent Biosciences Corporation, USAID (Environmental Health Program, Dar es Salaam Mission and the President's Malaria Initiative, all administered through Research Triangle International), and the Wellcome Trust. We are deeply thankful to all community members involved during different phases of the UMCP; to municipal, city, and ward officers; to local leaders and to community-owned resource persons; and to all interviewers, nurses, laboratory technicians, and data entry personnel who supported the assembly of the UMCP database. We are particularly thankful to Khadija Kannady, Abdullah Hemed, James Msami, and George Makanyadigo. We thank Francesca Dominici for valuable comments on an earlier version of this paper.

Author Contributions

Conceived and designed the experiments: MCC. Performed the experiments: MCC. Analyzed the data: MMG MCC. Contributed reagents/materials/analysis tools: MMG MCC. Wrote the paper: MMG MCC.

References

1. Keiser J, Singer BH, Utzinger J (2005) Reducing the burden of malaria in different eco-epidemiological settings with environmental management: a systematic review. Lancet Infect Dis 5: 695–708.
2. Shousha AT (1948) Species-eradication: The Eradication of *Anopheles gambiae* from Upper Egypt, 1942-1945. Bull World Health Organ 1: 309–352.
3. Soper F, Wilson D (1943) *Anopheles gambiae* in Brazil, 1930 to 1940. New York, NY: Rockefeller Foundation. 262 p.
4. Watson M (1953) African Highway: The Battle for Health in Central Africa. London, UK: John Murray. 294 p.
5. Ross R (1907) An address on the prevention of malaria in British possessions, Egypt, and parts of America. Lancet 2: 879–887.
6. Derryberry O, Gartrell F (1952) Trends in malaria control program of the Tennessee Valley Authority. American Journal of Tropical Medicine and Hygiene 1: 500–507.

7. Nájera JA, González-Silva M, Alonso PL (2011) Some lessons for the future from the Global Malaria Eradication Programme (1955–1969). PLoS Med 8: e1000412.
8. Baer F, McGahey C, Wijeyaratne P (1999) Summary of EHP Activities in Kitwe, Zambia, 1997–1999. Kitwe Urban Health Programs. Washington, DC: U.S. Agency for International Development. 48 p.
9. Lindsay S, Egwang T, Kabuye F, Mutambo T, Matwale G (2004) Community-Based Environmental Management Program for Malaria Control in Kampala and Jinga, Uganda. Washington, DC: U.S. Agency for International Development. 62 p.
10. Castro M, Yamagata Y, Mtasiwa D, Tanner M, Utzinger J, et al. (2004) Integrated urban malaria control: a case study in dar es salaam, Tanzania. Am J Trop Med Hyg 71: 103–117.
11. Chaki P, Dongus S, Fillinger U, Kelly A, Killeen G (2011) Community-owned resource persons for malaria vector control: enabling factors and challenges in

an operational programme in Dar es Salaam, United Republic of Tanzania. Human Resources For Health 9: 21.

12. Chaki P, Govella N, Shoo B, Hemed A, Tanner M, et al. (2009) Achieving high coverage of larval-stage mosquito surveillance: challenges for a community-based mosquito control programme in urban Dar es Salaam, Tanzania. Malaria Journal 8: 311.

13. Gu W, Utzinger J, Novak RJ (2008) Habitat-based larval interventions: a new perspective for malaria control. Am J Trop Med Hyg 78: 2–6.

14. Fillinger U, Lindsay SW (2011) Larval source management for malaria control in Africa: myths and reality. Malar J 10: 353.

15. Killeen GF, Fillinger U, Kiche I, Gouagna LC, Knols BG (2002) Eradication of Anopheles gambiae from Brazil: lessons for malaria control in Africa? Lancet Infect Dis 2: 618–627.

16. Worrall E, Fillinger U (2011) Large-scale use of mosquito larval source management for malaria control in Africa: a cost analysis. Malar J 10: 338.

17. Imbahale SS, Mweresa CK, Takken W, Mukabana WR (2011) Development of environmental tools for anopheline larval control. Parasit Vectors 4: 130.

18. Walker K, Lynch M (2007) Contributions of Anopheles larval control to malaria suppression in tropical Africa: review of achievements and potential. Med Vet Entomol 21: 2–21.

19. WHO (2012) Interim Position Statement - The role of larviciding for malaria control in sub-Saharan Africa. Geneva, Switzerland: World Health Organization - Global Malaria Programme. 21 p.

20. Fillinger U, Ndenga B, Githeko A, Lindsay S (2009) Integrated malaria vector control with microbial larvicides and insecticide-treated nets in western Kenya: a controlled trial. Bulletin of the World Health Organization 87: 655–665.

21. Geissbuhler Y, Kannady K, Chaki PP, Emidi B, Govella NJ, et al. (2009) Microbial larvicide application by a large-scale, community-based program reduces malaria infection prevalence in urban Dar es Salaam, Tanzania. PLoS One 4: e5107.

22. Majambere S, Pinder M, Fillinger U, Ameh D, Conway D, et al. (2010) Is mosquito larval source management appropriate for reducing malaria in areas of extensive flooding in The Gambia? A cross-over Intervention Trial. American Journal of Tropical Medicine and Hygiene 82: 176–184.

23. Geissbuhler Y, Chaki P, Emidi B, Govella NJ, Shirima R, et al. (2007) Interdependence of domestic malaria prevention measures and mosquito-human interactions in urban Dar es Salaam, Tanzania. Malar J 6: 126.

24. Griffin JT, Hollingsworth TD, Okell LC, Churcher TS, White M, et al. (2010) Reducing Plasmodium falciparum malaria transmission in Africa: a model-based evaluation of intervention strategies. PLoS Med 7: e1000324.

25. Ranson H, N'guessan R, Lines J, Moiroux N, Nkuni Z, et al. (2011) Pyrethroid resistance in African anopheline mosquitoes: what are the implications for malaria control? Trends Parasitol 27: 91–98.

26. Cuamba N, Morgan JC, Irving H, Steven A, Wondji CS (2010) High level of pyrethroid resistance in an Anopheles funestus population of the Chokwe District in Mozambique. PLoS One 5: e11010.

27. Morgan JC, Irving H, Okedi LM, Steven A, Wondji CS (2010) Pyrethroid resistance in an Anopheles funestus population from Uganda. PLoS One 5: e11872.

28. Munhenga G, Masendu HT, Brooke BD, Hunt RH, Koekemoer LK (2008) Pyrethroid resistance in the major malaria vector Anopheles arabiensis from Gwave, a malaria-endemic area in Zimbabwe. Malar J 7: 247.

29. Ferguson HM, Dornhaus A, Beeche A, Borgemeister C, Gottlieb M, et al. (2010) Ecology: a prerequisite for malaria elimination and eradication. PLoS Med 7: e1000303.

30. malERA Consultative Group on Vector Control (2011) A research agenda for malaria eradication: vector control. PLoS Med 8: e1000401.

31. WHO (2004) Global Strategic Framework for Integrated Vector Management. Geneva, Switzerland: World Health Organization. 15 p.

32. WHO (2012) Handbook for integrated vector management. Geneva, Switzerland: World Health Organization. 67 p.

33. UN-HABITAT (2010) The State of African Cities 2010 - Governance, Inequality and Urban Land Markets. United Nations Settlements Programme. 276 p.

34. Keiser J, Utzinger J, Castro CM, Smith TA, Tanner M, et al. (2004) Urbanization in sub-saharan Africa and implication for malaria control. Am J Trop Med Hyg 71: 118–127.

35. Robert V, Macintyre K, Keating J, Trape JF, Duchemin JB, et al. (2003) Malaria transmission in urban sub-Saharan Africa. Am J Trop Med Hyg 68: 169–176.

36. Fillinger U, Kannady K, William G, Vanek MJ, Dongus S, et al. (2008) A tool box for operational mosquito larval control: preliminary results and early lessons from the Urban Malaria Control Programme in Dar es Salaam, Tanzania. Malar J 7: 20.

37. UN (2007) World Population Prospects: The 2006 Revision and World Urbanization Prospects: The 2007 Revision. Population Division of the Department of Economic and Social Affairs of the United Nations Secretariat.

38. Yhdego M, Paul M (1988) Malaria Control in Tanzania. Environment International 14: 479–483.

39. Dongus S, Nyika D, Kannady K, Mtasiwa D, Mshinda H, et al. (2007) Participatory mapping of target areas to enable operational larval source management to suppress malaria vector mosquitoes in Dar es Salaam, Tanzania. International Journal of Health Geographics 6: 37.

40. Dongus S, Mwakalinga V, Kannady K, Tanner M, Killeen G (2011) Participatory Mapping as a Component of Operational Malaria Vector Control

in Tanzania Geospatial Analysis of Environmental Health. In: Maantay JA, McLafferty S, Jensen RR, editors: Springer Netherlands. pp. 321–336.

41. Vanek M, Shoo B, Mtasiwa D, Kiama M, Lindsay S, et al. (2006) Community-based surveillance of malaria vector larval habitats: a baseline study in urban Dar es Salaam, Tanzania. Bmc Public Health 6.

42. Filmer D, Pritchett LH (2001) Estimating wealth effects without expenditure data - or tears: an application to educational enrollments in states of India. Demography 38: 115–132.

43. Castro M, Tsuruta A, Kanamori S, Kannady K, Mkude S (2009) Community-based environmental management for malaria control: evidence from a small-scale intervention in Dar es Salaam, Tanzania. Malar J 8: 57.

44. Xie P, Arkin P (1996) Analyses of global monthly precipitation using gauge observations, satellite estimates, and numerical model predictions. Journal of Climate 9: 840–858.

45. Krefis AC, Schwarz NG, Krüger A, Fobil J, Nkrumah B, et al. (2011) Modeling the relationship between precipitation and malaria incidence in children from a holoendemic area in Ghana. Am J Trop Med Hyg 84: 285–291.

46. Zhou G, Minakawa N, Githeko AK, Yan G (2004) Association between climate variability and malaria epidemics in the East African highlands. Proc Natl Acad Sci U S A 101: 2375–2380.

47. Loevinsohn ME (1994) Climatic warming and increased malaria incidence in Rwanda. Lancet 343: 714–718.

48. Kristan M, Abeku TA, Beard J, Okia M, Rapuoda B, et al. (2008) Variations in entomological indices in relation to weather patterns and malaria incidence in East African highlands: implications for epidemic prevention and control. Malar J 7: 231.

49. Barbazan P, Baldet T, Darriet F, Escaffre H, Djoda DH, et al. (1998) Impact of treatments with Bacillus sphaericus on Anopheles populations and the transmission of malaria in Maroua, a large city in a savannah region of Cameroon. J Am Mosq Control Assoc 14: 33–39.

50. Cameron A, Gelbach J, Miller D (2008) Bootstrap-based improvements for inference with clustered errors. Review of Economics and Statistics 90: 414–427.

51. Schrodle B, Held L (2011) A primer on disease mapping and ecological regression using INLA. Computational Statistics 26: 241–258.

52. Natario I, Knorr-Held L (2003) Non-parametric ecological regression and spatial variation. Biometrical Journal 45: 670–688.

53. Congdon P (2006) Bayesian statistical modelling. Chichester, England ; Hoboken, NJ: John Wiley & Sons. xi, 573 p. p.

54. Spiegelhalter DJ, Best NG, Carlin BP, Linde Avd (2002) Bayesian measures of model complexity and fit. Journal of the Royal Statistical Society: Series B (Statistical Methodology) 64: 583.

55. Rue H, Martino S, Chopin N (2009) Approximate Bayesian inference for latent Gaussian models by using integrated nested Laplace approximations. Journal of the Royal Statistical Society Series B-Statistical Methodology 71: 319–392.

56. Beguin J, Martino S, Rue H, Cumming SG (2012) Hierarchical analysis of spatially autocorrelated ecological data using integrated nested Laplace approximation. Methods in Ecology and Evolution 3: 921–929.

57. R Development Core Team (2012) R: A Language and Environment for Statistical Computing. Vienna, Austria: R Foundation for Statistical Computing.

58. Rue H, Martino S, Lindgren F (2009) INLA: Functions which allow to perform a full Bayesian analysis of structured (geo-)additive models using Integrated Nested Laplace Approximaxion. R Package version 0.0 ed.

59. Miettinen OS (1985) Theoretical epidemiology : principles of occurrence research in medicine. New York: Wiley. xxii, 359 p. p.

60. WHO (2004) Global Strategic Framework for Integrated Vector Management. Geneva, Switzerland: World Health Organization. 12 p.

61. Howard SC, Omumbo J, Nevill C, Some ES, Donnelly CA, et al. (2000) Evidence for a mass community effect of insecticide-treated bednets on the incidence of malaria on the Kenyan coast. Trans R Soc Trop Med Hyg 94: 357–360.

62. Hawley WA, Phillips-Howard PA, ter Kuile FO, Terlouw DJ, Vulule JM, et al. (2003) Community-wide effects of permethrin-treated bed nets on child mortality and malaria morbidity in western Kenya. Am J Trop Med Hyg 68: 121–127.

63. Alonso PL, Brown G, Arevalo-Herrera M, Binka F, Chitnis C, et al. (2011) A research agenda to underpin malaria eradication. PLoS Med 8: e1000406.

64. WHO (2004) Global Strategic Framework for Integrated Vector Management. Geneva, Switzerland: World Health Organization. 14 p.

65. WHO (2008) WHO position statement on integrated vector management. Geneva, Switzerland: World Health Organization. 4 p.

66. Castro M, Kanamori S, Kannady K, Mkude S, Killeen G, et al. (2010) The importance of drains for the larval development of lymphatic filariasis and malaria vectors in Dar es Salaam, United Republic of Tanzania. PLoS Negl Trop Dis 4: e693.

67. Shililu JI, Tewolde GM, Brantly E, Githure JI, Mbogo CM, et al. (2003) Efficacy of Bacillus thuringiensis israelensis, Bacillus sphaericus and temephos for managing Anopheles larvae in Eritrea. J Am Mosq Control Assoc 19: 251–258.

68. Majambere S, Lindsay SW, Green C, Kandeh B, Fillinger U (2007) Microbial larvicides for malaria control in The Gambia. Malar J 6: 76.

69. Fillinger U, Knols BG, Becker N (2003) Efficacy and efficiency of new Bacillus thuringiensis var israelensis and Bacillus sphaericus formulations against Afrotropical anophelines in Western Kenya. Trop Med Int Health 8: 37–47.

9

New Integrated Strategy Emphasizing Infection Source Control to Curb *Schistosomiasis japonica* in a Marshland Area of Hubei Province, China: Findings from an Eight-Year Longitudinal Survey

Yan-Yan Chen[1,2], Jian-Bing Liu[2]*, Xi-Bao Huang[2], Shun-Xiang Cai[2], Zheng-Ming Su[2], Rong Zhong[1], Li Zou[1], Xiao-Ping Miao[1]*

1 Ministry of Education Key Lab of Environment and Health, Department of Epidemiology and Biostatistics, School of Public Health, Tongji Medical College, Huazhong University of Science and Technology, Wuhan, China, 2 Hubei Center for Disease Control and Prevention, Wuhan, China

Abstract

Background: Schistosomiasis remains a major public health problem in China. The major endemic foci are the lake and marshland regions of southern China, particularly the regions along the middle and lower reach of the Yangtze River in four provinces (Hubei, Hunan, Jiangxi, and Anhui). The purpose of our study is to assess the effect of a new integrated strategy emphasizing infection source control to curb schistosomiasis in marshland regions.

Methods: In a longitudinal study, we implemented an integrated control strategy emphasizing infection source control in 16 villages from 2005 through 2012 in marshland regions of Hubei province. The interventions included removing cattle from snail-infested grasslands, providing farmers with mechanized farm equipment, improving sanitation by supplying tap water, building lavatories and latrines, praziquantel chemotherapy, controlling snails, and environmental modification.

Results: Following the integrated control strategy designed to reduce the role of bovines and humans as sources of *Schistosoma japonicum* infection, the prevalence of human *S. japonicum* infection declined from 1.7% in 2005 to 0.4% in 2012 (*P*<0.001). Reductions were also observed in both sexes, across all age groups, and among high risk occupations. Moreover, the prevalence of bovine *S. japonicum* infection decreased from 11.7% in 2005 to 0.6% in 2012 (*P*<0.001). In addition, all the 16 villages achieved the national criteria of infection control in 2008.

Conclusion: Our findings indicate that the integrated strategy was likely effective in controlling the transmission of *S. japonicum* in marshland regions in China.

Editor: Sten H. Vermund, Vanderbilt University, United States of America

Funding: This study was funded by the National Science and Technology Support Program in China (2009BAI78B07), and the Provincial Research of Schistosomiasis Prevention of Health Department of Hubei, China (XF2010-5, XF2010-30, XF2012-24, XF2012-26). The funders had no role in study design, data collection and analysis, decision to publish, or preparation of the manuscript.

Competing Interests: The authors have declared that no competing interests exist.

* E-mail: miaoxp@gmail.com (XPM); hbcdcxf@126.com (JBL)

Introduction

Schistosomiasis is one of the most prevalent parasitic diseases in the world [1]. The global prevalence is currently estimated to be 207 million cases, with another 779 million people at risk of infection in 76 countries and territories [2,3]. Documented evidence indicates that *Schistosoma japonicum* has been endemic for a long time in China [4]. *S. japonicum* eggs were identified in a male corpse dating back to the Western Han dynasty some 2100 years ago that was exhumed in Jianglin Hsien, Hubei Province in 1975 [5]. Schistosome eggs were also found in another female corpse buried about the same time in Hunan Province [6].

Schistosomiasis is mainly endemic today in lake and marshland areas (Hubei, Hunan, Jiangxi, Anhui, and Jiangsu provinces) and in hilly and mountainous regions (Sichuan and Yunnan provinces)

in China [7]. The Chinese government has given high priority to the control of schistosomiasis since the founding of the People's Republic of China in 1949. A number of special bodies were established to manage control activities from the national to the township level [4,8,9]. Schistosomiasis control strategies in China have gone through three stages since the 1950s. Prior to the 1980s, schistosomiasis controls primarily comprised waterway management and snail control. From 1980s to 2004, control measures were primarily synchronous chemotherapy for humans and domestic animals. Since 2005, integrated strategy emphasizing infection source control and mollusciciding were added to chemotherapy of humans. Significant achievements on schistosomiasis control have been attained through ongoing national control programs over the past 50 years [10]. The third nationwide schistosomiasis sampling survey indicated that the

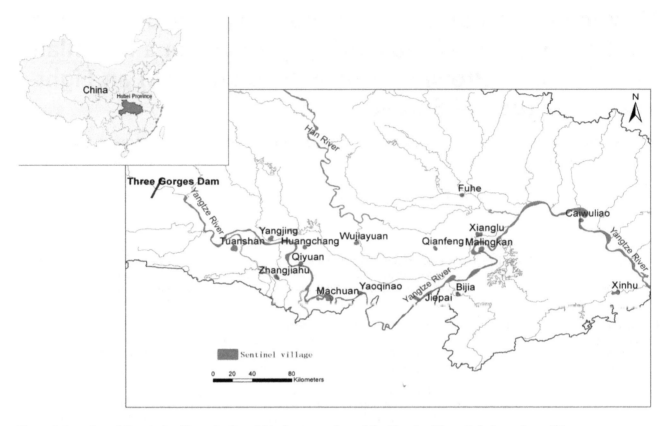

Figure 1. Location of the study villages in the mid-to-lower reaches of the Yangtze River, Hubei province, China.

number of schistosomiasis patients decreased by 55.7%, from 1,638,103 cases in 1989 to 726,112 cases in 2004 [11].

However, each phase also has limitations. In the first stage, the endemic areas decreased immensely due to environmental modification, but severe environment pollution ensued because of the molluscicide widely used in endemic areas. Eliminating snails in lake and marshland habitats, in contrast to rivers and irrigation canals, has proved difficult [12]. In the second stage, human and bovine schistosomiasis were reduced to relatively low level temporarily, but re-infection occurred and continuous chemotherapy based schistosomiasis control could not be maintained [13]. At the beginning of the 21st century, schistosomiasis was still of considerable economic and public health concern in China. The progress of schistosomiasis control has stalled since the termination of the World Bank Loan Project for schistosomiasis control at the end of 2001 [14,15]. Data suggest that 110 counties had not yet reached the criteria for transmission control in 2003 [16]. The third national survey conducted in 2004 showed that the prevalence of *S. japonicum* infection in humans had not substantially changed in lakes, marshlands, and other areas of Southern China since 1995 [11].

In China, most of the current uncontrolled schistosomiasis endemic areas are concentrated in four provinces (Hubei, Hunan, Jiangxi, and Anhui) along the middle and lower reach of the Yangtze River. The interruption of transmission has proven particularly difficult to achieve, thus finding an effective and sustainable strategy for schitosomiasis control in the lake and marshland regions of China has become critical [16,17,18]. In 2004, the Chinese government directed much attention to the control of schistosomiasis, which was placed on the top priority list for the control of communicable diseases in China. Therefore, two control targets were established: to reduce human infection rates in all endemic counties to <5% by 2008, and then to <1% by 2015 [19]. A comprehensive strategy was developed to reduce the sources of infection (bovines and humans) in endemic regions of China [20].

Hubei province is located in the middle reaches of the Yangtze River. Following the effect of flood along the upper reaches of the Yangtze River during the annual monsoon season, the marshlands along the Yangtze River operate in a "winter–land, summer–water" cycle, and vast grass-covered marshlands emerge as floodwaters recede. These are ideal breeding sites for *Oncomelania hupensis* snails [21,22]. *O. hupensis* is the unique intermediate host of *S. japonicum*, which has a key function during the transmission of schistosomiasis. Hubei province is a highly endemic area of schistosomiasis in China. Research has indicated a relatively high prevalence within the human population in Hubei province [18]. To achieve the two national control goals, an integrated control strategy aimed at reducing the roles of humans and cattle as the main sources of infection for snails was carried out in marshland areas of Hubei Province, China. The new strategy includes measures of improving water supply and sanitation, prohibiting the grazing of cattle in grasslands, captive feeding of livestock, and replacement of bovine with machines for farming [19].

In this paper, we examined this integrated control strategy which was carried out over an eight-year period in marshland areas of Hubei. Specifically we examined the effectiveness of this strategy in 16 study villages along the middle reaches of the Yangtze River, Hubei province from 2005 to 2012.

Table 1. *S. japonicum* prevalence and intensity of infection in human in the 16 villages at baseline in 2005.

	Sub-group	No. of people examined	Prevalence (95% CI)
All		12872	1.65%(1.43,1.87)
Sex			
	Male	6655	2.04 % (1.70,2.38)
	Female	6210	1.22 % (0.95,1.50)
Age			
	6–9	667	0.30 % (−0.12,0.71)
	10–19	2481	0.85 % (0.49,1.21)
	20–29	920	1.09 % (0.42.1.76)
	30–39	2476	2.18 % (1.61,2.76)
	40–49	2912	2.16 % (1.64,2.69)
	50–59	1979	2.22 % (1.57,2.87)
	60-	1437	1.25 % (0.68,1.83)
Occupation			
	Farmer	9662	1.88 % (1.61,2.15)
	Fisherman	73	8.22 % (1.92,14.52)
	Student	2882	0.69 % (0.39,1.00)
	Other	255	1.57 % (0.04,3.09)

Materials and Methods

Study area

The study was conducted in the middle and lower reaches of the Yangtze River, Hubei province, China. In 2004, Hubei province had 5,499 schistosomiasis-endemic villages and 292,059 cases of chronic schistosomiasis; the prevalence of schistosomiasis in humans and bovines was 3.9% and 6.2%, respectively [23].

We carried out a longitudinal survey in 16 villages from 16 counties in Hubei Province (Figure 1). The areas were selected through a three-stage random sampling procedure. The 16 counties were first randomly selected from 63 counties; then 16 towns were randomly selected from 16 counties; finally, 16 schistosomiasis-endemic villages were randomly selected from the 16 towns (each town selected a village). The 16 villages had approximately 23,835 people and 2,385 hm^2 of agricultural acreage. Most of the residents were farmers whose principal activity was cultivation of rice and cotton. The residents were exposed to contaminated water when they performed their agricultural or daily activities (i.e., cultivating crops, catching fish, and washing clothes).

The participants met the following inclusion criteria: a) must have been a resident of the village for more than 12 months; b) should be more than six years old; c) should continuously reside in the village for the study period; d) has no serious diseases, such as malignant cancer.

Integrated strategy emphasizing infection source control to curb schistosomiasis

The interventions were not only carried in study villages, but all the schistosomiasis-endemic villages in Hubei Province. The uptake of interventions in the study areas was no different than elsewhere in the province.

Interventions to control sources of S. japonicum infection. During the study period from 2005 to 2012, 575 cattle were replaced with small farm machines, representing 41.9% of the total cattle in the villages. The remaining bovines were not allowed to graze in marshlands containing known snail habitat and instead were confined to fences whenever possible. To reduce the potential of humans as a source of infection for snails, measures were implemented to attempt to reduce the transmission. Safe water was provided to 5,915 households by pipeline or well water supply, covering 97.6% of the total households. To deal with human feces, 5,696 home lavatories were constructed or repaired with three-cell septic tanks, representing 94.0% of the total households. Furthermore, fecal-matter containers were provided for fisherman to prevent them from excreting feces directly into river or lake freshwater.

Praziquantel chemotherapy. During the study period, all residents and bovines found positive for *S. japonicum* were treated with praziquantel (PZQ) (humans: 40 mg/kg; bovines: 25 mg/kg), in accord with World Health Organization recommendation [24]. For schistosomiasis japonica, cure rates of 70% to 90% have been recorded with these regiments. Those uncured also benefit because their egg counts fall to one fifth or less that of the pretreatment levels [25]. PZQ has also been used as a chemoprophylactic among high-risk groups, such as flood relief workers, tourists known to have been exposed recently and fishermen.

Comprehensive control of snail habitats. A comprehensive approach was employed to control snails by mollusciciding together with environmental modification. Mollusciciding using niclosamide was conducted after measuring snails twice a year in the 16 villages. Over eight years, 481.1 hm^2 of snail habitats in the study villages were treated with molluscicides (niclosamide), and 54.6 hm^2 were environmentally improved to control snails by ditch sclerosis, hardening river banks with concrete, sluice transformation, planting trees and constructing fish ponds.

Outcome measurement

During the study period, all residents more than 6 years old in the 16 villages were examined using indirect hemagglutination assay (IHA) [26] in October and November annually, and IHA-positive individuals were subjected to the quantitative Kato–Katz thick smear technique (KK, three slides based on a single stool sample) [27]. *S. japonicum* egg counts were expressed in eggs/gram of stool (EPG). Residents that were positive for both IHA and KK were defined as infected, and the prevalence of *S. japonicum* in the residents of all 16 villages was determined in autumn.

During the same period, the infection of *S. japonicum* in all bovines (water buffaloes and cattle) in the 16 villages was examined annually using miracidial hatching test with fecal samples [28].

The snail survey was performed yearly along the river banks, ditches, and marshlands around the villages (April or May) through systematic sampling combined with environmental sampling (0.11 m^2-sized frames, 20 m between frames) [29]. All snails within the square frames were collected and brought to the laboratory. The collected snails were then counted and crushed to examine microscopically for schistosome infections. The prevalence of infection of snails was calculated as the proportion of infected snails, and the density of infected snails as the number of infected snails/0.11 m^2.

Infected snail area calculations. If an isolated infected snail spot has been found, then spread to each direction for 50 meters, that's total of 10000 m^2. If more than one infected snail spots have been found within 50 meters, then add the interval between each spot, and spread to each direction for 50 meters, than calculate the area. If more than one infected snail spots have been found exceed 50 meters, calculate the area by isolated infected snail spots.

Table 2. *S. japonicum* prevalence in human in the 16 villages from 2005 to 2012.

Year	No. of people qualified	No. of people tested	No. of people infected	Prevalence (95% CI)	Geometric mean EPG (95% CI)
2005	14514	12872	212	1.65 % (1.43, 1.87)	51.53(9.13,93,93)
2006	13032	11552	215	1.86% (1.61, 2.11)	16.50(10.10, 22.90)
2007	12471	11378	158	1.39 % (1.17, 1.60)	20.90(11.52,30.29)
2008	12725	11225	120	1.07 % (0.88, 1.26)	16.09(6.43,25.75)
2009	12027	10574	103	0.97% (0.79, 1.16)	19.98(3.09,36.87)
2010	11296	9941	86	0.86% (0.68, 1.05)	16.22(1.36,31.07)
2011	10536	9689	38	0.39 % (0.27, 0.52)	20.82(2.09,39.55)
2012	10699	9778	35	0.36% (0.24, 0.48)	18.34(3.67,33.01)

Ethical approval

The study was approved by the Ethics Review Committee of Hubei Provincial Center for Disease Control and Prevention, and the National Institute of Parasitic Diseases, Chinese Center for Disease Control and Prevention. Written informed consents were obtained from all adults and from parents or guardians of minors before participation in the study. The participants had the opportunity to withdraw from the study at any time.

The snail survey for each location was approved by Health Department of Hubei Province.

All animal work was approved by Ethics Review Committee of Animal Experiments, Hubei Provincial Center for Disease Control and Prevention. Animal cure and sacrifice were carried out by Animal Husbandry Bureau according to the guidance of the Institute for Laboratory Animal Research.

Statistical analysis

Data were compiled in Microsoft Visual Foxpro database. Chi-square test or Fisher exact probability test were used to examine the proportions of differences. Spearman correlation was used to analyze the association between density and prevalence of snails.

All *P* values were two-tailed, with a significant level at 0.05. All statistical analyses were carried out in SPSS 16.0.

Results

Baseline

The human *S. japonicum* prevalence in the 16 villages in 2005 was 1.7% (95% CI: 1.4–1.9) (N = 12872) (Table 1).

The number of males (N = 6655) was greater than that of females (N = 6218) in 2005. Infection prevalence was higher in males (2.0%; 95% CI: 1.7–2.4) than in females (1.2%; 95% CI: 1.0–1.5) (Table 1).

S. japonicum prevalence varied according to age. People between 50 to 59 years of age had the highest prevalence (2.2%; 95% CI: 1.6–2.9).

Majority of the residents were farmers (N = 9662), followed by students (N = 2882) (Table 1). The highest prevalence was found in the fisherman group (8.2%; 95% CI: 1.9–14.5) and the lowest was in the student group (0.7%; 95% CI: 0.4–1.0) (Table 1).

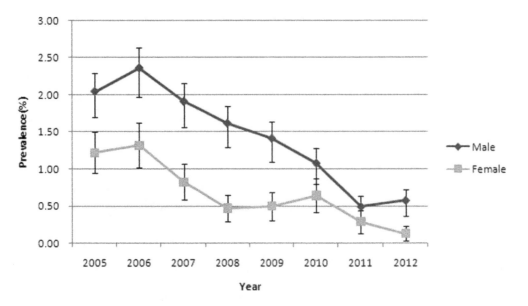

Figure 2. Human *S. japonicum* prevalence by sex and year (with 95% CI).

Table 3. Human *S. japonicum* prevalence by age in the 16 villages from 2005 to 2012.

Year	6-9		10-19		20-29		30-39		40-49		50-59		60-	
	No. tested	Prevalence %	No. tested	Prevalence %	No. tested	Prevalence %	No. tested	Prevalence %	No. tested	Prevalence %	No. tested	Prevalence %	No. tested	Prevalence %
2005	667	0.30 (-0.12, 0.71)	2481	0.85 (0.49,1.21)	920	1.09 (0.42,1.76)	2476	2.18 (1.61,2.76)	2912	2.16 (1.64,2.69)	1979	2.22 (1.57,2.87)	1437	1.25 (0.68,1.83)
2006	558	0.18 (-0.17,0.53)	1989	0.85 (0.45,1.26))	777	1.54 (0.68,2.41)	2136	2.81 (2.11,3.51)	2761	2.06 (1.53,2.59)	2091	2.10 (1.49,2.72)	1240	1.94 (1.17,2.70)
2007	502	0.20 (-0.19,0.59)	1887	0.26 (0.03,0.50)	734	0.41 (-0.05,0.87))	1912	1.46 (0.93,2.00)	2677	1.94 (1.42,2.47)	2231	2.06 (1.47,2.65)	1435	1.60 (0.95,2.25)
2008	357	0.00	1863	0.05 (-0.05,0.16)	838	0.48 (0.01,0.94)	1754	1.14 (0.64,1.64)	2712	1.62 (1.15,2.10)	2292	1.44 (0.95,1.93)	1409	1.28 (0.69,1.86)
2009	324	0.00	1492	0.00	652	0.77 (0.10,1.44)	1397	1.00 (0.48,1.52)	2677	0.86 (0.51,1.21)	2471	1.46 (0.98,1.93)	1561	1.60 (0.98,2.22)
2010	269	0.00	1105	0.00	577	0.17 (-0.17,0.51))	1119	0.80 (0.28,1.33)	2933	0.61 (0.33,0.90)	2284	1.31 (0.85,1.78)	1654	1.69 (1.07,2.31)
2011	239	0.00	881	0.00	840	0.12 (-0.11,0.35)	1184	0.25 (-0.03,0.54)	2908	0.34 (0.13,0.56)	2148	0.70 (0.35,1.05)	1489	0.60 (0.21,1.00)
2012	217	0.00	759	0.00	931	0.00	998	0.30 (-0.04,0.64)	2686	0.56 (0.28,0.84)	2270	0.44 (0.17,0.71)	1917	0.37 (0.10,0.64)

Participant flow

Over the eight-year period, about 90% (ranged from 88.0% to 92.0%) of the qualified residents participated, and more than 95% of the IHA-positive individuals submitted a fecal sample for examination. PZQ treatment coverage was high, with 100% of those found infected successfully treated (Table 2).

S. japonicum infection in humans

Human prevalence. Human prevalence of *S. japonicum* infection declined from 2005 to 2012, showing a significant difference ($\chi^2 = 204.2$, $P<0.001$). The prevalence decreased from 1.7% in 2005 to 0.4% in 2012, showing a decline of 78.2% ($P<0.001$). The geometric mean intensity of infection in the positives decreased from 51.5 EPG in 2005 to 18.3 EPG in 2012, showing a decline of 64.4 % (Table 2).

Human prevalence by sex, age, and occupation over eight years. The human prevalence of *S. japonicum* infection by sex and year is shown in Figure 2. Prevalence was higher in males than in females over eight years. Significant difference between males and females was observed during the study period ($P<0.05$). The prevalence of *S. japonicum* in males decreased from 2.0% in 2005 to 0.6% in 2012, showing a decline of 71.6% ($P<0.001$). The prevalence of *S. japonicum* in females decreased from 1.2% in 2005 to 0.1% in 2012, showing a decline of 89.3% ($P<0.001$).

The human prevalence of *S. japonicum* infection by age and year is shown in Table 3. Human prevalence increased with age, and most infected residents were 30 years of age and older. Individuals younger than 30 years were infrequently infected, and the prevalence rate was low.

The human prevalence of *S. japonicum* infection by occupation and year is shown in Figure 3 and Table 4. The majority of infected residents were fishermen and farmers, followed by students over eight years. *S. japonicum* prevalence decreased over the years. The prevalence in farmers decreased from 1.88% in 2005 to 0.4% in 2012, showing a decline of 77.1% ($P<0.001$). The prevalence in fishermen decreased from 8.2% in 2005 to 0% in 2012, showing a decline by 100% ($P=0.001$).

S. japonicum infection in bovines

S. japonicum prevalence in bovines is shown in Table 5. Significant difference ($\chi^2 = 298.8, P <0.001$) in prevalence was observed from 2005 to 2012. The prevalence decreased from 11.7% (95% CI: 9.5–13.9) in 2005 to 0.6% (95% CI: 0.0–1.1) in 2012, showing a decline of 95.0% ($P<0.001$).

S. japonicum infection in snails

The infected snail habitat decreased from 58.95 hm^2 in 2005 to 0 hm^2 in 2012, showing a decline of 100%. The snail infection decreased from 0.3% (95% CI: 0.2–0.3) in 2005 to 0% in 2012, showing a decline of 100% (Table 6).

The density of living snails decreased from 0.9 No./0.11 m^2 in 2005 to 0.2 No./0.11 m^2 in 2012, showing a decline of 76.5%. The density of infected snails decreased from 0.003 No./0.11 m^2 in 2005 to 0 No./0.11 m^2 in 2012, showing a decline of 100%. The prevalence of *S. japonicum* infection in the snails decreased from 0.3% in 2005 to 0% in 2012, showing a decline of 100% (Table 6).

Significant correlations were observed between the density of infected snails and the density of living snails (r = 0.881, $P = 0.004$).

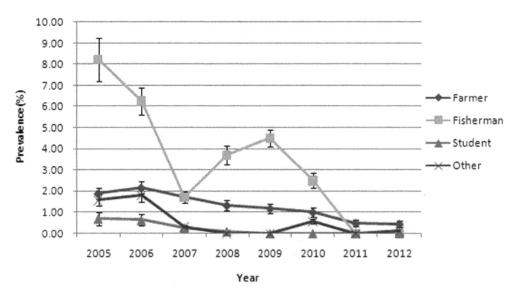

Figure 3. Human *S. japonicum* prevalence by occupation and year (with 95% CI).

Discussion

Investigations showed that more than 80% of *S. japonicum* patients in China were distributed in the lake and marshland areas of Hubei, Hunan, Jiangxi, Anhui, and Jiangsu provinces [30]. In 2005, 291,111 cases of *S. japonicum* patients, 14,816 heads of infected bovines, and 76,435 hm^2 of areas infested with *Oncomelania* snails were reported in Hubei province, accounting for the highest numbers in China [31]. Therefore, schistosomiasis control in Hubei province is especially important.

To achieve the compliance requirements of the national prevention plan, the Hubei Provincial government formulated a document called "Mid- and long-term plan on prevention and control of schistosomiasis in Hubei province (2005–2015)". Hubei province has conducted a series of comprehensive prevention and control measures based on blocking the transmission of *S. japonicum* from cattle/buffalos and humans to snails since 2005 [20]. We selected 16 villages and used longitudinal observation for eight years to evaluate *S. japonicum* transmission following comprehensive prevention and control measures.

The prevalence of human *S. japonicum* infection declined considerably in all surveyed villages over the course of the study.

In 2008, the prevalence of human *S. japonicum* infection decreased to below 5% and no acute schistosomiasis case was reported. The national criterion of infection control was achieved. In the following years, the prevalence continued to decline. The integrated control strategy had an active and important function in the decrease in human prevalence observed in the study villages. The decrease in human prevalence attributable to human fecal contamination of the environment was effectively controlled by sanitary latrine construction. Annual PZQ treatment, which on ethical grounds had to be administered to all infected individuals, may also have helped.

In the study, we found that infection prevalence was higher in males than females in the study villages. This finding may be related to the fact that males often take part in agricultural labor and have more exposure to infested water. To a far lesser extent, females are exposed while bathing or washing clothes. Male-dominated exposure has been reported elsewhere in China [32]. In Jiangxi Province, China, most of the reported contact for males occurred while washing or fishing [32]. The result showed that the infected population was mainly concentrated in ages 30 to 60. Infection prevalence was low in children, which may due to the

Table 4. Human *S. japonicum* prevalence by occupation in the 16 villages from 2005 to 2012.

Year	Farmer			Fisherman			Student			Other		
	No. tested	No. infected	Prevalence %	No. tested	No. infected	Prevalence %	No. tested	No. infected	Prevalence %	No. tested	No. infected	Prevalence %
2005	9662	182	1.88	73	6	8.22	2882	20	0.69	255	4	1.57
2006	8932	193	2.16	48	3	6.25	2350	15	0.64	222	4	1.80
2007	8780	150	1.71	58	1	1.72	2229	6	0.27	311	1	0.32
2008	8692	115	1.32	81	3	3.70	2108	2	0.09	344	0	0.00
2009	8416	99	1.18	89	4	4.49	1784	0	0.00	285	0	0.00
2010	8221	82	1.00	80	2	2.50	1296	0	0.00	344	2	0.58
2011	8104	38	0.47	76	0	0.00	1141	0	0.00	368	0	0.00
2012	7912	34	0.43	90	0	0.00	903	0	0.00	873	1	0.11

Table 5. *S. japonicum* prevalence in bovines in the 16 villages from 2005 to 2012.

Year	No. tested	No. infected	Prevalence (95% CI)
2005	821	96	11.69% (9.49, 13.89)
2006	771	93	12.06% (9.76, 14.36)
2007	834	43	5.16% (3.66, 6.66)
2008	990	24	2.42% (1.47, 3.38)
2009	1162	23	1.97% (1.18, 2.78)
2010	780	11	1.41% (0.58, 2.24)
2011	725	6	0.83% (0.17, 1.49)
2012	693	4	0.58 % (0.01, 1.14)

health education carried in schools every year. Health education plays an important role on schistosomiasis control as part of the integrated strategy[33]. Our previous research had found that heath education was effective on reducing the prevalence of *S. japonicum* infection in human especially students[34]. Moreover, some prevalent cases may be persons with long-lived *S. japonicum*[35]. Thus, most human exposures to schistosomiasis in a typical Chinese lake and marshland region are occupationally driven.

A number of reports have shown that bovines are the major transmission sources for human schistosomiasis in the lakes and marshland areas of Southern China, and that interventions targeting bovines can reduce the incidence of human infection [36,37]. Over the study period, bovine prevalence rates decreased in 16 villages, possibly due to measures including providing farmers with machines instead of bovines and prohibiting bovines in regions with snails. These measures have been shown to be effective elsewhere [19]. However, some challenges exist for the implementation of these measures in some rural areas of Hubei Province. Not all farms were suitable for tractor farming and bovines also bring additional economic benefits to villages beyond their use in farming [38].

The densities and rates of *S. japonicum* infection in snails were consistent with the prevalence of human *S. japonicum* infection. Our findings suggest that environmental management is very effective for snail control, and could result in a persistent reduction in the density of living snails. However, environmental management is limited by the floods. Major flooding of the lakes and marshland

areas downstream of the Three Gorges Dam (TGD) can drown adult snails, resulting in decreased transmission [7].

During the past years, the Hubei Provincial government placed a high priority on the control of schistosomiasis and carried out numerous control programs [39,40,41]. Our research suggested that the integrated control strategy implemented in our study area was effective. Earlier studies showed that the chemotherapy-based approach could not perennially reduce the prevalence of *S. japonicum* [42,43]. Given the decreasing drug efficacy after widespread and repeated use of PZQ, chemotherapy alone can only control the infection rates at a low level; it cannot eliminate re-infections, and therefore, cannot interrupt the transmission of schistosomiasis. However, the key for integrated control strategy is to manage feces of humans and animals, preventing the eggs from entering the water so the snails will not be infected, reducing human and animal re-infected [43].

However, some challenges still remain for schistosomiasis control in Hubei province, China. Schistosomiasis endemic may be infected by large-scale hydro-projects [7,44,45]. The ecology of central China will change substantially in ways that could affect the distribution and transmission of schistosomiasis due to the TGD [25,46]. It is a giant impoundment across the Yangtze River, which was close to a height of 135 m by 2003 and reached its full height of 175 m by 2009. TGD has been predicted to alter water and sand distributions downstream, which will have a significant effect on ecological systems; these include the canals of Hubei and Dongting and Poyang lakes, where *S. japonicum* transmission is generally projected to increase, although decreased transmission is projected for other locations [7,22,47]. Specifically, TGD is anticipated to result in large changes to the flow, depth, and sedimentation load of the Yangtze; thus, the distribution and number of schistosome-infected *Oncomelania* snails will be altered, increasing transmission of schistosomiasis in some areas and its re-introduction into places where the infection is currently under control [7,21,48,49]. Gray and colleagues believe that TGD has no immediate effect on schistosome transmission, and that the predicted changes may take longer to occur [50]. Continued surveillance should be undertaken to monitor the future ecological effects of the dam in Hubei province.

Large population movements taking place in southern China have exacerbated the schistosomiasis problem [47]. The downturn in the world economy in 2007 has also induced further population movements as businesses in urban centers close and employment opportunities recede. The foregoing is likely to affect schistosomiasis control efforts negatively, with millions of residents returning

Table 6. *S. japonicum* prevalence in snail in the 16 villages from 2005 to 2012.

Year	No. snails dissected	Density of living snails(No./0.11 m²)	No. infected snails	Infected snail habitat (hm²)	Density of infected snails (No./0.11 m²)	Prevalence (95% CI)
2005	46978	0.85	138	58.95	0.0025	0.29% (0.24, 0.34)
2006	35328	0.67	60	64.72	0.0011	0.17% (0.13, 0.21)
2007	22386	0.44	38	35.78	0.0008	0.17% (0.12, 0.22)
2008	14601	0.28	33	27.71	0.0006	0.23% (0.15, 0.30)
2009	15824	0.29	22	18.08	0.0004	0.14% (0.08, 0.20)
2010	16914	0.32	10	4.07	0.0002	0.06% (0.02, 0.10)
2011	13927	0.19	3	1.13	0.00004	0.02% (0.00, 0.05)
2012	15010	0.20	0	0	0	0

from cities to their home villages in areas of endemicity to seek employment.

In our research, the densities and rates of infected snails were both at a low level, possibly caused by the low sensitivity of the microscopy-based method; identifying snails that were infected early through this microscopy method is very difficult [51]. To assess the long-term effectiveness of the integrated strategy and to maintain the success of this strategy, using these new molecular tools to monitor snails in the future may be necessary. Molecular tools, such as polymerase chain reaction and loop-mediated isothermal amplification, can detect potential infection in snails with early sporocysts, and show promise for monitoring early infection rate in snails [51].

Schistosomiasis control is a systematic project involving sectors of water conservancy, agriculture, forestry, health, and others. External factors of equal importance, such as political, economic, cultural and social, should not be ignored [52]. With the development of the social economy, the living standards of residents in the endemic areas will be improved continually, so control of schistosomiasis will become easier. This study shows that an integrated strategy, predominantly involving the elimination of infection sources in the environment and snail control, combined with treatment of infected humans is achievable and effective[43]. Therefore, a multi-component, integrated control program will be continuously required to combat the spread of schistosomiasis [53] and to achieve the national criteria of transmission control to reduce the prevalence of S. japonicum to less than 1% in Hubei Province.

Our study had somewhat potential limitations. The interventions were carried province-widely, thus there are no available uncontrolled villages for comparison. Our study data could represent the basic schistosomiasis endemic of the province as a whole because of random selection and 8-years longitudinal survey were carried. Though there may have bias when selected 16 villages from 5502 villages in the province with random selection.

We did not investigate the status of S. japonicum infection in people who withdraw from the study. Thus, we did not know the differential dropout of infected persons. The study populations shrunk over the course of the study, though most of the dropouts were below 30 years and were not the high risk persons. Moreover, we tested about 90% of the eligible persons each year. So we consider that the limitations of study have little effect on the results.

Conclusions

An integrated control strategy aimed at controlling the roles of bovines and humans as sources of S. japonicum infection is suggested to be likely effective in controlling the transmission of S. japonicum in marshland regions of China.

Acknowledgments

We are very grateful to all staff at the Huangzhou, Yangxin, Honghu, Jianli, Jianglin, Jingzhou, Shashi, Shishou, Songzi, Caidian, Chibi, Jiayu, Huanchuan, Qianjiang, Xiantao County Institute for Schistosomiasis Control and the villagers who participated in this study. We acknowledge Qiu Juan for assistance in the preparation of the map.

Author Contributions

Conceived and designed the experiments: YYC JBL XBH SXC XPM. Performed the experiments: YYC JBL XBH SXC ZMS XPM. Analyzed the data: YYC JBL ZMS RZ LZ XPM. Contributed reagents/materials/ analysis tools: JBL XBH SXC XPM. Wrote the paper: YYC JBL XPM.

References

1. King CH, Dickman K, Tisch DJ (2005) Reassessment of the cost of chronic helmintic infection: a meta-analysis of disability-related outcomes in endemic schistosomiasis. Lancet 365: 1561–1569.

2. Steinmann P, Keiser J, Bos R, Tanner M, Utzinger J (2006) Schistosomiasis and water resources development: systematic review, meta-analysis, and estimates of people at risk. Lancet Infect Dis 6: 411–425.

3. Hotez PJ, Molyneux DH, Fenwick A, Kumaresan J, Sachs SE, et al. (2007) Control of neglected tropical diseases. N Engl J Med 357: 1018–1027.

4. Zhou XN, Wang LY, Chen MG, Wu XH, Jiang QW, et al. (2005) The public health significance and control of schistosomiasis in China—then and now. Acta Trop 96: 97–105.

5. Zhou D, Li Y, Yang X (1994) Schistosomiasis control in China. World Health Forum 15: 387–389.

6. Mao SP, Shao BR (1982) Schistosomiasis control in the people's Republic of China. Am J Trop Med Hyg 31: 92–99.

7. McManus DP, Gray DJ, Li Y, Feng Z, Williams GM, et al. (2010) Schistosomiasis in the People's Republic of China: the era of the Three Gorges Dam. Clin Microbiol Rev 23: 442–466.

8. Mao SB (1986) Recent progress in the control of schistosomiasis in China. Chin Med J (Engl) 99: 439–443.

9. McManus DP, Gray DJ, Ross AG, Williams GM, He HB, et al. (2011) Schistosomiasis research in the dongting lake region and its impact on local and national treatment and control in China. PLoS Negl Trop Dis 5: e1053.

10. Utzinger J, Zhou XN, Chen MG, Bergquist R (2005) Conquering schistosomiasis in China: the long march. Acta Trop 96: 69–96.

11. Zhou XN, Guo JG, Wu XH, Jiang QW, Zheng J, et al. (2007) Epidemiology of schistosomiasis in the People's Republic of China, 2004. Emerg Infect Dis 13: 1470–1476.

12. Balen J, Zhao ZY, Williams GM, McManus DP, Raso G, et al. (2007) Prevalence, intensity and associated morbidity of Schistosoma japonicum infection in the Dongting Lake region, China. Bull World Health Organ 85: 519–526.

13. Chen MG (2005) Use of praziquantel for clinical treatment and morbidity control of schistosomiasis japonica in China: a review of 30 years' experience. Acta Trop 96: 168–176.

14. Xianyi C, Liying W, Jiming C, Xiaonong Z, Jiang Z, et al. (2005) Schistosomiasis control in China: the impact of a 10-year World Bank Loan Project (1992–2001). Bull World Health Organ 83: 43–48.

15. Changsong S, Binggui Y, Hongyi L, Yuhai D, Xu X, et al. (2002) Achievement of the World Bank loan project on schistosomiasis control (1992-2000) in Hubei province and the challenge in the future. Acta Trop 82: 169–174.

16. Zhao GM, Zhao Q, Jiang QW, Chen XY, Wang LY, et al. (2005) Surveillance for schistosomiasis japonica in China from 2000 to 2003. Acta Trop 96: 288–295.

17. Li SZ, Luz A, Wang XH, Xu LL, Wang Q, et al. (2009) Schistosomiasis in China: acute infections during 2005–2008. Chin Med J (Engl) 122: 1009–1014.

18. Liu R, Dong HF, Jiang MS (2013) The new national integrated strategy emphasizing infection sources control for schistosomiasis control in China has made remarkable achievements. Parasitol Res 112: 1483–1491.

19. Wang LD, Chen HG, Guo JG, Zeng XJ, Hong XL, et al. (2009) A strategy to control transmission of Schistosoma japonicum in China. N Engl J Med 360: 121–128.

20. Wang LD, Guo JG, Wu XH, Chen HG, Wang TP, et al. (2009) China's new strategy to block Schistosoma japonicum transmission: experiences and impact beyond schistosomiasis. Trop Med Int Health 14: 1475–1483.

21. Zhu HM, Xiang S, Yang K, Wu XH, Zhou XN (2008) Three Gorges Dam and its impact on the potential transmission of schistosomiasis in regions along the Yangtze River. Ecohealth 5: 137–148.

22. Seto EY, Wu W, Liu HY, Chen HG, Hubbard A, et al. (2008) Impact of changing water levels and weather on Oncomelania hupensis hupensis populations, the snail host of Schistosoma japonicum, downstream of the Three Gorges Dam. Ecohealth 5: 149–158.

23. Zhu HG, Xiao Y, Huang XB, Zhang XF, Ding H, Liu JB, Su ZM (2008) Study on endemic situation of schistosomiasis in Hubei Province from 2004 to 2007. Chin J Schisto Control 20: 251–254.

24. WHO (2006) WHO preventive chemotherapy in human helminthiasis. World Health Organization, Geneva.

25. Ross AGP, Sleigh AC, Li Y, Davis GM, Williams GM, et al. (2001) Schistosomiasis in the People's Republic of China: Prospects and Challenges for the 21st Century. Clinical Microbiology Reviews 14: 270–295.

26. Zhu YC (2005) Immunodiagnosis and its role in schistosomiasis control in China: a review. Acta Trop 96: 130–136.

27. Katz N, Chaves A, Pellegrino J (1972) A simple device for quantitative stool thick-smear technique in Schistosomiasis mansoni. Rev Inst Med Trop Sao Paulo 14: 397–400.

28. Ross AG, Bartley PB, Sleigh AC, Olds GR, Li Y, et al. (2002) Schistosomiasis. N Engl J Med 346: 1212–1220.

29. World Bank Loan Program completion report on infectious and endemic disease control project: schistosomiasis control component. 1992–2001. Department of Disease Control & Foreign Loan Office MoH.

30. LD W (2006) The epidemic status of schistosomiasis in China: results from the third nationwide sampling survey in 2004 [in Chinese]. Shanghai: Shanghai Scientific and Technological Literature Publishing House. 36p.

31. Yang Hao X-hW, Gang Xia, Hao Zheng, Jia-Gang Guo, Li-Ying Wang, Xiao-nong Zhou (2006) Schistosomiasis situation in Peopel's Republic of China in 2005. Chin J Schisto Control 18: 321–324.

32. Wu Z, Shaoji Z, Pan B, Hu L, Wei R, et al. (1994) Reinfection with Schistosoma japonicum after treatment with praziquantel in Poyang lake region, China. Southeast Asian J Trop Med Public Health 25: 163–169.

33. Liu R, Dong HF, Jiang MS (2012) What is the role of health education in the integrated strategy to control transmission of Schistosoma japonicum in China? Parasitol Res 110: 2081–2082.

34. He Hui SZ, Tu Zuwu, Fan Hongping (2010) The Analysis of Surveillance Results of Schistosomiasis in Hubei Province in 2009. Journal of Tropical Medicine 10: 1243–1245.

35. Hall S.C KEL (1970) Prolongeol survival of Schistosoma Japonicum. California Med 113: 75.

36. Gray DJ, Williams GM, Li Y, Chen H, Forsyth SJ, et al. (2009) A cluster-randomised intervention trial against Schistosoma japonicum in the Peoples' Republic of China: bovine and human transmission. PLoS One 4: e5900.

37. Guo J, Li Y, Gray D, Ning A, Hu G, et al. (2006) A drug-based intervention study on the importance of buffaloes for human Schistosoma japonicum infection around Poyang Lake, People's Republic of China. Am J Trop Med Hyg 74: 335–341.

38. Hong X-C, Xu X-J, Chen X, Li Y-S, Yu C-H, et al. (2013) Assessing the Effect of an Integrated Control Strategy for Schistosomiasis Japonica Emphasizing Bovines in a Marshland Area of Hubei Province, China: A Cluster Randomized Trial. PLoS Neglected Tropical Diseases 7: e2122.

39. ZHU Hong HX-b, CAI Shun-xiang,TU Zu-wu,CHEN Yan-yan,LI Guo,XIA Jing,XIAO Ying,ZHOU Xiao-rong,CAO Mu-min,GAO Hua (2011) Evaluation of comprehensive measures to control schistosomiasis primarily through infection source control in Hubei Province. Journal of Pathogen Biology 6: 908–911.

40. ZHU Hong CD-e, HUANG Xi-bao,CAI Shun-xiang,TU Zu-wu,CHEN Yan-yan,LI Guo,GAO Hua,CAO Mu-min,DAI Ling-feng,ZHOU Xiao-rong,XIAO Ying (2010) Cost-effectiveness of Comprehensive Measures with Emphasis on Infection Source Control for Schistosomiasis in Marshland Endemic Regions. Journal Of Tropical Medicine 10: 982–985.

41. LIU Wei CC-l, CHEN Zhao, LI Shi-zhu, TANG Li, XIAO Ying, ZHANG Hua-ming, YANG Zhi-qiang, WANG Yi, SU Shang-yang, WANG Li-ying, WANG Qiang, XU Jun-fang, BAO Zi-ping, HUANG Xi-bao, ZHOU Xiao-nong (2013) Evaluation of the Comprehensive Schistosomiasis Control Measures with Emphasis on Infection Source of Replacing Cattle with Machine. Chinese Journal of Parasitology and Parasitic Diseases 31: 206–211.

42. Zhou YB, Zhao GM, Jiang QW (2007) Effects of the praziquantel-based control of schistosomiasis japonica in China. Ann Trop Med Parasitol 101: 695–703.

43. Zhou YB, Liang S, Chen GX, Rea C, He ZG, et al. (2011) An integrated strategy for transmission control of Schistosoma japonicum in a marshland area of China: findings from a five-year longitudinal survey and mathematical modeling. Am J Trop Med Hyg 85: 83–88.

44. LIAO Si-qi CY-y, HONG Zhi-hua, ZHANG Jian-hua, JIANG Yong, CAI Zong-da, WANG Yi, LI Ke-hua, HUANG Shui-sheng (2009) Impact of South-North Water Diversion Middle-Line Project on the Transmission of Schistoso-miasis in Hubei Province. Journal Of Tropical Medicine 9: 1288–1290.

45. Chen Yan-yan YY, Zhou Bin,Zhu Yun,Peng Xun,Xu Xing-jian (2010) Effect evaluation of large water conservancy project on control of schistosomiasis transmission. Chinese Journal of Schistosomiasis Control 22: 411–414.

46. Ross AG, Li Y, Williams GM, Jiang Z, McManus DP (2001) Dam worms. Biologist (London) 48: 121–124.

47. Li YS, Raso G, Zhao ZY, He YK, Ellis MK, et al. (2007) Large water management projects and schistosomiasis control, Dongting Lake region, China. Emerg Infect Dis 13: 973–979.

48. Stone R (2008) China's environmental challenges. Three Gorges Dam: into the unknown. Science 321: 628–632.

49. Zheng J, Gu XG, Xu YL, Ge JH, Yang XX, et al. (2002) Relationship between the transmission of schistosomiasis japonica and the construction of the Three Gorge Reservoir. Acta Trop 82: 147–156.

50. Gray DJ, Thrift AP, Williams GM, Zheng F, Li Y-S, et al. (2012) Five-Year Longitudinal Assessment of the Downstream Impact on Schistosomiasis Transmission following Closure of the Three Gorges Dam. PLoS Neglected Tropical Diseases 6: e1588.

51. Kumagai T, Furushima-Shimogawara R, Ohmae H, Wang TP, Lu S, et al. (2010) Detection of early and single infections of Schistosoma japonicum in the intermediate host snail, Oncomelania hupensis, by PCR and loop-mediated isothermal amplification (LAMP) assay. Am J Trop Med Hyg 83: 542–548.

52. Jiang MS LR, Zhao QP, Dong HF (2010) Social epidemiological thinking about schistosomiasis. Chin J Schisto Control 22: 201–205.

53. Gray DJ, McManus DP, Li Y, Williams GM, Bergquist R, et al. (2010) Schistosomiasis elimination: lessons from the past guide the future. Lancet Infect Dis 10: 733–736.

10

Field Efficacy of Vectobac GR as a Mosquito Larvicide for the Control of Anopheline and Culicine Mosquitoes in Natural Habitats in Benin, West Africa

Armel Djènontin[1]*, Cédric Pennetier[2], Barnabas Zogo[2], Koffi Bhonna Soukou[2], Marina Ole-Sangba[2], Martin Akogbéto[3], Fabrice Chandre[4], Rajpal Yadav[5], Vincent Corbel[6]*

1 Faculté des Sciences et Techniques/MIVEGEC (IRD 224-CNRS 5290-UM1-UM2), Université d'Abomey Calavi/Centre de Recherche Entomologique de Cotonou (CREC), Cotonou, Bénin, 2 MIVEGEC (IRD 224-CNRS 5290-UM1-UM2), Centre de Recherche Entomologique de Cotonou (CREC), Cotonou, Bénin, 3 Faculté des Sciences et Techniques/Centre de Recherche Entomologique de Cotonou (CREC), Université d'Abomey Calavi/Centre de Recherche Entomologique de Cotonou (CREC), Cotonou, Bénin, 4 MIVEGEC (IRD 224-CNRS 5290-UM1-UM2), Laboratoire de lutte contre les Insectes Nuisibles (LIN), Montpellier, France, 5 Department of Control of Neglected Tropical Diseases, World Health Organization, Geneva, Switzerland, 6 MIVEGEC (IRD 224-CNRS 5290-UM1-UM2)/Department of Entomology, Kasetsart University, Ladyaow Chatuchak Bangkok, Thailand

Abstract

Introduction: The efficacy of Vectobac GR (potency 200 ITU/mg), a new formulation of bacterial larvicide *Bacillus thuringiensis var. israelensis* Strain AM65-52, was evaluated against *Anopheles gambiae* and *Culex quinquefasciatus* in simulated field and natural habitats in Benin.

Methods: In simulated field conditions, Vectobac GR formulation was tested at 3 dosages (0.6, 0.9, 1.2 g granules/m^2 against *An. gambiae* and 1, 1.5, 2 g granules/m^2 against *Cx. quinquefasciatus*) according to manufacturer's product label recommendations. The dosage giving optimum efficacy under simulated field conditions were evaluated in the field. The efficacy of Vectobac GR in terms of emergence inhibition in simulated field conditions and of reduction of larval and pupal densities in rice fields and urban cesspits was measured following WHO guidelines for testing and evaluation of mosquito larvicides.

Results: Vectobac GR caused emergence inhibition of ≥80% until 21 [20–22] days for *An. gambiae* at 1.2 g/m^2 dose and 28 [27–29] days for *Cx. quinquefasciatus* at 2 g/m^2 in simulated field habitats. The efficacy of Vectobac GR in natural habitats was for 2 to 3 days against larvae and up to 10 days against pupae.

Conclusions: Treatment with Vectobac GR caused complete control of immature mosquito within 2–3 days but did not show prolonged residual action. Larviciding can be an option for malaria and filariasis vector control particularly in managing pyrethroid-resistance in African malaria vectors. Since use of larvicides among several African countries is being emphasized through Economic Community of West Africa States, their epidemiological impact should be carefully investigated.

Editor: Georges Snounou, Université Pierre et Marie Curie, France

Funding: The authors have no support or funding to report.

Competing Interests: The authors have declared that no competing interests exist.

* E-mail: armeldj@yahoo.fr (AD); vincent.corbel@ird.fr (VC)

Introduction

Malaria in Sub-Saharan Africa is a major public health problem accounting for 79% of global incidence of cases and 90% of deaths [1]. Lymphatic filariasis is a widely prevalent neglected vector-borne disease in Africa [2]. While chemotherapy for malaria control and mass drug administration against filariasis have been extensively used in disease endemic countries, vector control can complement strategies for prevention and control of these diseases [3]. Complementary vector control tools targeting exophagic and exophilic vectors or targeting another stage in the mosquito's lifecycle (e.g. the aquatic stage) are then needed to achieving the Millennium Development Goals for malaria control by 2015 [4].

Larval source management is an important component of an integrated vector management approach [5] and has extensively been used for the control of anophelines since the 1950s [6]. Recent studies in rural areas of Eastern Africa demonstrated that larval control by hand application of larvicides can reduce the abundance of malaria mosquito larvae and adults and transmission by 70–90% where the majority of aquatic mosquito larval habitats are accessible and relatively limited in number and size [7]. Larval source management offers the dual benefits of reducing numbers of house-frequenting mosquitoes and those that bite outdoors.

Larviciding is a commonly used method of mosquito control in different ecological patterns mostly in urban areas or some coastal areas where breeding sites are well identified. Currently 10 formulations are recommended by WHOPES for mosquito larval control, including microbial agents [8]. These bio-pesticides offer interesting prospects for the control of malaria vectors through varied and diverse groups of micro-organisms including viruses, bacteria and fungi which constitute an important part of the active ingredient arsenal for Integrated Vector Control [5].

Bacillus thuringiensis israelensis (Bti) and *Bacillus sphaericus (Bs)* have been extensively evaluated in the laboratory against anophelines and culicines larvae and also tested in a variety of environmental settings [9]. *Bti* is unlikely to pose any hazard to humans, other vertebrates and non-target invertebrates, provided that it is free from non-*Bt* microorganisms and biologically active products other than the insecticidal crystal proteins [10]. It was recently demonstrated that long-term use of *Bacillus thuringiensis israelensis* in coastal wetlands had no influence on the temporal evolution of the taxonomic structure and taxa abundance of non-target aquatic invertebrate communities [11]. In Benin, larviciding by the use of *Bti* was recently integrated as a part of vector management for malaria prevention [12].

While various *Bti* formulations are available as mosquito larvicides today, there has always been a need to improve them for better efficacy, ease of application and acceptability. In the present study in southern Benin, a new granular formulation of *Bti*, Vectobac GR of Valent BioSciences Corp, USA, was evaluated against *Anopheles gambiae* and *Culex quinquefasciatus* in simulated field experiments and in natural breeding habitats. The experimental procedures followed the WHO guidelines for testing and evaluation of mosquito larvicides [13]. The National Ethical Committee for Medical Research of Benin (N°006) cleared the study and the work was supervised by the WHO Pesticide Evaluation Scheme.

Methods

1. Ethics Statement

Ethics clearance for the study was obtained from the National Ethical Committee for Medical Research in Benin (ethics clearance N°006 of 28th April, 2011). The trial on *An. gambiae* was conducted after having received formal agreement from the president of local farmers named Lokossou Nestor. Concerning the trial on *Cx. quinquefasciatus*, permission from each owner of houses where cesspits were located was obtained before the trial was conducted.

2. Test Material

Vectobac GR is a new granular formulation of *Bacillus thuringiensis, subsp. israelensis*, strain AM65-52 developed by Valent BioSciences Corp., USA. The bio potency of this larvicide is 200 International Toxic Units (ITU)/mg product. Bio potency of products based on *Bti* is compared with a lyophilized reference powder (IPS82, strain1884) of this bacterial species using early fourth instar larvae of *A. aegypti* (strain Bora Bora). The potency of IPS82 has been arbitrarily designated as 15 000 ITU/mg powder against this strain of mosquito larva.

According to the manufacturer's Material Safety Data Sheet, Vectobac GR is non-toxic by ingestion, skin contact or inhalation. It has no adverse effect on birds, earthworms, fish, or numerous other non-target aquatic invertebrates.

3. Mosquito Species

The Kisumu strain of *An. gambiae* and the F1 progeny of local population of *Culex quinquefasciatus* were used for the simulated field trial. Kisumu strain of *An. gambiae* is a reference strain maintained

at the insectary of the Centre de Recherche Entomologique de Cotonou (CREC) and is free of any resistance mechanism.

4. Study Area

The simulated field trial was carried out in the Centre de Recherche Entomologique de Cotonou (CREC). The field trial with *An. gambiae* was conducted in a rice field located in Lélé, Cové district located in Department of Zou (7°13′ 8″ N, 2°20′ 22″ E). Concerning *Cx. quinquefasciatus*, the field trial was conducted in Cotonou, Department of Littoral (6°23N–2°25E).

5. Study Design

5.1. Simulated field studies. The main objective of simulated field studies were to test and determine the optimum field application dosage of Vectobac GR. Vectobac GR was tested at 3 dosages against *An. gambiae* (0.6, 0.9, 1.2 g granules/m^2) and *Cx. quinquefasciatus* (1, 1.5, 2 g granules/m^2) according to manufacturer's product label recommendations. Four replicates of the experiments were run for both treatments and control.

Experimental set up: Artificial cement containers (i.e. rectangular pits of 60 cm long × 30 cm width × 30 cm depth) were used to study the Vectobac GR dose-efficacy relation. Containers were half-filled with water and covered with a mosquito netting piece to prevent oviposition by wild female mosquitoes and the deposit of debris, and were placed under a shelter to prevent direct exposure of rain and sunlight.

Bti application: At t0, measured quantity of Vectobac GR was dispensed manually taking necessary safety precautions using gloves and facial masks.

Cohort monitoring: Larvicidal activity might last longer than the developmental period. In this context, cohort of 30 to 50 second instars larvae of *An. gambiae* or *Cx. qinquefasciatus* were released in each container every 7 to 10 days, depending on the larval development time frame. Each *An. gambiae* and *Cx. quinquefasciatus* larvae cohort was fed with 0.5 g and 1 g of cat food respectively when released. Each day after treatment, pupae were counted and removed from the containers and placed in plastic cups with water and covered with a netting piece. Temperature and pH of water in the containers were recorded daily and meteorological data were obtained from the National Meteorology Department. The studies were conducted between 8 June and 21 July 2011 with *An. gambiae* and between 10 August and 21 September 2011 with *Cx. quinquefasciatus*.

5.2. Field trials. The field trial was launched with a formal agreement with the president of local rice field farmers. For the study, thirty ponds of 8 m^2 (2 m×4 m) each were delimited with natural barrier made of local mud. Thirty cesspits with a surface ranging from 0.14 to 3.46 m^2 housing *Cx. quinquefasciatus* larvae were selected in Cotonou and geo-referenced. All breeding sites were checked to confirm the presence of larvae before applying Vectobac GR.

One dose that provided the optimum efficacy in the simulated field studies was tested in natural habitats. Fifteen breeding sites of each type were treated with Vectobac while the remaining ones were left untreated and served as controls. Vectobac GR was uniformly applied manually on the water surface. Three replicates were run for each treatment or control corresponding to 45 treated habitats and 45 of untreated habitats.

Before treatment, each breeding site was sampled twice to determine the density of mosquito larvae and pupae. After treatment, sampling was done on days 1, 2, 3, and 7, and thereafter every third day until the density of larvae in the treated habitats reached to that of the control. The larval sampling method consisted of 3 dips using a ladle (350 ml). Sampling was

Table 1. Emergence and Emergence Inhibition Rates (EIR) of *An. gambiae* larvae according to treatments.

N day post treatment		Control	0.6 g/m^2	0.9 g/m^2	1.2 g/m^2
11	N	200	200	200	200
	NE	179	2	1	1
	ER (%) [95%CI]	90 [86–94]	1 [0–2]	1 [0–2]	1 [0–1]
	EIR (%) [95%CI]	–	**99**	**99**	**99**
19	N	200	200	200	200
	NE	180	31	27	17
	ER (%) [95%CI]	90 [86–94]	16 [11–21]	14 [9–19]	9 [5–13]
	EIR (%) [95%CI]	–	**83**	**85**	**91**
26	N	200	200	200	200
	NE	191	107	86	71
	ER (%) [95%CI]	96 [93–99]	54 [47–61]	43 [36–50]	36 [29–43]
	EIR (%) [95%CI]	–	**44**	**55**	**63**
35	N	120	120	120	120
	NE	114	105	104	72
	ER (%) [95%CI]	95 [91–99]	88 [82–94]	87 [81–94]	60 [51–69]
	EIR (%) [95%CI]	–	**8**	**9**	**37**
43	N	120	120	120	120
	NE	112	107	105	87
	ER (%) [95%CI]	93 [89–97]	89 [83–95]	88 [82–94]	73 [65–81]
	EIR (%) [95%CI]	–	**5**	**6**	**22**

N = Number of larvae; NE = Number of larvae emerged; ER = Emergence Rate; EIR = Emergence Inhibition Rate.

Table 2. Emergence and Emergence Inhibition Rates (EIR) of *Cx. quinquefasciatus* larvae according to the treatments.

N days post treatment		Control	1 g/m^2	1.5 g/m^2	2 g/m^2
11	N	200	200	200	200
	NE	191	0	0	0
	ER [95%CI]%	96 [93–99]	0	0	0
	EIR (%) [95%CI]%	–	**100**	**100**	**100**
19	N	200	200	200	200
	NE	198	51	30	3
	ER [95%CI]%	99 [98–100]	26 [20–32]	15 [10–20]	02 [0–4]
	EIR (%) [95%CI]%	–	**74**	**85**	**99**
26	N	200	200	200	200
	NE	180	96	80	28
	ER [95%CI]%	90 [86–94]	48 [41–55]	40 [33–47]	14 [9–19]
	EIR (%) [95%CI]%	–	**47**	**56**	**84**
34	N	160	160	160	160
	NE	152	93	80	65
	ER [95%CI]%	95[92–98]	58[50–66]	50[42–58]	41[33–49]
	EIR (%) [95%CI]%		**38**	**47**	**57**
42	N	180	180	180	180
	NE	171	157	154	134
	ER [95%CI]%	95[92–98]	87[82–92]	86[81–91]	74[68–80]
	EIR (%) [95%CI]%		**8**	**10**	**22**

N = Number of larvae; NE = Number of larvae emerged; ER = Emergence Rate; EIR = Emergence Inhibition Rate.

Table 3. Mean number of larvae and pupae per dip and density reduction (DR) after treatment Vectobac in natural habitats.

N day post treatment	Parameters	An. gambiae								Cx. quinquefasciatus							
		Control				Treatment (1.2 g/m^2)				Control				Treatment (2 g/m^2)			
		L1+L2	L3+L4	Pupae	Total	L1+L2	L3+L4	Pupae	Total	L1+L2	L3+L4	Pupae	Total	L1+L2	L3+L4	Pupae	Total
0	N larvae/dip	2.93	0.56	0.07	3.56	3.02	0.73	0.04	3.80	7.2	8.2	0.7	16.1	16.6	7.8	2.2	26.6
1	N larvae/dip	1.69	0.70	0.02	2.41	1.21	0.04	0.02	1.27	8.0	7.2	1.3	16.5	1.4	2.2	0.5	4.1
	DR (%)					31	95	0	53					92	68	88	85
2	N larvae/dip	3.36	0.90	0.08	4.33	2.0	0.2	0.00	2.3	6.5	7.4	0.9	14.8	1.9	1.2	0.3	3.4
	DR (%)					42	82	100	53					88	83	88	86
3	N larvae/dip	4.00	1.38	0.05	5.43	1.93	0.51	0.00	2.45	6.2	5.8	0.8	12.8	2.8	1.3	0.3	4.4
	DR (%)					54	73	100	60					81	76	90	79
7	N larvae/dip	3.02	2.12	0.06	5.20	3.68	3.03	0.17	6.89	5.5	6.6	0.7	12.8	4.3	2.6	0.2	7.1
	DR (%)					0	0	0	0					66	59	90	66
10	N larvae/dip	2.54	2.09	0.29	4.92	3.77	3.26	0.31	7.34	5.2	4.7	1.6	11.5	4.7	3.4	0.8	8.9
	DR (%)					0	0	0	0					61	26	83	53
13	N larvae/dip	–	–	–	–	–	–	–	–	5.9	5.4	1.5	12.8	7.7	5.4	1.7	14.9
	DR (%)					–	–	–	–					44	0	62	29
16	N larvae/dip	–	–	–	–	–	–	–	–	4.3	4.9	1.4	10.5	10.8	6.9	1.8	19.5
	DR (%)					–	–	–	–					0	0	57	0

RD = Density reduction.

done by the same operator. The larval instars as well as pupae were counted separately. Temperature and pH were monitored at each sampling day in each mosquito breeding sites. The field trials were conducted between 10 November and 23 December 2011 with *An. gambiae* and between 14 September and 5 November 2011 with *Cx. quinquefasciatus*.

6. Data Analysis

The data analyses were performed using R software (version 2.11.1). Data from the simulated field trial were used to estimate the Emergence Inhibition Rates (% EIR) for each treatment according to the following formula:

$$\% \, EIR = ((CT)/C) \times 100$$

where C is the emergence rate in the control and T is the emergence rate in the treated containers at the same period of time [13].

A logistic regression model with a logit link was fitted to the data to investigate the effect of the treatments on the emergence rate. The influence explanatory covariables on the emergence rate was investigated by including in the models the dose, the number of day post-treatment and the replicates. The number of day after which the emergence rates significantly increased to more than 20% with 95% Confidence Intervals was estimated for each treatment based on the logistic regression model.

Concerning data from the field trials, the mean number of larvae and pupae collected (i.e. density) per sampling day was calculated for both treated and control groups. The first and second instars larvae (L1+L2) were pooled as early instars and the third and fourth instars (L3+L4) as late instars. Density Reduction (DR) of early and late instars larvae as well as pupae was estimated post-treatment using Mulla's formula as follows:

$$DR = 100 - (C_1/T_1) \times (T_2/C_2) \times 100$$

where C_1 is the average number of larvae or pupae in control breeding sites prior to treatment and C_2 is the average number of larvae in control breeding sites at each day of sampling. T_1 is the average number of larvae or pupae in breeding sites to be treated with Vectobac GR and T_2 is the average number of larvae or pupae in treated breeding sites for each sampling day [13]. When DR was negative (i.e. densities were higher in the treated group than the control group), the value was taken as zero.

Then, a linear regression model was fitted to the data to investigate the effect of the treatment on the density reduction. The influence of the time as explanatory covariable on the density reduction was investigated by including the time in the models. The number of day after which the density reduction reached 80% and 50% was then estimated.

Results

1. Simulated Field Studies

The average temperature recorded in containers during trials with *An. gambiae* was 26.5°C (ranging from 24.0°C to 27.8°C) and 26.5°C (25.0°C to 27.7°C) with *Cx. quinquefasciatus*. The water pH was 7.5 (6.8 to 8.7) and 6.9 (6.6 to 8.0) for *An. gambiae* and *Cx. quinquefasciatus* containers, respectively.

Three thousands three hundred and sixty (3,360) larvae of *An. gambiae* were released in the containers for the trial. Emergence rates (ER) and Emergence Inhibition Rates (EIR) for each treatment are shown in Table 1. Emergence rates in the control ranged from 90% [86–94] to 96% [93–99]. The EIR were >80% for all dosages up to day 19 post-treatment. After day 26, the EIR was 44%, 55% and 63% at 0.6 g/m^2, 0.9 g/m^2 and 1.2 g/m^2 doses of Vectobac GR, respectively. According to the logistic

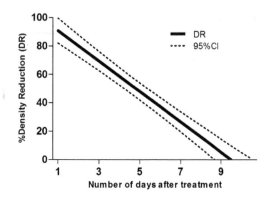

Figure 1. Density reduction (DR) of *An. gambiae* old instars larvae estimated by the regression model according to the number of days after treatment.

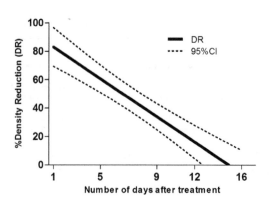

Figure 3. Density reduction (DR) of *Cx. quinquefasciatus* old instars larvae estimated by the regression model according to the number of days after treatment.

regression model, the estimated period of effectiveness (i.e. emergence rates <20%) was 15 days [14–17], 17 days [16–18] and 21 days [20–22] at the doses of 0.6, 0.9 and 1.2 g granule/m², respectively.

Three thousands seven hundred and sixty (3,760) larvae of *Cx. quinquefasciatus* were released in the containers during the simulated studies. Emergence and EIR for each treatment are shown in Table 2. Emergence rates in the control ranged from 90% [86–94] to 99% [98–100]. The EIR were 100% at day 11 regardless of the doses, and then decreased to 80% after 19 days of treatment with 1 g/m² dose, 26 days at 1.5 g/m² and 34 days at 2 g/m². According to the logistic regression model, the estimated period of effectiveness (i.e. emergence rates <20%) was 19 days [18–20], 22 days [20–23] and 28 days [27–29] days at 1, 1.5 and 2 g granules/m², respectively.

2. Field Trials

Based on the results of simulated studies, doses that showed highest efficacies and residual activities against *An. gambiae* (1.2 g granules/m²) and *Cx. quinquefasciatus* (2 g granules/m²) were selected for the field trials.

The average temperature recorded in the breeding sites through the trial was 35.1°C (ranging from 28°C to 41.7°C) and 27.1°C (ranging from 25.1°C to 32.2°C) for *An. gambiae* and *Cx. quinquefasciatus*, respectively. The water pH was 6.6 (ranging from 5.1 to 8.8) and 6.8 (ranging from 5.7 to 8.1) in habitats with *An.*

gambiae and *Cx. quinquefasciatus*, respectively. No rain was recorded during the trial on *An. gambiae* as the study was conducted during the dry season. The aerial average temperature recorded in the rice fields was 27.8°C, (23.2°C to 35.0°C). During the trial on *Cx. quinquefasciatus*, there was 345.90 mm rainfall while the average temperature was 27.3°C (24.6°C to 30.0°C).

The mean number of *An. gambiae* larvae sampled per dip and the density reduction (DR) at each sampling day are shown in Table 3. Before treatment, mosquito larvae densities in the control ponds were 2.93 per dip, 0.56 per dip and 0.07 per dip for early instars larvae, late instars larvae and pupae respectively. In the ponds to be treated these densities were 3.02 per dip, 0.73 per dip and 0.04 per dip for early instars larvae, late instars larvae and pupae respectively. The highest efficacy of Vectobac GR in terms of reduction of early instars larvae of *An. gambiae* was observed three days post-treatment but was below 60% reduction. Vectobac GR reduced late instars larvae density by >80% up to 2 days post-treatment. The DR decreased to 73% after 3 days and to nil after day 7. The number of pupae was too low to make any comparisons between control and treated ponds. According to the logistic regression model, the estimated period for which the density of late instars larvae would be reduced by 80% (DR$_{80}$) and 50% (DR$_{50}$) was 2 days (1–3) and 5 days (4–6), respectively (Figure 1). The highest DR induced by Vectobac GR against early instars larvae was about 50%. Consequently The DR$_{80}$ and DR$_{50}$ values could not be estimated.

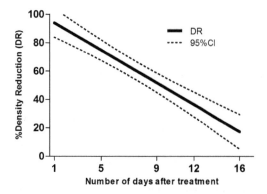

Figure 2. Density reduction (DR) of *Cx. quinquefasciatus* young instars larvae estimated by the regression model according to the number of days after treatment.

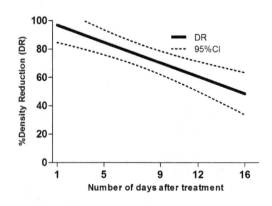

Figure 4. Density reduction (DR) of *Cx. quinquefasciatus* pupae estimated by the regression model according to the number of days after treatment.

The densities of early instars, late instars and pupae of *Cx. quinquefasciatus* in control cesspits before treatment were 7.2 per dip, 8.2 per dip and 0.7 per dip respectively. In the cesspits to be treated the densities were 16.6 per dip, 7.8 per dip and 2.2 per dip respectively. After treatment with Vectobac GR, >80% reduction was observed until day 3 in early instars, until 2 days in late instars and until day 10 in pupae (Table 3). According to the linear regression model, the estimated numbers of days after which the density of late instars larvae would be reduced by 80% (DR_{80}) and 50% (DR_{50}) were 2 days (0–4) and 7 days (5–8), respectively. DR_{80} and DR_{50} were 4 days (2–5) and 10 days (8–11), respectively for early instars larvae and 6 days (4–9) and 16 days (11–20), respectively for pupae (Figures 2, 3 and 4).

Discussion and Conclusions

The efficacy of Vectobac GR, a new formulation of *Bacillus thuringiensis var. israelensis* Strain AM65-52, was evaluated against *An. gambiae* and *Cx. quinquefasciatus* in both simulated and natural conditions.

Under simulated field conditions, Vectobac GR caused emergence inhibition of ≥80% up to 21 days (20–22) post-treatment for *An. gambiae* (at 1.2 g/m^2) and 28 days (27–29) for *Cx. quinquefasciatus* (at 2 g/m^2). The longer efficacy of Vectobac GR against *Cx. quinquefasciatus* can be explained by the higher dosage of Vectobac GR used during the trial (as per manufacturer's recommendation) and/or by a higher susceptibility of *Cx. quinquefasciatus* larvae to *Bti* as reported elsewhere [14,15].

In the field, Vectobac GR formulation, designed for deep penetration of overgrown vegetation after application, induced a ~50% reduction of *An. gambiae* density in rice fields 3 days after application. The reduction of *Cx. quinquefasciatus* larvae was about 80% in urban cesspits 3 days after application. The short residual efficacy against both *Anopheles* and *Culex* mosquitoes in open water bodies may be due to its low ITU content (200 ITU/mg compared with previous *Bti* formulations with 3000 ITU/mg) or a faster degradation or sequestration of *Bti* toxins in natural habitats as previously reported [16]. This short residual efficacy of Vectobac GR in natural habitats inevitably rose the question about the bioavailability of the *Bti* toxins in the rice field ponds and highly polluted habitats such as cesspits. *Bti* toxins are known to sediment in the breeding sites and this is also true with the Vectobac GR (granules were found at the bottom of the cement containers in simulated field conditions). In rice fields, the targeted species population (*An. gambiae s.s.*) was exclusively made of the M molecular form (recently renamed *Anopheles coluzzii*) [17]. Compared to the S molecular (now *Anopheles gambiae*) form, the M form larvae of are known to spend significantly more time at the bottom of the water column in breeding sites to collect food than the S form larvae [18]. Consequently, it is unlikely that the sedimen-

tation of the Vectobac GR toxins might have caused lower control of M form *An. gambiae*. In contrast, we observed that in the water ponds, the granules were sometimes found buried in the mud. Between the granule sp, thus were not fully available for mosquito control.

In addition, it is likely that the direct sunlight exposure of the larval habitats contributed to reduce the residual efficacy of Vectobac GR. Regarding *Culex quinquefasciatus*, results obtained in this present study are consistent with previous trials conducted in polluted (stagnant) waters in Africa [19] and India [20]. The presence of debris and heavy load of organic materials in the cesspits are known to absorb the *Bt* toxins and hence reduce the performance of *Bti*-based products.

With the development and rapid spread of insecticide resistance in malaria vectors [21] and increased proportion of malaria vectors that feed outdoors in response to the implementation of vector control intervention such treated nets and indoor residual spraying [22,23], there is a urgent need for complementary vector control strategies that could better impact vector density and malaria transmission. As suggested by Corbel et al. [24], the use of larvicide products could be complementary tool in the context of an integrated vector management for malaria transmission reduction and vector resistance management.

The use of larvicidal products for malaria control has long history in Africa with however more or less success [6]. In Gambia, hand application of water-dispersible granular formulations of *Bti* (Valent BioSciences, USA) to water bodies was associated with a 88% reduction in larval densities but had no effect on adult mosquito density and clinical malaria [25]. It is essential to better assess the impact of larviciding on mosquito density, malaria transmission and malaria morbidity. The opportunity to reinforce the use of larviciding in public health is currently under the spotlights among African countries through Economic Community of West Africa States (ECOWAS) including Benin [26]. This can also be an option to manage the spread of pyrethroid-resistance in African malaria vectors, as well as complement control of lymphatic filariasis in Africa south of Sahara. Nevertheless this complementary tool is highly dependent of the larval breeding site dynamics. The cost-effectiveness of such vector control strategy should be also carefully investigated. The present results emphasize the crucial need to improve basic knowledge on mosquito ecology as well as precise identification, mapping and monitoring of larval habitats in order to enhance the public health benefit to implement larval control programs.

Author Contributions

Conceived and designed the experiments: VC AD FC RY MA. Performed the experiments: BZ MO AD. Analyzed the data: KBS AD VC. Contributed reagents/materials/analysis tools: AD VC CP RY.

References

1. WHO (2012) World Malaria Report.
2. WHO (2013) Sustaining the drive to overcome the global impact of neglected tropical diseases. Second WHO report on neglected tropical diseases. WHO/HTM/NTD/2013.1. Geneva, World Health Organization.
3. Townson H, Nathan MB, Zaim M, Guillet P, Manga L, et al. (2005) Exploiting the potential of vector control for disease prevention. Bull World Health Organ 83: 942–947.
4. Matthew TB, Godfray HCJ, Read AF, Berg Hvd, Tabashnik BE, et al. (2012) Lessons from Agriculture for the Sustainable Management of Malaria Vectors. PLoS Med 9.
5. WHO (2008) WHO position statement on integrated vector management. Weekly epidemiological record 20: 177–184.
6. Fillinger U, Lindsay SW (2011) Larval source management for malaria control in Africa: myths and reality. Malaria J 10: 353.

7. Geissbühler Y, Kannady K, Chaki P, Emidi B, Govella N, et al. (2009) Microbial Larvicide Application by a Large-Scale, Community-Based Program Reduces Malaria Infection Prevalence in Urban Dar Es Salaam, Tanzania. PLoS ONE 4.
8. WHO (2011) WHO position statement on integrated vector management to control malaria and lymphatic filariasis. *Weekly Epidemiological Record* Number 13: 120–127.
9. Lacey LA (2007) *Bacillus thuringiensis* serovariety *israelensis* and *Bacillus sphaericus* for mosquito control. Amer Mosq Cont Assoc 23, Supplement to No.2: 133–163.
10. WHO (2006) Pesticides and their application for the control of vectors and pests of public health importance. Sixth edition. 1–125 p.
11. Lagadic L, Roucaute M, Caquet T (2013) *Bti* sprays do not adversely affect non-target aquatic invertebrates in French Atlantic coastal wetlands. Journal of Applied Ecology doi: 10.1111/1365-2664.12165.

12. Kinde-Gazard D, Baglo T (2012) Assessment of microbial larvicide spraying with *Bacillus thuringiensis israelensis*, for the prevention of malaria. Med Mal Infect 42: 114–118.
13. WHO (2005) Guidelines for laboratory and field testing of mosquito larvicides. WHO/CDS/WHOPES/GCDPP/200513.
14. Porter A, Davidson E, Liu J (1993) Mosquitocidal toxins of bacilli and their genetic manipulation for effective biological control of mosquitoes. Microbiological Reviews 57: 838–861.
15. Charles JF, Nielsen-LeRoux C (2000) Mosquitocidal bacterial toxins : diversity, mode of action and resistance phenomena. Mem Inst Oswaldo Cruz 95 Suppl 1: 201–206.
16. Madliger M, Gasser CA, Schwarzenbach RP, Sander M (2010) Adsorption of transgenic insecticidal Cry1Ab protein to silica particles. Effects on transport and bioactivity. Environ Sci Technol 45: 4377–4384.
17. Coetzee M, Hunt RH, Wilkerson R, Della Torre A, Mamadou B, et al. (2013) *Anopheles coluzzii* and *Anopheles amharicus*, new members of the *Anopheles gambiae* complex. *Zootaxa* 3619 246–274.
18. Gimonneau G, Pombi M, Dabiré RK, Diabaté A, Morand S, et al. (2012) Behavioural responses of *Anopheles gambiae* sensu stricto M and S molecular form larvae to an aquatic predator in Burkina Faso. Parasites & Vectors 5: 65.
19. Skovmand O, Sanogo E (1999) Experimental formulations of *Bacillus sphaericus* and *B. thuringiensis israelensis* against *Culex quinquefasciatus* and *Anopheles gambiae* (Diptera: Culicidae) in Burkina Faso. Journal of Med Entomol 36: 62–67.
20. Haq S, Bhatt R, Vaishnav K, Yadav R (2004) Field evaluation of biolarvicides in Surat city, India. J Vector Borne Dis 41: 61–66.
21. Ranson H, N'Guessan R, Lines J, Moiroux N, Nkuni Z, et al. (2011) Pyrethroid resistance in African anopheline mosquitoes: what are the implications for malaria control? Trends Parasitol 27: 91–98.
22. Okumu F, Moore SJ (2011) Combining indoor residual spraying and insecticide-treated nets for malaria control in Africa: a review of possible outcomes and an outline of suggestions for the future. *Malar J* 10: 208.
23. Moiroux N, Gomez MB, Pennetier C, Elanga E, Djènontin A, et al. (2012) Changes in Anopheles funestus Biting Behavior Following Universal Coverage of Long-Lasting Insecticidal Nets in Benin. J Infect Dis 206: 1622–1629.
24. Corbel V, Akogbéto M, Damien GB, Djènontin A, Chandre F, et al. (2012) Combination of malaria vector control interventions in pyrethroid resistance area in Benin: a cluster randomised controlled trial. Lancet Infect Dis 12: 617–626.
25. Majambere S, Lindsay SW, Green C, Kandeh B, Fillinger U (2007) Microbial larvicides for malaria control in The Gambia. Malar J 6: 76.
26. Economic Community of West African States (ECOWAS) (2011) ECOWAS Report of technical meeting on malaria vector control in ECOWAS region, Cotonou, Republic of Benin 28–30 nov. 2011.

11

Sustained Reduction of the Dengue Vector Population Resulting from an Integrated Control Strategy Applied in Two Brazilian Cities

Lêda N. Regis[1]*, Ridelane Veiga Acioli[2], José Constantino Silveira Jr.[1], Maria Alice Varjal Melo-Santos[1], Wayner Vieira Souza[3], Cândida M. Nogueira Ribeiro[4], Juliana C. Serafim da Silva[5], Antonio Miguel Vieira Monteiro[6], Cláudia M. F. Oliveira[1], Rosângela M. R. Barbosa[1], Cynthia Braga[7], Marco Aurélio Benedetti Rodrigues[8], Marilú Gomes N. M. Silva[8], Paulo Justiniano Ribeiro Jr.[9], Wagner Hugo Bonat[9], Liliam César de Castro Medeiros[10], Marilia Sa Carvalho[11], André Freire Furtado[12]

1 Departameto de Entomologia, Fundação Oswaldo Cruz-Fiocruz-Pe, Recife-PE, Brazil, 2 Secretaria Estadual de Saúde, Recife-PE, Brazil, 3 Departameto de Saúde Coletiva, Fundação Oswaldo Cruz-Fiofruz-PE, Recife-PE, Brazil, 4 Secretaria Municipal de Saúde, Santa Cruz do Capibaribe-PE, Brazil, 5 Secretaria Municipal de Saúde, Ipojuca-PE, Brazil, 6 Divisão de Processamento de Imagens, Instituto Nacional de Pesquisas Espaciais-INPE, São José dos Campos-SP, Brazil, 7 Departameto de Parasitologia, Fundação Oswaldo Cruz-Fiocruz-PE, Recife-PE, Brazil, 8 Departameto de Eletrônica e Sistemas, Universidade Federal de Pernambuco, Recife-PE, Brazil, 9 Departameto de Estatística, Universidade Federal do Paraná, Curitiba-PR, Brazil, 10 Centro de Ciência do Sistema Terrestre, Instituto Nacional de Pesquisas Espaciais-INPE, São José dos Campos-SP, Brazil, 11 Fundação Oswaldo Cruz-Fiocruz, Rio de Janeiro-RJ, Brazil, 12 Departameto de Virologia, Fundação Oswaldo Cruz-Fiocruz-PE, Recife-PE, Brazil

Abstract

Aedes aegypti has developed evolution-driven adaptations for surviving in the domestic human habitat. Several trap models have been designed considering these strategies and tested for monitoring this efficient vector of Dengue. Here, we report a real-scale evaluation of a system for monitoring and controlling mosquito populations based on egg sampling coupled with geographic information systems technology. The SMCP-Aedes, a system based on open technology and open data standards, was set up from March/2008 to October/2011 as a pilot trial in two sites of Pernambuco -Brazil: Ipojuca (10,000 residents) and Santa Cruz (83,000), in a joint effort of health authorities and staff, and a network of scientists providing scientific support. A widespread infestation by Aedes was found in both sites in 2008–2009, with 96.8%–100% trap positivity. Egg densities were markedly higher in SCC than in Ipojuca. A 90% decrease in egg density was recorded in SCC after two years of sustained control pressure imposed by suppression of >7,500,000 eggs and >3,200 adults, plus larval control by adding fishes to cisterns. In Ipojuca, 1.1 million mosquito eggs were suppressed and a 77% reduction in egg density was achieved. This study aimed at assessing the applicability of a system using GIS and spatial statistic analysis tools for quantitative assessment of mosquito populations. It also provided useful information on the requirements for reducing well-established mosquito populations. Results from two cities led us to conclude that the success in markedly reducing an Aedes population required the appropriate choice of control measures for sustained mass elimination guided by a user-friendly mosquito surveillance system. The system was able to support interventional decisions and to assess the program's success. Additionally, it created a stimulating environment for health staff and residents, which had a positive impact on their commitment to the dengue control program.

Editor: Pedro Lagerblad Oliveira, Universidade Federal do Rio de Janeiro, Brazil

Funding: This work received financial support from grants from CNPq, http://www.cnpq.br (PQ-301277/2005-2 and APQ-479214/2010-7); FACEPE, http://www.facepe.br (APQ-0692-2.13/08); and PDTSP-FIOCRUZ, http://www.fiocruz.br (RDVE-03). The State Health Department of Pernambuco offered ground transport for people and materials related to the project from Recife to Santa Cruz do Capibaribe and Ipojuca, during 2008–2011. The funders had no role in the study design, data collection and analysis, decision to publish, or preparation of the manuscript.

Competing Interests: The authors have declared that no competing interests exist.

* E-mail: leda@cpqam.fiocruz.br

Introduction

Aedes aegypti Limnaeus 1762 (Diptera:Culicidae) populations appear to be currently well established in most households at almost every tropical urban setting and are also established in some sub-tropical areas. The main consequence is that two fifths of the world's population is potentially exposed to four infections by the dengue viruses, resulting in 50 million infections annually, as estimated by the World Health Organization. Dengue fever is currently considered as one of the fastest spreading diseases in the world [1]. Even if an efficient dengue fever vaccine becomes available, *A. aegypti and A. albopictus* mosquitoes established in urban environments will remain as a matter of concern as these Culicids are highly efficient vectors of other arbovirusis such as, Yellow Fever, West Nile viruses and Chikungunya. For Brazil and other South American countries, widespread infestation by *A. aegypti* is a relatively recent scenario: a few decades ago villages and cities were considered free of dengue vectors. Rapid spread and stable installation of the main DENV vector *A. aegypti* in urban and semi-urban territories are greatly favored by evolution-driven adaptations to the human host [2] and to unstable aquatic habitats commonly found in the human house. Some of these adaptations are the mosquito's ability to colonize a very wide variety of water holding containers, to spread eggs from the same batch on different sites, and the high resistance of the egg chorion. Due to these biological characteristics, and to the high number of diverse objects that hold water often found in modern urban environments, the classical method of visual inspection to detect larvae/pupae is extremely difficult and time-consuming, resulting in an ineffective way to monitor urban mosquito populations. Moreover, it is unlikely that classical control interventions based mainly on the application of larvicides in aquatic habitats will succeed on reducing dengue vector populations. Nevertheless, this is still the method used in most dengue endemic countries and despite the large amounts of financial resources spent in vector control for over a decade, most urban territories in those countries remain heavily infested by *Ae. aegypti*, as is the case of Brazil [3].

Several experimental and practical evidences obtained from studies developed in different countries show that the use of traps is a more appropriated strategy to monitor *Ae. aegypti* and *Ae. albopictus* than classical surveillance methods. Different models of ovitraps have been clearly shown as effective monitoring tools when integrated to entomological surveillance systems able to generate quantitative information on mosquito presence and densities [4–14]. Such systems are essential for directing control actions and measuring their impact on the vector abundance. A great potential of traps as a mass attract-and-kill strategy for reducing mosquito population has been revealed in some studies [7,10,15–17,18]. Applying a sensitive surveillance system (SMCP-*Aedes*) in a district of Recife City, Brazil, we have shown that *Aedes* population boosts resulting from massive eggs hatching promoted by rainfall that follows a dry season, can be prevented through this control strategy by eliminating, at the dry season, more than 6 million eggs collected in 4000 traps throughout 4 months [7].

The SMCP-*Aedes* - Monitoring System and Population Control for urban *Aedes*, is an entomological surveillance framework to provide baseline data for dengue epidemiological surveillance. It consists of an all-integrated approach supported by the intensive use of the web and free software to collect, store, analyze and disseminate information on the spatial-temporal distribution of the estimated density of *Aedes*, based on data that is systematically collected with the use of ovitraps and integrated through the use of open geospatial technologies such as GIS, Remote Sensing Images and Spatial Statistics analytical tools [19]. This system was developed based on a 3-year longitudinal experiment using 460 georeferenced ovititraps in seven districts of Recife [7]. It was subsequently (2008–2011) settled as a pilot program in two cities and since 2010 it has been set up in an Oceanic Island in Pernambuco, all endemic areas for dengue.

In this article, we report the results of a strategy based on integrated control interventions guided by mosquito surveillance applied by the local health authorities and staff coupled with scientific team support and community participation. This strategy was successfully able to reduce mosquito populations in Santa Cruz do Capibaribe-PE and Ipojuca-PE, Brazil. Our ultimate goals were to evaluate at real-scale and to improve the applicability of an environmentally friend vector monitoring and control system aiming to contribute to reduce dengue virus transmission.

Methods

Ethics Statement

This study was reviewed and approved by the Ethics Committee of the CPqAM-Fiocruz-PE, Brazil (No. 14/04). Field works were carried out by the Municipal Health personnel, as part of the Municipal Dengue Control Program in agreement with the Brazilian rules for dengue control in endemic areas. In areas where the presence of *Aedes aegypti* is confirmed, a written consent for house visits and mosquito control interventions by the municipal government is not required. With the consent of the municipal health authorities the scientific team had full access to all data generated by the use of the SMCP-*Aedes*, as well as the teams from the Health Secretary had access to all the results of analyzes of the data made by the scientific team.

The SMCP-*Aedes* was evaluated from March 2008 to October 2011, in a joint effort of the state health authorities, local health staff and a scientific team organized in a network.

Implementation of the Program

1. Site selection. The system was evaluated in two municipalities chosen by the State Health Department of Pernambuco based on dengue fever incidence, different geographic characteristics, and commitment level of the health staff: Ipojuca is located on the coast at 50.2 km from Recife, the capital of the state; Santa Cruz do Capibaribe is located in the Semi-Arid region of the state, at 194.3 km from Recife (Figure 1).

Santa Cruz do Capibaribe (7° 57′ 27″ S and 36°12′17″ W, 438 m above sea level) has annual mean temperature of 26.9°C (min 21° max 32.5° C) and 360.3 mm of annual average rainfall, which occurs from January to July. The municipality, here referred as SCC, has a total area of 335 km² with 87582 inhabitants [20]. From 2001 to 2008 SCC notified 2394 dengue cases (1124 during the Denv-3 epidemics in 2002), with most of them occurring from January to April. The Premise Index (PI, percentage of inspected premises found to have containers positive for *Ae. aegypti* larvae/pupae), used as an indicator of dengue risk by the PNCD-National Program for Dengue Control, ranged from 3.8 to 12.9 in 2007 and 2008, with annual means of 4.8 and 9.4, respectively. The present study was conducted in an urban area of 5.67 km² where 83161 residents live. Access to water is not regular and residents often store water in underground cisterns (approximately 20000 cisterns with an average capacity of 10000 L), which are potential breeding places for culicids. The field work team of the dengue control program was composed of 32 endemic agents (EA), three supervisors and one coordinator. Ipojuca (08°24′00″ S and 35°03′45′ W) has a population of 80637 inhabitants [20] and a 532 Km² area. The annual mean temperature is 26.1°C (min 21.8°C - max 30.7°C), with a 1719.4 mm mean annual rainfall concentrated from March to August. Ipojuca notified 3060 dengue cases (1586 confirmed cases) from 2000 to 2008, with approximately 80% of them occurring in January-June. The PI was 0.8 in 2007 and 0.4 in 2008. The study was conducted in the Municipal Seat, referred as IpojucaMS, an area of 0.8 km² with 10037 people living at. The local field team was composed by eight EA and one supervisor.

For administrative reasons the PNCD procedures for mosquito control, implemented since 1997, were maintained during this study in both study areas. PNCD vector control activities comprise routine bimonthly application of temephos (larvicide); an annual campaign for source elimination - the Dengue Day; and application of organophosporous (OP) or piretroids adulticides through ULV in the case of epidemics. Following a decision by the PNCD, the OP temephos was replaced by chitin synthesis

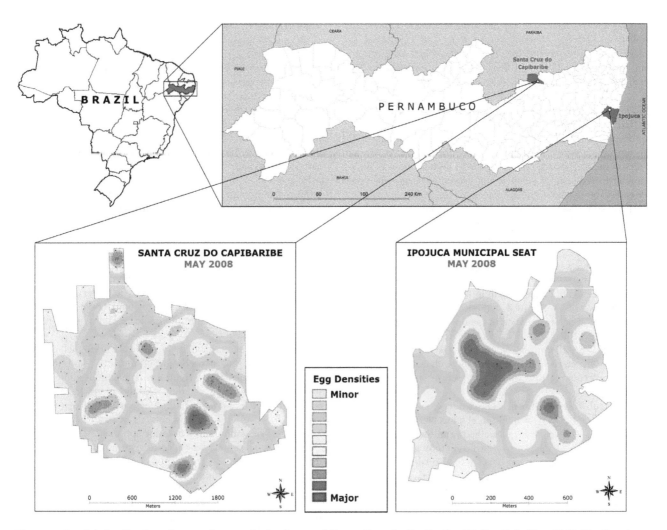

Figure 1. Spatial distribution of mosquito egg for Ipojuca and Santa Cruz do Capibaribe-PE, Brazil, in May 2008. The Kernel maps showing smoothed egg densities are based on the number of eggs deposited in each sentinel-ovitrap during one month. Data from the sentinel network encompassing 262 ovitraps distributed over a 5.6 km² urban area in Santa Cruz do Capibaribe and 75 ovitraps over a 0.8 km² urban area of Ipojuca municipal seat.

inhibitors in January/2011in IpojucaMS and in May/2011 in SCC.

2. Pre-trial. Prior to implementation, the procedures proposed by the SMCP-*Aedes* were approved by the management and technical staff of the municipal health department, and then discussed in open meetings (April 2008) with the local authorities of Education, Urban Environment Management and Social Actions departments, the Municipal Health Council, and citizens. Training and workshops were offered to the local dengue program staff by the scientific research team in April 2008 and in September 2008, providing theoretical and practical knowledge on the strategies, methods and tools to be adopted by the SMCP-*Aedes* for mosquito monitoring and control.

3. Sentinel network set up. The number of traps for surveying each site was calculated based on a logistic function according to house density, as described in Regis et al (2009). For ovitrap distribution within the study area, grids with 40 m×40 m cells were overlaid in the study area map. The grids were made of approximately 3500 cells in SCC and 490 in IpojucaMS. Out of these, a sample of 262 and 75 cells were systematically and randomly chosen in SCC and IpojucaMS, respectively. The sentinel networks were installed in May 2008 by the field team of

each local health service equipped with maps indicating the quadrants for geo-referenced sentinel-ovitraps (S-ovt) location, a GPS, Bti-larvicide and labeled traps. The networks were composed of 262 S-ovt in SCC and 75 in IpojucaSM. Each S-ovt was installed at a fixed sampling station, hung 1 m above ground level outdoor of the residence, in the shade and protected from rainfall. Geographical coordinates of trap location were taken using a Global Positioning System (GPS) and entered into the SMCP-*Aedes* geographical database. This database is based on open technology, open protocols and open data and was designed to make available reports and data analysis [19]. The S-ovt model used for sampling *Aedes* eggs, described in Regis et al [7], consists of a black plastic cup filled with two liters of water, two 5×15 cm wood paddles as egg substrates and two or four drops of a Bti-based product as larvicide, to prevent the trap from becoming a mosquito breeding site [6]. Liquid Bti-products in 30 or 50 ml bottles (Bt-horus, Btheck, Brasilia DF, Brazil or Killarv-Bti, Caruaru-PE, Brazil) were chosen instead of granular formulations, because they are more practical to transport and do not deposit residues on the paddles. Each ovitrap was labeled with an identification code, the set up date and a telephone number. The first household visit was aimed at explaining the project to and obtaining the agreement of the householder, followed by trap installation. Monthly checking of the S-ovt was conducted to replace paddles, water and Bti. The paddles taken each month from Ipojuca (150 units) and SCC (524) were delivered to the Department of Entomology/Fiocruz-PE in Recife for egg counting.

Semi-automatic egg counting: A system for acquiring digital images, the SDP-Egg Counting System, was developed to hasten and improve the process of counting the eggs laid on the paddles [21]. The system consists of an equipment to perform optical scanning images of the trap paddles and of software installed on a desktop computer to assist egg counting. Image acquisition starts when the user enters the paddle in the system. *Aedes* eggs are recognized by the operator on the digitalized image. The counting is based on mouse click over the picture and the number of identified eggs is registered and transmitted via web to the SMCP-*Aedes* Geographic Database [19]. Egg counting through the SDP system allowed analysis of paddles to be done around two times faster than the conventional microscopic method.

Aedes species identification. Fourth instar larvae (L4) reared from field-collected eggs were used for species identification. Egg samples from at least 70 different sentinel stations were used. For sample size calculation, it was considered relevant to detect *Aedes* species other than *A. aegypti*, with at least 0.1% prevalence. Assuming a hyper-geometric distribution with p = 1/1000, the sample size required per village was set as 1000 identified individuals.

4. Control Interventions. The strategy proposed in the SMCP-*Aedes* combines: i) the diffusion of information on mosquito biology and on control strategy, methods and tools; ii) interventions focused on mechanical mass destruction through incineration of eggs laid in large amounts of ovitraps loaded with Bti; and (iii) indoor collections of adults using aspirators, targeting places considered as highly important for virus transmission, such as health units, schools and premises located within hotspots of mosquito density.

For the control activity of mass-trapping eggs, a control-ovitrap (C-ovt) was designed and made of recycled PET bottle washed and paint in black. Bti was applied as a larvicide in 1.5 litter of water per trap. A 40×20 cm cotton fabric, covering the trap's inner wall, was used as the egg support, according to Lenhart et al. [22], instead of wood paddles. Ten thousand C-ovt were produced at a cost of R$ 0.97 (0.6 US$) each. An adhesive label with information on trap function and a telephone number for contact was put on the house wall near the C-ovt. For security reasons, each municipal government was responsible for supplying Bti-based larvicides, inspecting traps, and incinerating the egg supports.

Each municipal health department designed its own intervention plan according to the local characteristics and resources available. Members of the scientific team provided expertise in entomology.

SCC: starting on October 1st 2009, 2 C-ovt per building were installed in 17 health care units (HCU - hospitals, clinics, etc); 1203 traps were distributed, 2 per premise, over hotspot areas of mosquito density located in different neighborhoods; 4473 traps were concentrated in Rio Verde and Santa Tereza, neighborhoods considered as being critical for dengue cases. Overall, around 5680 C-ovt remained installed for two years. As an initiative of the EA team, larvivorous fishes taken from local sources were introduced as larval predators into cisterns, the main local water reservoirs. Fishes popularly known as "piabas", belonging to a group restrict to South America, the genus *Leporinus* Spix 1829 (Characiformes: Anostomidae), were used. Two to four fishes per cistern were added to 4690 cisterns in January–February/2010 in the

neighborhoods of Dona Lica and Palestina, and to 2415 cisterns in the neighborhoods of Cruz Alta, Malaquias, São Miguel, Centro, Nova Sta. Cruz and Bela Vista, in January-February 2011, totaling 7105 treated cisterns. According to the EA, treated sites are inspected at every two months during the regular house visits, and fishes are replaced when mosquito larvae are seen alive. However, no method was employed to evaluate the reduction/elimination of mosquito larvae by this process. *IpojucaMS*: during the first week of October 2009, about 2700 C-ovt were set up, two per household, in the hotspots of the neighborhoods of Centro and Campo do Avião, the latter showing the highest incidence of Dengue cases.

Overall, approximately 8.4 thousands C-ovt were set up in SCC and IpojucaMS. The fabric of each ovitrap used as a support for egg was taken by the EA team at every two months, appropriately incinerated and replaced with new ones. The EA also replaced water and Bti in the traps. Over 25 months, from October 2009 to October 2011, 12 mass-trapping cycles were carried out. Information was offered through a public exhibition, radio, television, banners, posters and leaflets, in order to make the public aware that the fewer available potential breeding sites in a house, the higher the amount of eggs to be laid by mosquitoes in the ovitraps.

Due to the limited availability of field work personnel, mosquito aspirations were not carried out in the scale and frequency expected to potentially contribute for reducing mosquito population size. For collection of adult mosquitoes inside buildings, a light-weight, battery-powered aspirator was used (Horst Armadilhas, SP, Brazil). With the householder's help, an EA carried out the catching of resting mosquitoes for approximately 20 minutes. Caught mosquitoes were identified to genus/species, counted and destroyed. Aspirations were carried out in SCC (*i*) monthly or fortnightly in 17 Health Care Unities (HCU) throughout 34 months starting on January 2009 (*ii*) and eventually in premises within mosquito density hotspots. Data from aspirations were used for mosquito surveillance. However, in IpojucaMS indoor collections of adult mosquitoes were restricted to a few aspiration sessions when only four *Ae. aegypti* specimens were collected, besides 1236 *Culex quinquefasciatus*.

Impact of the control interventions. Baseline data on the presence and density of *Aedes* eggs gathered for 16 months (May/2008 to Sept/2009) before the beginning of the control actions were used to assess changes in population density during the control pressure phase launched in the dry season that lasted from Oct/2009 to Oct/2011 (Table 1). Additionally, data from adult mosquito collections from places considered critical for virus transmission, acquired in the pre-control and control phases, provided additional data on the mosquito population.

It is noteworthy that the entomological routine activities of the official program, PNCD, implemented since 1997– larval search to estimate the entomological indices PI and BI, as well as the use of larvicides - were not interrupted during the study period (2008–2011).

Data Analysis

A spatial smooth kernel density estimator (KDE) [23] was used to produce surface maps of eggs collected in the S-ovt network, over successive counting cycles, in order to identify hotspots of vector density. Although KDE represents a simple alternative to analyze focal behavioral patterns, it does so by estimating the intensity of the point process throughout the study region. It is based on observed data, from which an unobservable density function thought of as the density according to which a large population is distributed, is estimated. The eggs collected in this

Table 1. Schematic representation of the study design in both study sites SCC and IpojucaMS, Pernambuco, Brazil.

Activities	2008				2009						2010						2011					
	3	4	5	6	1	2	3	4	5	6	1	2	3	4	5	6	1	2	3	4	5	6
Mosquito surveillance																						
Sentinel-ovitrap network	+	+	+	+	+	+	+	+	+	+	+	+	+	+	+	+	+	+	+	+	+	+
Mosquito aspiration					+	+	+	+	+	+	+	+	+	+	+	+	+	+	+	+	+	+
Integrated control measures																						
Mass elimination of eggs									+	+	+	+	+	+	+	+	+	+	+	+	+	+
Larvivorous fishes*										+	+	+	+	+	+	+	+	+	+	+	+	+
Education for source reduction									+	+	+	+	+	+	+	+	+	+	+	+	+	+

*only applied in Santa Cruz do Capibaribe.

case are usually thought of as a random sample from that unknown population. It is also a non-parametric estimation technique, which is good once there is no known parametric model for *Aedes* eggs distribution over urban spaces. And, most importantly, the outputs of the data analysis should be readily readable and easily understood by health staff at all levels (managers, field agents, planners, etc) and by city residents. The Density Maps of *Aedes* eggs produced by KDE procedure fits quite well this need. For the experiments in the SCC and IpojucaMS the KDE parameters were a quadratic function for the KDE kernel and a bandwidth of 300 m and 150 m. This radius covered approximately 12 S-ovt in SCC and 7 S-ovt in IpojucaMS considering each stage of the smoothening phase. Thus, the selected bandwidth allowed the smoothening to be performed with an appropriate number of traps and was chosen based on entomological parameters tested and adjusted over data collected on the local sites. The same grid and color definitions were applied for all cycles, in order to compare results across time. Analyses were performed with TerraView (www.dpi.inpe.br/terraview), an open GIS application.

Due to non-normality and overdispersion distribution of the eggs per ovitrap we have represented these distributions through their mean, minimum and maximum observed. The nonparametric Mann-Whitney test based on ranks was used to compare egg densities (eggs per trap per month). To evaluate the results achieved in reducing the intensity of mosquito eggs, we compared the series representing the number of eggs collected per ovitrap in the twelve mosquito surveys by plotting the median and the 25th and 75th percentiles for each period and each city.

The number of eggs collected per C-ovt was estimated in a sample of 100 oviposition supports (cotton fabric) from SCC and 50 from IpojucaMS, and used for calculating the amount of eggs suppressed at every cycle. According to a method developed by E.V.G. Silva & M.A.V Melo-Santos (unpublished data), the following categories were used for egg number estimation: zero, 1–100, 101–500, 501–1000, >1000. To estimate the total number of eggs suppressed at each mass-trapping cycle, the geometric mean of eggs from the C-ovt sample was calculated taking the mean points of the categories described above, and multiplying the mean by the total number of C-ovt.

Results

Operational Features of the Monitoring System

To set up the sentinel network, 32 health agents spent 9 h to install 262 S-ovt in SCC, and 16 agents spent 4 h to install 75 traps in IpojucaMS. Overall, one hour was required for each sentinel-station, including the first visit formalities, trap installation and geographical coordinate record. There was 100% acceptance by the householders. Monthly checking of the traps was carried out in one or, more often, two working days per month by the local EA. Trap maintenance was integrated as a monthly routine activity from May 2008 to October 2011, without significant operational surcharge for the Municipal Dengue Program. However, trap paddles containing eggs were not read every month. At first, due to the large amount of eggs (>400 thousands) collected in the first survey in SCC and IpojucaMS, three full-time technicians spent one month counting eggs on 674 paddles (two per trap) under stereoscopic microscopes, an average of 26 min per paddle. Egg counting was then discontinued until a prototype of the counting support system SDP was set up in the Department of Entomology/Fiocruz-PE in Recife where the paddles were read. Based on egg counting of 6600 paddles using the SDP system, a technician was able to read an average of 6.25 paddles per hour spending approximately 9.6 min per paddle.

Infestation Intensity before SMCP-*Aedes* Control Actions

The 1[rst] survey, undertaken in May 2008, showed wide distribution of *Aedes* population over the two territories: in IpojucaMS trap positivity was 92.6% and in SCC it was 100%. Marked differences (p<0.1%) in infestation intensity were however observed amongst sites: densities were 542.9 (0–2207) and 1587.2 (67–6027) eggs/trap/month in IpojucaMS and SCC respectively. Furthermore, the spatial distribution of egg densities revealed evident intra-site heterogeneity and high density clusters, as shown in kernel maps (Figure 1). The subsequent surveys carried out before implementation of the new control strategy, showed the persistence of widespread infestation by *Ae. aegypti* over the study territories, as indicated by 95.8% ovitrap positivity in IpojucaMS in June 2009, slightly higher than that observed one year before (92.6%) but with a lower density: 283.9 (0–1191) eggs/trap/month (p = 0.009). In SCC 100% trap positivity was recorded in the four surveys carried out in 2009 (January, April, June and July). Temporal fluctuations in egg densities showed growing active population densities from January (997.6 (7–5889) eggs/trap/month) to June (1461.1 (6–8209)) (p<0.1%), followed by fast decay in the following month to 762.9 (5–4732) (p<0.1%). The density in June/2009 was similar to that recorded the same month one year before (1587.2 eggs/trap/month). In addition to temporal variations, heterogeneous spatial distribution of densities was observed in each survey (Figure 2).

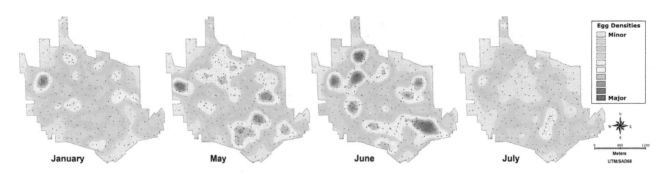

Figure 2. Spatial distribution of *Aedes* eggs in Santa Cruz do Capibaribe-PE, Brazil, in 2009. Each kernel smoothed map of egg densities is based on the number of eggs deposited throughout one month in 262 sentinel-ovitraps placed over a 5.6 km² urban area in Santa Cruz do Capibaribe, at different times along 2009.

Aedes Species

Ae. aepypti was the only species identified in a sample of 1569 larvae reared from eggs laid in 190 S-ovt in SCC, while in IpojucaMS 2.2% of 1812 specimens reared from eggs deposed in 36 ovitraps were *Ae. albopictus* and the other 97.8% were *Ae. aegypti.*

Infestation of the Health Care Unities

In SCC, 47 cycles of indoor mosquito aspiration performed in 17 HCU and encompassing 893 events resulted in the collection of 3280 *Ae. aegypti*, of which 62.2% were females, and 9336 *C. quinquefasciatus*. Every HCU has been found to be infested by *Aedes* in several rounds over the study period (Feb/2009 - Oct/2011). More than 10 (11–16) of them were found positive in most rounds along the pre-intervention period and such a high proportion of infested HCU persisted until August 2010 (Figure 3). A lower number of unities (6–11) remained positive in the following rounds from September 2010-May 2011, and none or very few (0–4) HCU were found to be positive for *Aedes* in the last months (June-October 2011). The mean number of *Aedes* per 17 HCU per aspiration round was approximately three to four, rarely reaching more than 10 (Figure 3). It is noteworthy that only in the last six cycles a decreasing trend in mosquito infestation could be observed, indicating the need for a continuous surveillance.

Control Actions and Social Engagement

The field workers showed proficiency in understanding data on mosquito density distribution expressed through the kernel maps. The health staff, supervisors and field workers continuously demonstrated great interest in discussing the surveys' results and planning the next actions aiming at further reducing the mosquito population. More than 2000 people in each site visited the public exhibition on the materials related to the proposed control strategy. The activities related to PET bottles collection and the production of C-ovt proved to be a motivating process that involved engaged groups from governmental and non-governmental organizations: health workers of the Dengue and other programs, local school segments - students and schoolteachers, and social organizations as volunteers; moreover, these activities stimulated people to participate in the distribution and set up processes of the traps. The installation of approximately 8000 C-ovt in a few days was in part due to the social engagement.

Egg Suppression

The continuous use of thousands C-ovt throughout 26 months starting in the dry season 2009 resulted in the incineration of millions of eggs in both cities. In IpojucaMS, where 2760 C-ovt were concentrated in the neighborhoods of Centro and Campo do Avião, it was estimated that more than 111,000 eggs were incinerated at the end of the 1st bimonthly trapping cycle (Figure 4A). An estimated total of 1.17 million eggs were incinerated over 12 trapping cycles. In SCC the amount of eggs trapped in about 5.7 thousands C-ovt decreased markedly from the 1st (2.26 million eggs) to the 2nd cycle (1.18 million eggs), and starting at the 5th cycle 300–400 thousand eggs were suppressed at every cycle (Figure 4B). A total of 7.5 million eggs were estimated to be incinerated in SCC over the whole period.

The Impact of Integrated Control Actions on Mosquito Population

The median and 75 and 25 percentiles of the egg-counting series before and during the control interventions in SCC are shown in Figure 5A, where a marked decline in the median egg density values can be observed. An estimated 90.5% reduction in egg density was observed in 2011 when the median number of eggs in the period of high mosquito abundance (May-July) was compared with that of 2008. Moreover, a relevant drop in the 75 percentile was recorded, where less than 62 eggs were found in 75% of traps in the last counting cycle. The reduction in infestation intensity can also be clearly seen when comparing smoothed kernel maps before and during interventions (Figure 6). Lastly, considering that the main aim of the present study was to

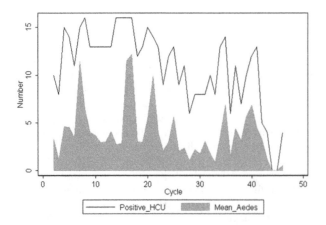

Figure 3. Results of indoor collection of mosquitoes in health care unities in Santa Cruz do Capibaribe-PE. Number of positive sites out of 17 surveyed health care unities (HCU), and mean number of adult *Aedes aegypti* collected through indoor aspirations carried out fortnightly or monthly in HCU in Santa Cruz do Capibaribe-PE, from February 2009 to October 2011.

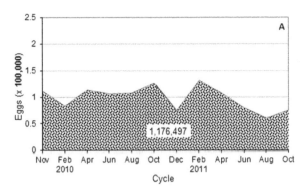 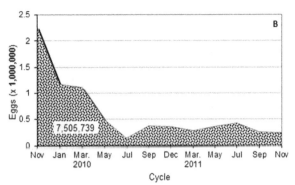

Figure 4. *Aedes* eggs trapped in bimonthly cycles. Estimated amounts of eggs suppressed at every two months through continuous mass-trapping using approximately 2700 control-traps with Bti in the Municipal Seat of Ipojuca (A) and 5700 in Santa Cruz do Capibaribe (B), from October 2009 to October 2011.

serve as a guide for public health activities, self-scale density maps were constructed for each period in order to facilitate the identification of priority areas by the field team (for instance, the map at the bottom of Figure 6).

However, while egg densities decreased progressively during the control intervention period, the trap positivity index remained 100% for at least one year. This index decreased to 98.3% in Jan/2011, to 95% in July and to 82.1% in the last survey, October 2011.

Data from IpojucaMS also showed important reduction in mosquito population (Figures 5B and 7), with a 77.1% decrease in the median egg density, as estimated comparing June 2011 to May 2008. In fact, the median value dropped from 413 in May 2008 to 94.5 in June 2011 notwithstanding the value of the 75th percentile remained above 130 eggs in the last survey. Data from February/2010 to October/2011, within the control intervention period, reported significantly lower densities than before interventions.

Discussion

The results of the present study showed that the SMCP-*Aedes*, evaluated at real scale in two different municipalities, was effective for monitoring dengue vectors and that it work as a decision support system. The system effectively indicated, in both localities, a wide spread infestation with moderate to high mosquito densities persisting for more than one year pre-intervention, followed by a

progressive and marked decrease in mosquito densities throughout the two years of implementation of integrated control measures.

The infestation scenario of the study areas by *Ae aegypti* found in May 2008 was typical of well-established mosquito populations, as demonstrated by high to moderate amounts of eggs laid in almost all ovitraps, which attests the presence of reproductively active females over the whole territory. Infestation levels remained very high in SCC and IpojucaMS in the following year: 96 and 100% trap positivity. Similar widespread infestation was previously observed in another city of Pernambuco State, when the overall positive ovitrap index remained close to 100% over 5616 trap observations throughout 12 months [7]. More recently, using the same monitoring system we observed an equally widespread infestation by *Ae. aegypti* in Fernando de Noronha Island, an oceanic island 545 km far from Recife (unpublished data).

While the SMCP-*Aedes* showed that *Ae. aegypti* was installed in almost all premises in both study areas, the Premise Index (based on visual detection of larvae) indicated very low infestation levels in the same territories and periods, estimating no more than 3.8 to 12.9% positive houses in SCC and 0.4 to 0.8% in Ipojuca. Such partial site occupation would, if true, characterize recent colonization by the mosquito, a hypothesis incompatible with the annual occurrence of many dengue cases in both cities over the last 12 years. While recent invasions by urban mosquitoes are characterized by limited space occupation, translated into low

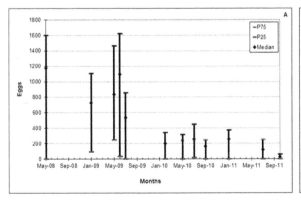 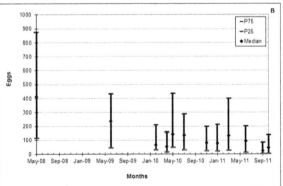

Figure 5. *Aedes* eggs collected in sentinel-ovitraps in Santa Cruz do Capibaribe (A) and Ipojuca MS (B). Median, and 75 and 25 percentiles for the quantitative series of collected eggs before (May/2008 to Sept/2009) and during (Oct/2009 to Oct/2011) the SMCP-*Aedes* control interventions.

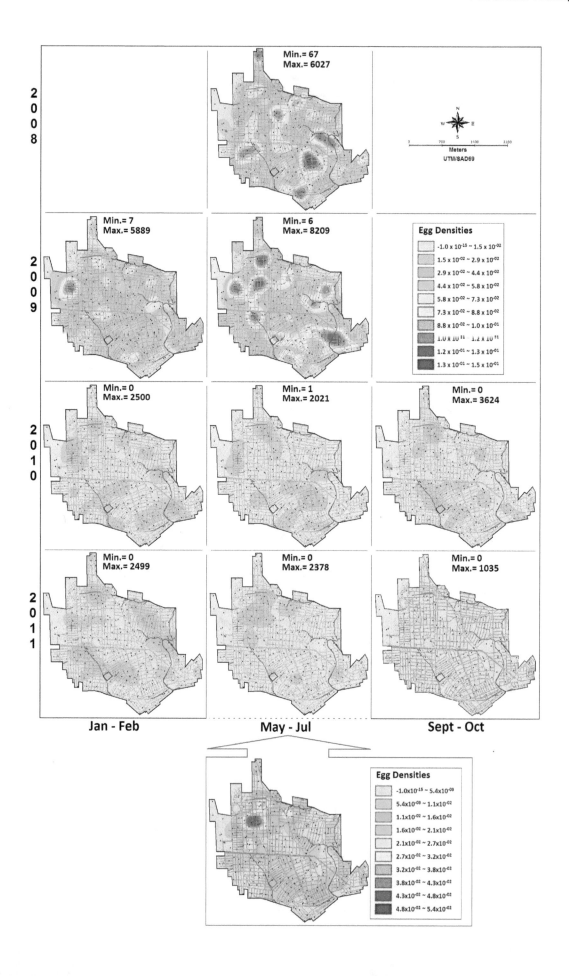

Figure 6. Kernel smoothed egg distribution in Santa Cruz do Capibaribe-PE, Brasil, 2008 to 2011. Maps are based on eggs laid in 262 sentinel-ovitraps during low and high egg density seasons, before (May/2008 to Sept/2009) and during (Oct/2009 to Oct/2011) integrated control measures including mass-suppression of eggs using 5700 control-ovitraps, and larvivorous fishes added to around 7 thousand cisterns. The same scale was used for all compared kernel maps. For each survey, a self-scale kernel map was also constructed to highlight hotspots, as that shown on the bottom of the middle column, for July 2011.

premise index, a widespread dispersion is conversely considered as an indicator of long-lasting mosquito establishment.

The present results provide a good example of the ability of trap-based quantitative methods on distinguishing different degrees of infestation intensity among sites presenting similar levels of mosquito spatial dispersion: the first mosquito survey showed similar rate of infested houses, while mosquito population size based on egg density was three-fold higher in SCC than in IpojucaMS. Local rainfall characteristics cannot explain the higher density of *Ae. aegypti* in SCC, where the average of annual rainfall was only 360.3 mm, against 1719 mm in Ipojuca. Furthermore, mosquito densities recorded in SCC in 2008–2009 were near as high as those previously observed in Recife [7], with high levels of rainfall (2455 mm per year). This apparent contradiction is possibly due to other factors such as the habits of domestic water storage and spatial distribution of rainfall associated with the terrain conditions, more relevant for *Aedes* establishment and proliferation than rainfall intensity taken as an isolated variable. Another factor possibly contributing for mosquito abundance in SCC was the great number of large cisterns (at least one per house), which in spite of being well covered, were not mosquito-proof. These underground potential breeding sites are the main reason for the large amount of larvicide used in the city. It is possible that an exaggerated confidence on the use of larvicides for controlling mosquito larvae in those habitats account for the maintenance of the high infestation levels observed, since (i) a wide diversity of other types of breeding site are available for mosquito oviposition as is the case in most tropical urban areas, and (ii) *Ae. aegypti* population became resistant to temephos, the organophosphorus larvicide regularly applied mainly in cisterns for more than one decade (Melo-Santos, unpublished data).

The mosquito surveys carried out throughout 2009 indicated that April-June was a period of high mosquito reproduction in SCC. It is reasonable to speculate that the existence of a period with higher mosquito abundance is partly due to other available larval habitats besides the cisterns, as these are located indoors and filled with tap water, being unaffected by rainfall. A few longitudinal studies employing trap-based survey methods to quantify mosquito abundance allowed the assessment of spatial

and seasonal patterns of *Ae. aegypti/Ae. albopictus* populations density influenced by climatic factors and by the availability of water-filled objects affected by rainfall [7,8,11,24–30].

Investigating possible factors associated to fluctuations in *Ae. aegypti* egg densities along two years in Recife, we observed [31] that although variations in temperature were subtle throughout the year in Pernambuco, egg densities were lower in months of lower temperature, supporting the assumption that temperature possibly also contributed for higher mosquito densities in SCC where annual mean temperature was ~1°C higher than in Ipojuca.

The spatial distribution of eggs displayed in smoothed maps for both villages showed to be a sensitive indicator of spatial concentration spots of vector reproduction activity (blood-fed active females), thus being able to indicate priority areas for control. This pattern of density hot spots was observed in all areas where the SMCP-*Aedes* was applied - Santa Cruz do Capibaribe, a district of Ipojuca, seven Recife neighborhoods and the Fernando de Noronha Island, showing a clear focal distribution of the *Ae. aegypti* population. This is consistent with a highly clustered spatial pattern observed in several studies on the distribution of *Ae aegypti* carried out in urban areas of different countries and using different sampling methods [32–37].

The scale at which *Aedes* reproduction concentrations were evident reinforced the well-known fact that the household environment is a determinant factor for vector breeding activity. This is in agreement with the unique habits of *Ae. aegypti*, a species that seldom moves out from the household with adequate conditions for blood feeding, resting, eggs laying and post-embryonic development [2,24,38]. The maps worked as an easy-to-use decision-support tool, as they were easily read and discussed by the health staff including the field workers, helping to decide how and where to direct further control efforts.

The results of this study stress the importance of using an efficient surveillance system, not only to know the real scenario of infestation, but specially to understand its local determinants to define appropriated control strategies.

It is worth mentioning that the scenario of high infestation by the dengue vector, which indicated a mature step of colonization

May 2008   **June 2009**   **May 2010**   **June 2011**

Egg Densities
Minor
Major

0   200   400   600
Meters
UTM/SAD69

Figure 7. Smoothed spatial distribution of *Aedes* eggs at the high density period (May-June) in Ipojuca, Brazil, 2008–2011. Kernel maps based on the number of eggs deposed in one sentinel-ovitrap during one month. Data from a sentinel network of 75 ovitraps distributed over a 0.8 km² urban area. Mass suppression of eggs using 2700 control-ovitraps started in October 2009.

(long-lasting establishment), was observed in areas that have been under systematic bimonthly use of larvicide for fourteen years. While monitoring mosquito population densities for two years in different Brazilian cities subjected to regular temephos-larviciding, as Santa Cruz and Ipojuca (2008–2009) and Recife (2004–2005), no reduction on mosquito population was observed. These findings, together with those performed in other Brazilian cities (Salvador-BA, Manaus-AM, and Rio de Janeiro-RJ) also showing high infestation by *Ae. aegypti* [8,39,40] after a long period of regular larviciding, indicate a lack of effectiveness of the classical larvicide-based *Aedes* control strategy applied in Brazil since 1998.

Integrated control actions targeting mosquito eggs and larvae with social participation impacted the mosquito population and caused an overall reduction of around 77% and 90% in egg densities, in the study sites, as compared to the pre-intervention period. This important decrease in population size was only reached at the end of two years of sustained control pressure imposed mainly through continuous eggs trapping, which resulted in more than 8.5 million eggs burned out. In both locations a clear decrease in egg density after the first mass-trapping cycle was efficiently detected by the sentinel ovitrap network in the following month. A sustained gradual reduction was subsequently observed through the sentinel traps. However, dispersion of the mosquito population remained surprisingly unchanged in SCC and slightly oscillated in Ipojuca for at least one year after the beginning of the control actions. Even at the end of two years, more than 80% of the traps remained positive for *Aedes* eggs in both sites, although many of them received just a few eggs per month. These observations clearly indicate that the percentage of positive houses is not a sensitive indicator either for vector population size or for host-vector exposure.

It should be mentioned that the contribution of this form of massive removal of eggs for the success in reducing vector population goes beyond the physical destruction of eggs laid in C-ovt traps. Constructed from plastic bottles the C-ovt embodied a symbolic value in addition to its didactic role and mobilized the community in all stages of the process. This included the active participation of people, mainly students and housewives, (i) in collecting bottles, (ii) in the handmade construction, distribution and installation of traps, (iii) in explaining how they work and persuading householders to adhere and (iv) to visually observe, after each cycle, the eggs deposited in the traps in their houses. With small capital investment, the process contributed to the knowledge of mosquito biology and gave people a real opportunity to work directly in the project with access of visible results and thus to contribute to prevent the proliferation of mosquitoes in their home environment, more so than the large sums spent on publicity in general.

In addition to this "lure and kill" strategy, larvae elimination through predacious fishes added to thousands of cisterns in SCC certainly contributed for reducing mosquito population and, very importantly, prevented the water stored in those cisterns from being contaminated with organophosphorous insecticides.

Although mosquito aspirations inside the houses were not performed at the frequency and length as planned, more than 3200 adult *Ae aegypti* were collected along three years from 17 health facilities in SCC. Considering that public locations could be the main source of DENV spread [41], data on the presence of mosquito females in Health Care Unities, a place where people with DENV infection suspicion proceed to, is of high epidemiological significance. The presence of *Ae. aegypti* in those places was very persistent, with the vector being found in most of the HCU in almost all surveys along the first (preintervention) year of aspirations. Importantly, 100% clearance was observed only at the end of the third year. The results from indoor aspirations: *i)* provided an additional evidence of the high infestation level in SCC, and the sharp reduction in mosquito population after two years of sustained control pressure, as shown by the sentinel-ovitrap network; *ii)* very possibly contributed to limit the number of dengue infection cases; and *iii)* contributed to improve the welfare of patients by suppressing a large number of *Culex* mosquitoes in addition to *Aedes*.

It is known that for *r*-strategist species, like most mosquitoes, a very high proportion of individuals have to be constantly killed in order to impose a strong and sustained control pressure to reduce the target population, as these species are able to recover quickly even after catastrophic mortality [42]. The ways to reach the required control pressure seems to be quite different for the main urban disease vectors *Culex* and *Aedes*. For the lymphatic filariais vector *Cx. quinquefasciatus* (Say) which lay grouped eggs in well-defined sites and use an oviposition aggregation pheromone, continuous pressure can be obtained through larval control, as shown in field trials [43–47]. Differently, for the dengue vector *Ae aegypti*, its unique oviposition habits result in a wide spread of its progeny in a very large amount of several breeding sites, and the integration of methods directed against at least two mosquito developmental stages seems to be essential for an effective vector control. In spite of enormous efforts to control the dengue vectors in recent decades, very few successful real-scale experiences have been reported in the scientific literature. Rare examples are the elimination of dengue transmission through community-based biological control of the vector in Vietnamese villages [48,49]; integrated control actions including sticky and lethal ovitraps, source reduction and residual spraying directed at dengue cases on Thursday Island/Australia [15]; and the integration of source reduction, copepods, Bti and lethal ovitraps in rural villages in Thailand [17]. Social engagement and the use of control methods targeting more than one mosquito developmental stage are the main features of these examples.

We evaluate that the success in markedly reducing *Ae. aegypti* populations in SCC and IpojucaMS can be attributed to the appropriate choice of control measures for sustained mass elimination guided by a sensitive and easy-to-use surveillance system, which was capable to motivate the health staff. Despite a much higher infestation degree in the pre-intervention period, the impact of interventions on mosquito population was greater in SCC. Although it is difficult to measure the contribution of people's involvement to the program's success, one factor seemed evident: the municipal authorities and health managers demonstrated to be more intensely involved and thus to create a greater commitment of the health staff in SCC than in Ipojuca. One of the factors possibly contributing to the greater commitment of the SCC population was that the actions covered an urban area where over 97% of the municipal inhabitants live, improving the notion of their help in reducing dengue transmission. As a consequence, control measures were more intense and residents possibly received more stimuli for positive actions to avoid mosquito breeding. Furthermore, social segments such as school communities participated more actively in control actions in SCC.

In practical terms, the observed reduction in egg density can be expressed as a decrease in the number of *Ae aegypti* active females ovipositing over a thousand eggs per month in each house, as observed in SCC, to only one or few females laying approximately one hundred eggs per month in 80% of the houses. The association of DENV transmission with the abundance of *Ae aegypti* has been demonstrated [50], however the threshold of vector density under which no virus transmission would occur is yet to be defined. This threshold is supposedly very low because by feeding preferentially and frequently on human blood *Ae. aegypti* exponentially boosts the basic reproduction rate of virus transition [2]. Historical series on dengue incidence comparing SCC and other cities within the same geographical region in Pernambuco are being analyzed in order to investigate whether the marked decrease in vector population size has any impact on DENV transmission.

Several trap models have been evaluated in regards to their potential to be used in large scale strategies to promote mass destruction of adult mosquitoes or eggs, and the development of promising genetic vector control techniques [51–54] enhances the options of appropriate approaches for an integrated management of dengue vector populations. In our opinion, the mass elimination of *Aedes* eggs should be considered as an efficient option because it may prevent mosquito population boosts resulting from sudden hatching of large amounts of eggs that have been maintained viable in the environment during dry season [7], and because ovitraps handling is simple and their use favor social engagement to the control program. Furthermore, by mass destructing *Aedes* eggs from endemic areas, DENV viruses present in the eggs may also be eliminated. Although it has not been experimentally demonstrated, it is possible that besides being the biological vectors of DENV, *Ae. aegypti* and *Ae. albopictus* could also play a role

as the reservoirs of dengue viruses in urban environments resulting in the persistence of the virus in the environment between periodic transmission periods through infected eggs.

Acknowledgments

We thank the Health staff of Santa Cruz do Capibaribe and Ipojuca City Halls for the precious collaboration. The contributions of Claudio Duarte, Health Secretary of Pernambuco State, made this study possible and we are particularly grateful to him also for his helpful suggestions during the progress of this work. We also thank the researchers C. Codeço, N. Honorio and C. Barcellos from Fiocruz-RJ for their contributions in discussions on the study design. Special thanks also to Mércia C. Santana for technical support.

Author Contributions

Conceived and designed the experiments: LNR JCS RVA MAVMS WVS CMNR AMVM CB MSC AFF PJR. Performed the experiments: LNR RVA JCS MAVMS CMNR JCSS CMFO RMRB. Analyzed the data: LNR WHB RVA JCS WVS CMNR JCSS AMVM CMFO RMRB PJR LCCM MSC AFF MABR MGNMS. Contributed reagents/materials/analysis tools: MABR MGNMS WHB AMVM WVS. Wrote the paper: LNR WVS AMVM AFF MSC JCS CB LCCM PJR WHB. Conceived and developed the system to help count Aedes eggs: MABR MGNMS.

References

1. WHO (2011) *Comprehensive Guidelines for the Prevention and Control of Dengue and Dengue Haemorrhagic Fever.* World Health Organization, Regional Office for South-East Asia. SEARO Technical Publication Series No.60.

2. Scott TW, Takken W (2012) Feeding strategies of anthropophilic mosquitoes result in increased risk of pathogen transmission. Trends Parasitol 28: 114–121.

3. Barreto ML, Teixeira MG, Bastos FI, Ximenes RAA, Barata RB, et al. (2011) Successes and failures in the control of infectious diseases in Brazil: social and environmental context, policies, interventions, and research needs. The Lancet 377 Issue 9780: 1877–1889.

4. Bellini R, Carrieri M, Burgio G, Bacchi M (1996) Efficacy of different ovitraps and binomial sampling in *Aedes albopictus* surveillance activity. JAMCA 12: 632–636.

5. Polson KA, Curtis C, Seng CM, Olson JG, Chantha N, et al (2002) The Use of Ovitraps Baited with Hay Infusion as a Surveillance Tool for *Aedes aegypti* Mosquitoes in Cambodia. Dengue Bull 26: 178–184.

6. Santos SRA, Melo-Santos MAV, Regis L, Albuquerque CMR (2003) Field evaluation of ovitraps consociated with grass infusion and *Bacillus thuringiensis* var. *israelensis* to determine oviposition rates. Dengue Bull 27: 156–162.

7. Regis L, Monteiro AM, Melo-Santos MAV, Silveira JC, Furtado AF, et al (2008) Developing new approaches for detecting and preventing *Aedes aegypti* population outbreaks: basis for surveillance, alert and control system. Mem Inst Oswaldo Cruz 103: 50–59.

8. Honório NA, Codeço CT, Alves FC, Magalhães MAFM, Lourenço-de-Oliveira R (2009) Temporal distribution of *Aedes aegypti* in different districts of Rio de Janeiro, Brazil, measured by two types of traps. J Med Entomol 46: 1001–1014.

9. Obenauer PJ, Kaufman PE, Allan SA, Kline DL (2009) Infusion-Baited Ovitraps to Survey Ovipositional Height Preferences of Container-Inhabiting Mosquitoes in Two Florida Habitats. J Med Entomol 46: 1507–1513.

10. Ritchie SA, Rapley LP, Williams C, Johnson PH, Larkman M, et al (2009) A lethal ovitrap-based mass trapping scheme for dengue control in Australia: I. Public acceptability and performance of lethal ovitraps. Med Vet Entomol 23(4): 295–302.

11. Albieri A, Carrieri M, Angelini P, Baldacchini F, Venturelli CR, et al (2010) Quantitative monitoring of *Aedes albopictus* in Emilia-Romagna, Northern Italy: cluster investigation and geostatistical analysis. Bull Insectol 63: 209–216.

12. Kaplan L, Kendell D, Robertson D, Livdahl T, Khatchikian C (2010) *Aedes aegypti* and *Aedes albopictus* in Bermuda: extinction, invasion, invasion and extinction. Biol Invasions 12: 3277–3288.

13. Khatchikian C, Sangermano F, Kendell D, Livdahl T (2011) Evaluation of species distribution model algorithms for fine-scale container-breeding mosquito risk prediction. Med Vet Entomol 25: 268–275.

14. Carrieri M, Albieri A, Angelini P, Baldacchini F, Venturelli C, et al (2011) Surveillance of the chikungunya vector *Aedes albopictus* (Skuse) in Emilia-Romagna (Northern Italy): organizational and technical aspects of a large scale monitoring system. J Vector Ecol 36: 108–118.

15. Montgomery BL, Ritchie SA, Hart AJ, Long SA, Walsh ID (2005) Dengue intervention on Thursday Island (Torres Strait) 2004: a blueprint for the future? Arbovirus Research in Australia 9: 268–273.

16. Williams CR, Ritchie SA, Richard RC, Zborowski P (2005) Development and application of 'lure and kill' strategies for the dengue vector *Aedes aegypti* in Australia. Arbovirus Research in Australia, 9: 397–402.

17. Kitayapong P, Yoksan S, Chansang C, Bhumiratana A (2008) Suppression of dengue transmission by application of integrated vector control strategies at seropositive GIS-based foci. Am J Trop Med Hyg 78: 70–76.

18. Rapley LP, Johnson PH, Williams CR, Silcock RM, Larkman M, et al (2009) A lethal ovitrap-based mass trapping scheme for dengue control in Australia: II. Impact on populations of the mosquito *Aedes aegypti*. Med Vet Entomol 23: 303–316.

19. Regis L, Souza WV, Furtado AF, Fonseca CD, Silveira JC Jr, et al (2009) An entomological surveillance system based on open spatial information for participative dengue control. An Acad Bras Ciências 81: 655–662.

20. Instituto Brasileiro de Geografia e Estatística (2010) Available from: http://www.ibge.gov.br/censo2010 (accessed 12.07.2012).

21. Silva MGN, Rodrigues MAB, Araujo RE (2011) *Aedes aegypti* egg counting system. *In:* 33rd Annual International Conference of the IEEE Engineering in Medicine and Biology Society (EMBC 11) 2011, Boston.

22. Lenhart AE, Walle M, Cedillo H, Kroeger A (2005) Building a better ovitrap for detecting *Aedes aegypti* oviposition. Acta Trop 96: 56–59.

23. Bailey TC, Gatrell AC (1995) Interactive spatial data analysis. Essex: Longman Scientific. 413 pp.

24. Scott TW, Morrison AC, Lorenz LH, Clark GG, Strickman D, et al. (2000) Longitudinal studies of *Aedes aegypti* (L.) (Diperta: Culicidae) in Thailand and Puerto Rico: Population dynamics. J Med Entomol 37: 77–88.

25. Stein M, Oria GI, Almirón WR, Willener JA (2005) Fluctuación estacional de *Aedes aegypti* in Chaco Province, Argentina. Rev Saude Publ 39: 559–564.

26. Facchinelli L, Valerio L, Pombi M, Reiter P, Costantini C, et al (2007) Development of a novel sticky trap for container breeding mosquitoes and evaluation of its sampling properties to monitor urban populations of *Aedes albopictus*. Med Vet Entomol 21: 183–195.

27. Gama RA, Silva EM, Silva IM, Resende MC, Eiras AE (2007) Evaluation of the sticky MosquiTRAP for detecting *Aedes* (Stegomyia) *aegypti* (L.) (Diptera: Culicidae) during the dry season in Belo Horizonte, Minas Gerais, Brazil. Neotrop Entomol 36: 294–902.

28. Vezzani D, Carbajo AE (2008) *Aedes aegypti, Aedes albopictu*, and dengue in Argentina: current knowledge and future directions. Mem Inst Oswaldo Cruz 103: 66–74.

29. Bellini R, Albieri A, Balestrino F, Carrieri M, Porretta D (2010) Dispersal and Survival of *Aedes albopictus* (Diptera: Culicidae) Males in Italian Urban Areas and Significance for Sterile Insect Technique Application. J Med Entomol 47: 1082–1091.

30. Giatropoulos A, Emmanouel N, Koliopoulos G, Michaelakis A (2012) A study on distribution and seasonal abundance of *Aedes albopictus* (Diptera: Culicidae) population in Athens, Greece. J Med Entomol 49: 262–269.

31. Bonat WH, Ribeiro PJ Jr, Dallazuanna HS, Regis LN, Monteiro AM, et al (2009) Investigando fatores associados a contagens de ovos de *Aedes aegypti* coletados em ovitrampas em Recife/PE. Rev Bras Biom 27: 519–537.

32. Getis A, Morrison AC, Gray K, Scott TW (2003) Characteristics of the spatial pattern of the dengue vector, Aedes aegypti, in Iquitos, Peru. Am J Trop Med Hyg 69: 494–505.

33. Chansang C, Kittayapong P (2007) Application of mosquito sampling count and geospatial methods to improve dengue vector surveillance. Am J Trop Med Hyg 76: 820–826.

34. Williams CR, Long SA, Webb CE, Bitzhenner M, Geier M, et al (2007) *Aedes aegypti* population sampling using BG-Sentinel traps in North Queensland Australia: statistical considerations for trap deployment and sampling strategy. J Med Entomol 44: 345–350.

35. Fernandes MT, Da Costa Silva W, Souza-Santos R (2008) Identification of key areas for *Aedes aegypti* control through geoprocessing in Nova Iguaçu, Rio de Janeiro State, Brazil. Cad Saude Publica 24: 70–80.

36. Jeffery JAL, Yen NT, Nam VS, Nghia LT, Hoffmann AA, et al (2009) Characterizing the *Aedes aegypti* population in a Vietnamese village in preparation for a Wolbachia-based mosquito control strategy to eliminate Dengue. PLoS NTD 3 (11): e0000552.

37. Barrera R (2011) Spatial stability of Adult *Aedes aegypti* populations. Am J Trop Med Hyg 85: 1087–1092.

38. Harrington LC, Scott TS, Lerdthusnee K, Coleman RC, Costero A, et al (2005) Dispersal of the dengue vector *Aedes aegypti* within and between rural communities. Am J Trop Med Hyg 72: 209–220.

39. Morato VCG, Teixeira MG, Gomes AC, Bergamaschi DP, Barreto M (2005) Infestation of *Aedes aegypti* estimated by oviposition traps in Brazil. Rev Saúde Pub 39: 553–558.

40. Ríos-Velasquez CM, Codeço CT, Honório NA, Sabrosa PS, Moresco M, et al. (2007) Distributions of dengue vectors in neibghborhoods with different urbanization types of Manaus, state of Amazonas, Brazil. Mem Inst Oswaldo Cruz 102: 617–623.

41. Medeiros LCC, Castilho CAR, Braga C, Souza WV, Regis L, et al (2011) Modeling the Dynamic Transmission of Dengue Fever: Investigating Disease Persistence. PLoS Negl Trop Dis 5(1): e942. doi:10.1371/journal.pntd.0000942.

42. Schofield C (1991) Vector population responses to control intervention. Ann Soc Belg Med Trop 71: 201–217.

43. Hougard J-M, Mbentengam R, Lochouam L, Escafhe H, Darriet F, et al (1993) Lutte contre *Culex quinquefasciatus* par *Bacillus sphaericus*: resultats d'une campagne pilote dans une grande agglomeration urbaine d'Afrique equatoriale. Bull WHO 1: 367–375.

44. Kumar A, Sharma VP, Thavaselvam D, Sumodan PK, Kamat RH, et al (1996) Control of *Culex quinquefasciatus* with *Bacillus sphaericus* in Vasco City, Goa. JAMCA 12: 409–413.

45. Barbazan P, Baldet T, Darriet F, Escafhe H, Haman Djoda D, et al (1997) Control of *Culex quinquefasciatus* (Diptera: Culicidae) with *Bacilus sphaericus* in Maroua, Cameroon. JAMCA 13: 263–269.

46. Silva-Filha MH, Regis L, Oliveira CF, Furtado A (2001) Impact of a 26-month *Bacillus sphaericus* trial on the pre-imaginal density of *Culex quinquefasciatus* in an urban area of Recife, Brazil. JAMCA, 17: 45–50.

47. Regis L, Oliveira CMF, Silva-Filha MH, Silva SB, Maciel A, et al (2000) Efficacy of *Bacillus sphaericus* in control of the filariasis vector *Culex quinquefasciatus* in an urban area of Olinda, Brazil. Trans R Soc Trop Med Hyg 94: 488–492.

48. Kay B, Nam VS (2005) New strategy against *Aedes aegypti* in Vietnam. Lancet 365: 613–17.

49. Kay BH, Tuyet Hanh TT, Le NH, Quy TM, Nam VS, et al (2010) Sustainability and Cost of a Community-Based Strategy Against *Aedes aegypti* in Northern and Central Vietnam. Am J Trop Med Hyg 82: 822–830.

50. Mammen MP, Pimgate C, Koenraadt CJM, Rothman AL, Aldstadt J, et al (2008) Spatial and temporal clustering of dengue virus transmission in Thai villages. PLoS Med 5: 1605–1616.

51. Beech C, Vasan S, Quinlan M, Capurro ML, Alphey L, et al (2009) Deployment of innovative genetic vector control strategies: Progress on Regulatory and Biosafety Aspects, Capacity Building and Development of Best Practice Guidance. Asia-Pac J Mol Biol 17: 75–85.

52. De Barro P, Murphy B, Jansen C, Murray J (2011) The proposed release of the yellow fever mosquito, *Aedes aegypti*, containing a naturally occurring strain of *Wolbachia pipientis*, a question of regulatory responsibility. Journal für Verbraucherschutz und Lebensmittelsicherheit (Journal for Consumer Protection and Food Safety) 6 (suppl 1): S33–S40.

53. Harris A, Nimmo D, McKemey A, Kelly N, Scaife S, et al (2011) Field performance of engineered male mosquitoes. Nature Biotechnol 29: 1034–1037.

54. Wise De Valdez M, Nimmo D, Betz J, Gong H, James AA, et al (2011) Genetic elimination of dengue vector mosquitoes. PNAS 108: 4772–4775.

12

Pleiotropic Impact of Endosymbiont Load and Co-Occurrence in the Maize Weevil *Sitophilus zeamais*

Gislaine A. Carvalho[1], Juliana L. Vieira[2], Marcelo M. Haro[2], Alberto S. Corrêa[3], Andrea Oliveira B. Ribon[1], Luiz Orlando de Oliveira[1], Raul Narciso C. Guedes[2]*

1 Departamento de Bioquímica e Biologia Molecular, Universidade Federal de Viçosa, Viçosa, MG, Brazil, **2** Departamento de Entomologia, Universidade Federal de Viçosa, Viçosa, MG, Brazil, **3** Departamento de Entomologia e Acarologia, Escola Superior de Agricultura "Luiz de Queiroz", Universidade de São Paulo, Piracicaba, São Paulo, Brazil

Abstract

Individual traits vary among and within populations, and the co-occurrence of different endosymbiont species within a host may take place under varying endosymbiont loads in each individual host. This makes the recognition of the potential impact of such endosymbiont associations in insect species difficult, particularly in insect pest species. The maize weevil, *Sitophilus zeamais* Motsch. (Coleoptera: Curculionidae), a key pest species of stored cereal grains, exhibits associations with two endosymbiotic bacteria: the obligatory endosymbiont SZPE ("*Sitophilus zeamais* Primary Endosymbiont") and the facultative endosymbiont *Wolbachia*. The impact of the lack of SZPE in maize weevil physiology is the impairment of nutrient acquisition and energy metabolism, while *Wolbachia* is an important factor in reproductive incompatibility. However, the role of endosymbiont load and co-occurrence in insect behavior, grain consumption, body mass and subsequent reproductive factors has not yet been explored. Here we report on the impacts of co-occurrence and varying endosymbiont loads achieved via thermal treatment and antibiotic provision via ingested water in the maize weevil. SZPE exhibited strong effects on respiration rate, grain consumption and weevil body mass, with observed effects on weevil behavior, particularly flight activity, and potential consequences for the management of this pest species. *Wolbachia* directly favored weevil fertility and exhibited only mild indirect effects, usually enhancing the SZPE effect. SZPE suppression delayed weevil emergence, which reduced the insect population growth rate, and the thermal inactivation of both symbionts prevented insect reproduction. Such findings are likely important for strain divergences reported in the maize weevil and their control, aspects still deserving future attention.

Editor: Wolfgang Arthofer, University of Innsbruck, Austria

Funding: The financial support was provided by the National Council of Scientific and Technological Development (CNPq, Brazilian Ministry of Science and Technology), CAPES Foundation (Brazilian Ministry of Education), and Minas Gerais State Foundation of Research Aid (FAPEMIG). The funders had no role in study design, data collection and analysis, decision to publish, or preparation of the manuscript.

* Email: guedes@ufv.br

Introduction

Symbiosis is the result of intricate ecological relationships. Such intricacy may lead to shifts in the selection pressure over an organism, which may result in advantage or disadvantage to at least one of the interacting organisms of different species [1–3]. Intracellular bacteria are common endosymbionts of arthropods, either in obligatory or facultative associations, that live within the cells of their hosts [4,5]. Not only nutrition-involved obligatory endosymbionts, such as *Buchnera* and *Wigglesworthia*, are of recognized importance in arthropods but also facultative endosymbionts, such as *Wolbachia*, *Hamiltonella*, and *Serratia*, among others [2,3,5–8]. Approximately 10% of insect species exhibit a primary (i.e., obligatory) endosymbiont, while an estimated 40% of insect species host some *Wolbachia* strain [9,10].

The specialized and unbalanced diets of several arthropod species is an indication of the potential importance of their endosymbionts, which frequently play a fundamental role in complementing nutrition in their host, allowing host survival in novel environments and under alternate food source [2,3,6–8,9]. Although such a role is likely a pivotal innovation in arthropod evolution, the specific roles of the majority of their endosymbionts remains unknown [2,3,9]. The suppression or inactivation of endosymbionts shed some light on this matter, as exemplified by the *Wolbachia*-mediated fitness increase and parasitism protection of whiteflies [11], and high temperature tolerance and parasitoid resistance provided by *Serratia* and *Hamiltonella* [12–15].

Understanding the role of endosymbionts in the behavioral, ecological and evolutionary processes of arthropods is no easy task. This is so not only because of individual trait variation within an arthropod population [16] but also because an arthropod may host varying loads of more than one endosymbiont, confounding and/or masking their impact and importance in the host individual. Weevils in the genus *Sitophilus*, which encompasses three grain weevil species of key importance for stored grain protection (*Sitophilus granarius*, *S. oryzae*, and *S. zeamais*), host both primary (obligatory) and secondary (facultative) endosymbionts, making then suitable models to study the roles of co-existing symbionts and their eventual relevance for pest control [17–23].

Grain weevils exploit a restrictive food source, cereal grains, and must complete their development within the grain kernel. The association between grain weevils and their primary endosymbiont SPE (*Sitophilus* Primary Endosymbiont) is hypothesized to be an important requirement allowing survival under such conditions [19,23,24]. However, physiological differences do exist among weevil strains, allowing strain variation in how well they are able to cope with cereal amylase inhibitors and insecticide exposure [25,26]. SPE was initially detected in the rice weevil (*S. oryzae*), where it is referred to as SOPE (*Sitophilus oryzae* Primary Endosymbiont; = *Candidatus* Sodalis pierantonius str. SOPE), and subsequently in the granary and maize weevils (*S. granarius* and *S. zeamais*), where it is referred to as SGPE (*Sitophilus granarius* Primary Endosymbiont; = *Candidatus* Sodalis pierantonius) and SZPE (*Sitophilus zeamais* Primary Endosymbiont; = *Candidatus* Sodalis pierantonius str. SZPE) [11–12,18–19,27], respectively.

SPE seems to provide vitamins to its weevil hosts, assisting in their amino acid metabolism, in addition to interacting with mitochondrial oxidative phosphorylation, thus enhancing respiration and mitochondrial enzyme activity in the host insect [17,28–30]. Such effects of SPE may affect development, immune response and flight activity in their weevil hosts [17,31]. Curiously, however, the focus of previous SPE studies has remained on the genetics and molecular biology of these endosymbionts [18,32–35], and not on their behavioral or physiological consequences in the weevil hosts. However, the co-occurrence of SPE and *Wolbachia* in cereal weevils [18–20], raises questions regarding their interaction and potential impact on this host species. Here we recognized the presence of both SZPE and *Wolbachia* in the maize weevil, subjected the colonized weevil hosts to different treatments for endosymbiont inactivation/suppression, assessed the impacts of endosymbiont loads of either one or both symbionts, and analyzed how they affect host reproductive fitness following a structured hierarchical approach. Past studies focused on the simultaneous presence/absence of such endosymbionts [17,28–30], while here presence was quantified and associated with behavioral and physiological traits potentially affecting the insect reproductive output.

Materials and Methods

Ethic Statement

This study did not involve any endangered or protected species. The insect species studied is a cereal pest species from a colony maintained in laboratory, where the experiments were performed, and no specific permission was required.

Insects

The insects were obtained from an insecticide-susceptible laboratory colony of the maize weevil (*S. zeamais*) that has been maintained in whole maize kernels free of insecticide residues since the mid-1980s [16,36,37]. The insects are maintained under controlled conditions of $27\pm2°C$ temperature, $70\pm10\%$ relative humidity, and a 12 h photoperiod, the same conditions employed in our bioassays.

Endosymbiont Quantification and Inactivation/ Suppression

PCR amplification for endosymbiont load quantification. The endosymbiont load in individual adult weevils was quantified using quantitative polymerase chain reaction (qPCR) after individual DNA extraction. The total genomic DNA of adult maize weevils (> one-week old) was extracted following Clark [38]. PCR amplification was performed in a total volume of 12 μL, and consisted of 1 μL DNA, 6.0 μL SYBR Green Master Mix (2x; Applied Biosystems, Foster City, CA, USA), and 200 nM of each primer (forward [F] and reverse [R] primers for the 16S rRNA gene of SZPE and for the 16S rRNA gene of *Wolbachia*). The following sets of primers were used: (1) 5′-AGACTCTAGCCTGCCAGTTT-3′ (F primer) and 5′-AGCTGTAATACAGAAAGTAAA-3′ (R primer) for the 16S rRNA of SZPE, generating a 145 bp DNA fragment; and (2) 5′-CGGGGGAAAAATTTATTGCT-3′ (F primer) and 5′-TAG-GAGTCTGGACCGTATCT-3′ (R primer) for the 16S rRNA of *Wolbachia*, generating a 198 bp DNA fragment. The design of the oligonucleotide pairs was performed using Primer3 Plus software [39], following the requirements of real time PCR. No-template controls, containing nuclease-free water, were included in each run.

The PCR was performed on an ABI Prism 7500 Sequence Detection System (Applied Biosystems, Foster City, CA, USA). The PCR cycles used the following conditions: 2 min at 50°C followed by 10 min denaturation at 95°C, 40 cycles of 45 s denaturation at 95°C, and annealing and extension at 60°C for 30 s. After the 40 cycles of amplification, all of the samples were subjected to gradual denaturation to elaborate the dissociation curve. The samples were heated at 1°C increments every 30 s from 60 to 94°C. The melting curves (65°C to 97°C) were obtained at the end of each reaction to ascertain the specificity of the PCR product. The standard curve was plotted using the following eight dilutions of the corresponding plasmids of each gene fragment: 2.92×10^1, 2.92×10^2, 2.92×10^3, 2.92×10^4, 5.03×10^4, 1.08×10^6, 2.08×10^6, and 3.08×10^6 copies/μL. The number of fragment copies of each gene was estimated using the standard curve, and the amount of total DNA (host+endosymbionts) in each sample was used to standardize the results of the number of copies of the 16S rRNA gene fragments [40]. The one-point calibration method (OPC) was used to correct the obtained values, minimizing the differences between plasmid and total DNA [41]. The results were presented as number of copies per ng of DNA. Three independent biological samples were analyzed in triplicate, and their endosymbiont load was quantified in independent amplifications. The same methods for quantifying endosymbiont load were performed on the adult progeny of adult weevils subjected to each of the different endosymbiont suppression treatments, in addition to adult weevils without such suppression (control).

Endosymbiont reduction. Two approaches were used for endosymbiont load reduction: inactivation via thermal treatment and suppression via antibiotic ingestion. The thermal treatment was based on the exposure of adult weevils to high temperature and humidity [19]. For this purpose, maize weevil adults (over one week old) were transferred to transparent plastic containers (250 mL) half-filled with whole maize grains and maintained for 21 days in an environmental chamber under controlled conditions of $37\pm2°C$ and $90\pm5\%$ relative humidity.

Endosymbiont suppression in adult weevils was performed by providing antibiotics through ingested water to insects subjected to 24 h of hydric stress. Hydric stress was achieved by individually containing one-week old adults in perforated Eppendorf tubes placed within glass desiccators (3,000 cm^3) at 1% relative humidity ($27\pm2°C$ and 12 h photoperiod) for 24 h, after which they avidly ingest water from water droplets [42]. The weevils were subsequently transferred individually to Petri dishes (9 cm diameter) containing a 5 μL droplet of water-diluted antibiotic (either amoxicillin, ciprofloxacin, rifamycin, or tetracycline, at

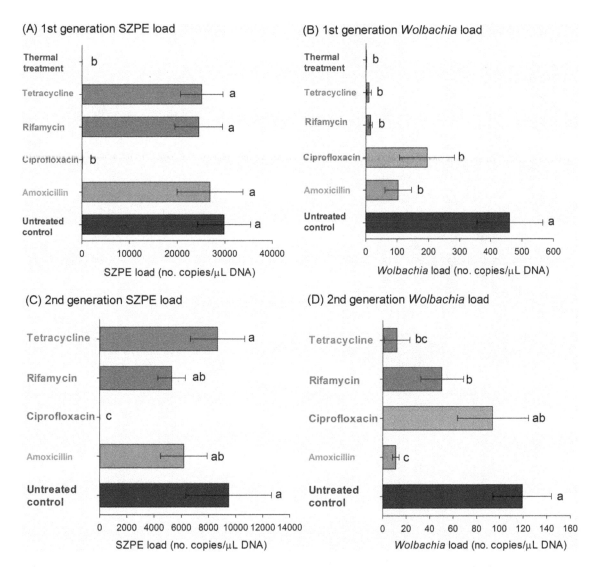

Figure 1. Load (± SE) of the endosymbionts SZPE (A, C) and *Wolbachia* (B, D) in F₁ (A, B) and F₂ progenies (C, D) of maize weevils (*Sitophilus zeamais*) exposed to different endosymbiont-reducing treatments. Means followed by the same letter in a histogram are not significantly different by Tukey's HSD test ($P<0.05$).

25 mg/mL). The antibiotics were obtained from Medley (Campinas, SP, Brazil), Genfar (Bogotá, Colômbia), Legrand (Campinas, SP, Brazil), and Bristol-Myers Squibb (São Paulo, SP, Brasil), respectively, at their available commercial formulations (Amoxicilina 250 mg, Ciprofloxacino 500 mg, Rifamicina 10 mg/ml, Tetrex 500 mg). The antibiotic concentration used was established after preliminary concentration-response bioassays using the following range of concentrations: 0, 1, 5, 10, 25, 50 and 100 mg/mL.

The insects were maintained for 40 min in the Petri dishes with the desired water-diluted antibiotic and subsequently transferred to maize contained in Petri dishes for 24 h; this procedure was repeated six times for each individual insect. The progeny of the treated insects was also subjected to the same antibiotic treatment. Therefore the antibiotic-treated insects were from the parental (P) generation when the F₁ progeny was assessed, and from the P and F₁ generations when the F₂ progeny was assessed. Only the progenies of the insects treated for one or two generations were used in the endosymbiont quantification and subsequent bioassays in order to eliminate the eventual deleterious effects of the

antibiotics themselves on insect performance. This was not possible for the thermal treatment because the treated (parental) weevil generation was unable to reproduce and the treated insects themselves were therefore used in the subsequent bioassays.

Behavioral Bioassays

Four batches of 10 adult weevils (> one week old) from each endosymbiont inactivation/suppression treatment were subjected to six behavioral bioassays assessing overall insect activity, walking activity, flight activity (take-off and free-fall flight), body righting, and death-feigning. The methods for determining overall insect activity were adapted from Tomé et al. [43], while those for the remaining bioassays were adapted from Morales et al. [16]. All methods are briefly described below.

Overall group insect activity. The batches of 10 adult weevils were transferred to a Petri dish arena (9 cm diameter) with its bottom covered with filter paper (Whatman no. 1), allowing for better traction when walking and contrast for activity determination, and its inner walls were coated with Teflon PTFE (DuPont, Wilmington, DE, USA) to prevent the insects from escaping. The

Table 1. Behavioral traits (± SE) of maize weevils (*Sitophilus zeamais*) exposed to different endosymbiont-reducing treatments.

Treatment	Overall group activity (Δ pixels/s$\times10^{-2}$)	Walking activity		Flight activity			Duration of death-feigning (s)	Length of time to body righting (s)
		Walking velocity (cm/s)	Resting time (s)	Horizontal Dislocation upon fall (cm)	No. taking off for flight	Flight height reached on take-off (cm)		
Untreated control	45.85±2.86 a	0.41±0.01 a	227.11±9.71	40.41±6.17 a	3.25±0.25 a	13.29±1.7 a	5.61±0.89 bc	3.38±0.35 b
Amoxicillin	27.81±2.13 b	0.37±0.02 bc	246.92±11.84	7.00±0.24 b	0.51±0.29 bc	0.75±0.05 bc	6.41±0.56 bc	4.41±0.41 b
Ciprofloxacin	17.42±0.73 cd	0.34±0.01 c	265.06±9.55	5.35±0.24 b	0.00±0.00 c	0.00±0.00 c	15.93±1.17 a	7.56±0.52 a
Rifamycin	25.37±1.12 bc	0.35±0.02 bc	268.63±11.94	9.53±1.01 b	1.25±0.48 b	4.38±1.72 b	5.42±0.40 bc	3.58±0.25 b
Tetracycline	22.48±3.75 bc	0.33±0.01 c	269.48±7.66	7.17±0.39 b	0.25±0.25 bc	0.25±0.25 c	3.33±0.19 c	3.41±0.24 b
Thermal treatment	11.54±1.51 d	0.31±0.01 c	280.26±11.17	5.75±0.26 b	0.00±0.00 c	0.00±0.00 c	7.86±0.95 b	6.65±0.46 a
$F_{5,18}$	26.77	4.65	2.61	160.34	21.51	18.80	19.44	25.27
P	<0.001*	0.007*	0.06	<0.001*	<0.001*	<0.001*	<0.001*	<0.001*

Means followed by the same letter in a column are not significantly different by Tukey's HSD test ($P<0.05$). Asterisks indicate significant differences among treatments by Fisher's F test from the (univariate) analyses of variance for each behavioral trait.

Table 2. Canonical loadings (between canonical structure) of the canonical axes for the behavioral traits of maize weevils (*Sitophilus zeamais*) exposed to different endosymbiont-reducing treatments.

Behavioral traits		Canonical axes				
		1st	2nd	3rd	4th	5th
Overall group activity (Δ pixels/s$\times 10^{-2}$)		**0.90**	−0.09	0.33	−0.17	0.03
Walking activity	Walking velocity (cm/s)	0.67	0.02	0.39	−0.14	0.29
	Resting time (s)	−0.55	**0.67**	−0.31	0.24	−0.28
Flight activity	Horizontal dislocation upon fall (cm)	**0.91**	0.17	−0.10	−0.02	0.01
	No. taking off for flight	**0.92**	0.01	0.08	0.23	−0.01
	Flight height reached on take-off (cm)	**0.87**	−0.04	0.15	0.39	−0.05
Duration of death-feigning (s)		−0.37	**0.82**	0.34	0.02	−0.20
Length of time to body righting (s)		−0.58	**0.75**	−0.10	0.04	0.25
$F_{appr.}$		8.29	3.83	1.69	1.14	1.02
P		<0.001*	<0.001*	0.09	0.23	0.43
Eigenvalue		79.47	14.22	1.40	0.77	0.27

Bold type indicates the main contributors of each axis and asterisks indicate the significant axes.

overall insect activity in each Petri dish arena, including walking behavior, insect interactions, and body part movements, were recorded for 10 min and digitally transferred to a computer using a video tracking system equipped with a digital CCD camera

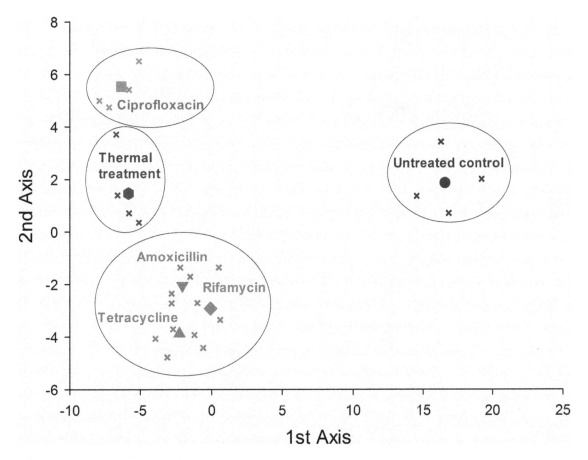

Figure 2. Ordination (CVA) diagram showing the divergence in behavioral traits of maize weevils (*Sitophilus zeamais*) exposed to different endosymbiont-reducing treatments (see Table 2). Both canonical axes are significant and account for 97.45% of the total variance explained. The solid symbols are centroids of treatments representing the class mean canonical variates and the smaller symbols of the same color represent the individual replicates. The large circles indicate clusters of treatments that are not significantly different by the approximated F-test (P< 0.05), based on the Mahalanobis (D²) distance between class means.

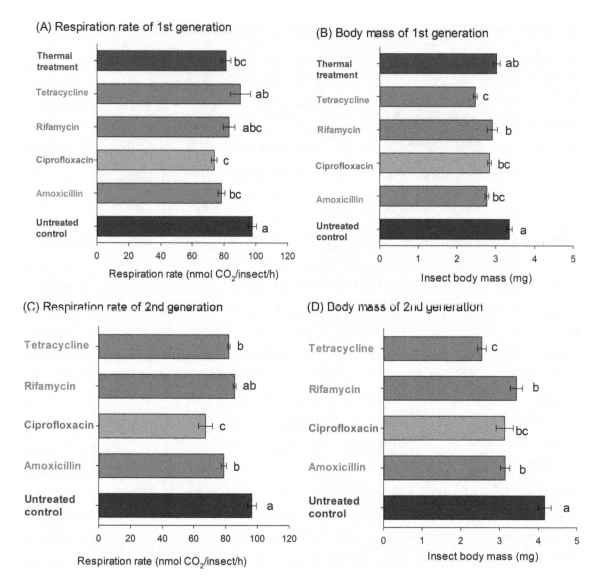

Figure 3. Respiration rate (A, C) and body mass (B, D) (± SE) of F₁ (A, B) and F₂ progenies (C, D) of maize weevils (*Sitophilus zeamais*) exposed to different endosymbiont-reducing treatments. Means followed by the same letter in a histogram are not significantly different by Tukey's HSD test ($P < 0.05$).

(ViewPoint LifeSciences, Montreal, QC, Canada). The overall insect activity was recorded as changes in pixels/s×10⁻².

Walking activity. Walking activity was recorded for individual insects for 10 min following their release into Petri dish arenas prepared as previously described. A single insect was released in the center of the arena and its movement was recorded using the same tracking system used in the assessment of overall group activity. The following characteristics were evaluated: distance walked (cm), walking velocity (cm/s), and resting time (s).

Take-off flight. A hand-made wooden square box (18 cm wide, 18 cm deep, 30 cm high) covered with a 2 mm steel frame was used. Groups of 10 adult insects were placed at the central bottom of the box within an open Petri dish (5 cm diameter) with its bottom covered with a piece of filter paper (Whatman no. 1) and its inner walls coated with Teflon PTFE. The length of time for the insects to take off for flight, the number of insects entering flight, and heights reached in flight during 10 min trials were recorded.

Free-fall flight. A hand-made wooden square box (44 cm wide, 44 cm deep, 88 cm high) with its top covered with organza tissue with a 5 cm-diameter hole in the top center was used. A chalk-covered funnel was inserted in the central hole at the top of the wooden box. The box was placed on a marked sheet of paper with concentric circles spaced 3 cm apart for one another. Each adult weevil was placed in the upper central funnel of the wooden box, and its landing site was recorded by determining its distance from the center. Each insect was released three times and the average distance of flight was determined.

Body righting. Each adult weevil was placed on its dorsum and the time taken to recover its regular ventral posture was recorded. The procedure was replicated three times, and the average determination was recorded.

Death-feigning. Death-feigning induction was performed by dorsally prodding the adult weevil with a fine-haired brush and recording the time taken for the insect to start moving after reaching its typical death-feigning (or thanatosis) posture. The

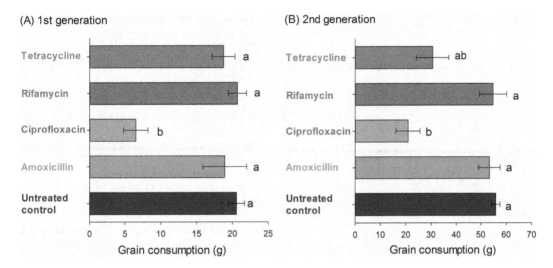

Figure 4. Grain consumption (± SE) of F₁ (A) and F₂ progenies (B) of maize weevils (*Sitophilus zeamais*) exposed to different endosymbiont-reducing treatments. Means followed by the same letter in a histogram are not significantly different by Tukey's HSD test ($P<$ 0.05).

procedure was replicated three times and the average determination was used as the duration of the death-feigning behavior.

Respiration Rate and Body Mass

The respirometry bioassays were carried out in a TR3C respirometer equipped with a CO_2 analyzer (Sable Systems International, Las Vegas, NV, USA), as detailed elsewhere [44,45]. Briefly, four replicates of 10 adult weevils from each endosymbiont-suppression treatment were gathered, and the insect body mass was determined with an analytical balance (Shimadzu AUW220D, Kyoto, Japan). The groups of 10 insects were subsequently contained in 25 mL glass respirometric chambers connected to a completely closed system. The CO_2 produced by the insects (μL CO_2/h) was determined by injecting CO_2-free air into the chambers and directing the insect-produced CO_2 to an infrared reader connected to the system. The CO_2 production in a control chamber without insects was also determined.

Developmental Rate and Grain Consumption

The experiment was performed using 1.0 L glass jars containing 300 g of whole maize. Ten adult couples of the maize weevil were released in each jar and removed 30 days later following methods by Trematerra et al. [46] and Fragoso et al. [47]. The daily and cumulative progeny emergence was assessed every other day, with four replicates (i.e., jars with ten couples and 300 g maize) for each endosymbiont-suppression treatment. The mass of grain consumption in each jar (i.e., replicate) was also determined at the end of the experiments when no more progeny emerged, 70 days after the experiment began.

Statistical Analyses

Endosymbiont load, respiration rate, adult weevil body mass, and grain consumption were subjected to analyses of variance and Tukey's HSD test when appropriate (PROC GLM; SAS v. 9) [48]. A canonical variate analysis (CVA) of the behavioral traits of weevils subjected to the different endosymbiont-suppression treatments was performed to recognize their eventual differences and the main contributing traits for observed differences (PROC CANDISC with Distance statement; SAS v. 9) [48]. Such

behavioral results were subsequently subjected to complementary analysis of variance for the individual traits assessed and Tukey's HDS test, if appropriate (PROC GLM; SAS v. 9) [48]. The normality and homoscedasticity assumptions were checked (PROC UNIVARIATE; SAS v. 9) [48], and log (x+1) transformation was necessary to stabilize the variance for the height of the flight take-off bioassay.

The daily and cumulative emergence results of weevils whose parental generation was subjected to the different endosymbiont-suppression treatments were subjected to non-linear regression analysis using the curve-fitting procedure of TableCurve 2D (Systat, San Jose, CA, USA). The significant regression models ($P<0.05$) were tested from the simplest (linear and quadratic) to more complex (peak and asymptotic) models basing the model selection on parsimony, high F-values (and mean squares), and a steep increase in R^2 with model complexity. Residual distribution was also checked for each analysis to validate parametric assumptions.

Path analysis was used to test the hypothesized relationships between endosymbiont load in host weevils of the F₁ progeny of the antibiotic-treated insects and potential direct and indirect consequences (including behavioral traits, respiration rate, body mass, grain consumption) potentially contributing to their progeny production. Only the data from endosymbiont suppression with antibiotic treatments was used in this analysis because the thermal treatment prevented assessment of the progeny of treated insects. This analysis was performed used the procedures PROC REG and PROC CALIS from SAS v. 9 [48], following guidelines provided by Mitchell [49].

Results

Endosymbiont Load and Reduction

The adult weevils subjected to the thermal treatment for endosymbiont inactivation and the F₁ and F₂ progenies of weevils subjected to antibiotic treatment for endosymbiont suppression were used to detect and quantify symbiont load based on the quantification of copy numbers of 16S rRNA gene fragments from SZPE and *Wolbachia*. The thermal treatment and the antibiotic ciprofloxacin were particularly effective in reducing the load of SZPE, while all antibiotics and the thermal treatment led to

Table 3. Summary of the non-linear regression analyses of the daily emergence curves (Fig. 5) of the F_1 and F_2 progenies of adult maize weevils (*Sitophilus zeamais*) exposed to different endosymbiont-suppression treatments via water-ingested antibiotics.

Generation	Model	Treatment	Parameter estimates (± SE)			df_{error}	F	P	R^2
			a	b	c				
1st	Gaussian (3-parameter) $y = a \exp(-0.5((x-b)/c)^2)$	Untreated control	36.73±1.76	33.05±0.62	11.13±0.62	93	154.51	<0.001	0.76
		Amoxicillin	36.68±1.56	27.84±0.52	10.55±0.52	93	203.18	<0.001	0.81
		Ciprofloxacin	16.60±1.11	31.32±0.69	8.93±0.69	93	89.03	<0.001	0.65
		Rifamycin	29.13±1.75	34.30±0.79	11.3)±0.79	93	94.97	<0.001	0.67
		Tetracycline	37.95±1.65	26.17±0.54	10.67±0.54	93	196.25	<0.001	0.81
2nd	Gaussian (3-parameter) $y = a \exp(-0.5((x-b)/c)^2)$	Untreated control	46.20±2.57	32.35±0.99	15.43±1.02	93	72.25	<0.001	0.61
		Amoxicillin	45.12±1.80	38.89±0.52	11.22±0.51	93	201.32	<0.001	0.80
		Ciprofloxacin	16.88±1.05	37.85±0.91	12.54±0.91	93	77.11	<0.001	0.62
		Rifamycin	61.21±2.32	38.47±0.48	10.87±0.47	93	223.09	<0.001	0.83
		Tetracycline	23.09±1.35	29.76±0.80	11.7)±0.80	93	84.06	<0.001	0.64

All parameter estimates were significant at $P<0.01$ by Student's t-test.

similar and significant reduction of *Wolbachia* loads in the F_1 weevil progeny (Figs. 1A and 1B).

The adult weevils that were subjected to the thermal treatment, and consequently full inactivation of both SZPE and *Wolbachia*, were unable to reproduce. Therefore, the endosymbiont load in the subsequent progeny was not determined for this endosymbiont-inactivation treatment. In the treatments with antibiotics, ciprofloxacin obtained complete suppression of SZPE, tetracycline did not obtain significant suppression, and amoxicillin and rifamycin exhibited intermediate results (Fig. 1C). Amoxicillin obtained significantly higher levels of *Wolbachia* suppression (i.e., lower load of *Wolbachia*) followed by tetracycline, rifamycin, and ciprofloxacin, which exhibited similar levels of *Wolbachia* suppression (Fig. 1D).

Behavioral Consequences of Endosymbiont Load

A multivariate analysis of variance performed with the CVA protocol from SAS indicated a significant overall effect of the endosymbiont-reducing treatments on the behavior of the F_1 progeny of antibiotic-exposed weevils and thermally treated weevils (Wilks' lambda = 0.0002, $F = 0.29$, d.f.$_{num/den}$ = 40/50, $P<0.001$). Subsequent (univariate) analyses of variance performed for each behavioral trait assessed indicated that symbiont reduction affected all behavioral traits except resting time (Table 1). The behavioral alterations caused by ciprofloxacin and thermal endosymbiont reduction are particularly noteworthy (Table 1).

The multidimensional behavioral construct obtained with the CVA analysis representing the behavioral consequences of endosymbiont reduction in the maize weevil provided significant overall results. The CVA ordination generated five axes, of which the two first (1st and 2nd) were significant ($P<0.001$), explaining 97.45% of the observed variance (Table 2). The number of insects taking off for flight and the horizontal dislocation upon free-fall flight followed by the flight height exhibited the greatest canonical loads for the 1st axis accounting for most of the observed divergence among endosymbiont-reducing treatments, followed by the duration of death-feigning and the length of time to upturn, which accounted for most of the divergence on the 2nd axis (Table 2). The CVA diagram derived from the CVA representing the maximum divergence in behavior among endosymbiont-reducing treatments emphasizes the differences in the thermal and ciprofloxacin treatments, with the other antibiotics exhibiting similar intermediate differences relative to the untreated weevils retaining their regular endosymbiont load (Fig. 2).

Respiration Rate, Body Mass, and Grain Consumption

Weevil respiration rate varied significantly among the endosymbiont-reducing treatments ($F_{5,18} = 5.72$, $P = 0.002$), with all treatments except rifamycin and tetracycline leading to significant reduction relative to the control (Fig. 3AC). Body mass followed a trend similar to respiration rate ($F_{5,18} = 12.81$, $P<0.001$), but the F_1 progeny of antibiotic-treated insects exhibited lower body mass (Fig. 3BD). Grain consumption also differed significantly among F_1 progeny weevils of endosymbiont-reduced parental insects ($F_{4,15} = 10.15$, $P<0.001$), with the F_1 progeny of ciprofloxacin-treated parents exhibiting the lowest levels of grain consumption (Fig. 4A). The results obtained with the F_2 progeny of the endosymbiont-reduced insects were also significant ($P<0.003$) and largely congruent with the results from the F_1 progeny (Fig. 4B).

Daily and Cumulative Emergence

The profile of daily adult emergence of the F_1 and F_2 progenies of parental weevils subjected to endosymbiont suppression

(A) 1st Generation

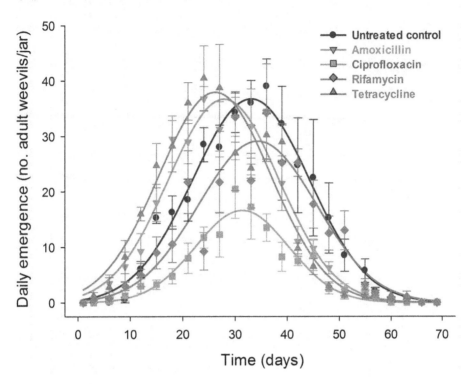

(B) 2nd Generation

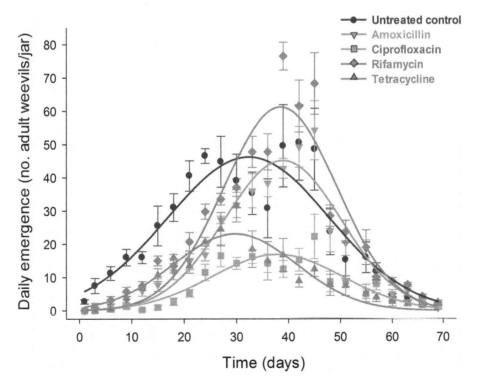

Figure 5. Daily emergence of F$_1$ (A) and F$_2$ progenies (B) of maize weevils (*Sitophilus zeamais*) exposed to different endosymbiont-reducing treatments. The symbols and vertical bars represent the means and standard errors of four replicates and the equation parameters are exhibited in Table 3.

Table 4. Summary of the non-linear regression analyses of the cumulative emergence curves (Fig. 5) of the F_1 and F_2 progenies of adult maize weevils (*Sitophilus zeamais*) exposed to different endosymbiont-suppression treatments via water-ingested antibiotics.

Generation	Model	Treatment	Parameter estimates (± SE)			df_{error}	F	P	R^2
			a	b	c				
1st	Sigmoid (3-parameter) $y = a/(1+\exp(-(x-b)/c))$	Untreated control	337.92±12.75	31.65±1.18	5.40±0.99	93	255.96	<0.001	0.85
		Amoxicillin	319.91±8.08	26.67±0.85	5.07±0.72	93	407.98	<0.001	0.89
		Ciprofloxacin	127.35±6.81	30.06±1.65	5.51±1.40	93	105.11	<0.001	0.69
		Rifamycin	275±9.44	32.61±1.07	5.70±0.89	93	339.27	<0.001	0.88
		Tetracycline	331.98±6.05	25.14±0.63	5.05±0.53	93	715.02	<0.001	0.94
2nd	Sigmoid (3-parameter) $y = a/(1+\exp(-(x-b)/c))$	Untreated control	587.37±15.65	31.25±0.93	3.92±0.74	93	679.42	<0.001	0.93
		Amoxicillin	437.49±10.19	36.96±0.68	7.12±0.55	93	1,046.67	<0.001	0.96
		Ciprofloxacin	173.51±8.64	36.85±1.46	7.15±1.17	93	234.68	<0.001	0.83
		Rifamycin	581.50±8.78	36.39±0.44	7.06±0.36	93	2,359.23	<0.001	0.98
		Tetracycline	228.41±7.93	29.41±1.16	7.00±0.97	93	283.58	<0.001	0.86

All parameter estimates were significant at P<0.01 by Student's t-test.

markedly differed among treatments, and followed the three-parameter Gaussian model used to describe the trend and selected as previously described (Table 3). Amoxicillin and tetracycline, although advancing the emergence peak of F_1 progenies, compromised adult emergence in the F_2 progeny but not in the F_1 progeny (Fig. 5). In contrast, rifamycin slightly delayed the peak of adult emergence relative to the control, reducing it for the F_2 progeny, but increasing the peak of emergence for the F_2 progeny when compared with the control (Fig. 5). Additionally, the ciprofloxacin endosymbiont-reduced progeny of treated parental weevils exhibited longer delays the for F_2 progeny, and reduced peaks of adult emergence for both F_1 and F_2 progenies (Fig. 5).

The cumulative emergence profiles of endosymbiont-suppressed weevils are a direct consequence of the daily emergence. Ciprofloxacin again exhibited a consistent, substantial reduction in emergence for both F_1 and F_2 progenies, and intermediate results were observed with rifamycin for the F_1 progeny. A trend reversal took place for amoxicillin and tetracycline, which exhibited reduced emergence only for the F_2 weevil progeny (Table 4, Fig. 6).

Endosymbiont Load and Consequences

The consequences of the reduction of endosymbiont load in adult weevils (by providing antibiotics to the parental insects) were tracked using a hierarchical approach structured as a path diagram subjected to path analysis (Fig. 7). We expected that the endosymbiont load (SZPE and *Wolbachia*) would potentially influence respiration rate and body mass in addition to potential direct effects on reproduction, and lead to potential indirect effects on weevil behavior and reproduction. No significant departures from expected covariance matrices were observed in the hypothesized path diagram ($\chi^2 = 9.37$, df = 9, $P = 0.40$), indicating that the path model used is valid (Fig. 7).

The loads of SZPE and *Wolbachia* are not correlated, and only SZPE exhibited significant direct effects on both weevil respiration rate and grain consumption (Fig. 7, Table 5). The contribution of the *Wolbachia* load in both traits was negligible, although it exhibited a significant direct effect in insect fertility (Fig. 7, Table 5). Grain consumption exhibited a significant direct effect in weevil body mass (Fig. 7, Table 5). Weevil behavior, represented by the main behavioral trait (i.e., number of insects taking off for flight), differed among endosymbiont-reducing treatments and was significantly affected by body mass and respiration rate with indirect contributions by endosymbiont load and grain consumption (Table 5, Fig. 7). Among the traits assessed for the F_1 progeny of endosymbiont-suppressed weevils subjected to antibiotic ingestion, only *Wolbachia* load and behavior had significant direct effects on the weevil reproductive output, but the whole of direct and indirect hypothesized effects in reproduction suggested in the path diagram were not significant (Fig. 7, Table 5).

Discussion

Individual traits vary within a population, and the co-existence of varying loads of different endosymbiont species within an individual host makes understanding the impact of such associations in insect species even more difficult. The SPE association with weevils was recognized as early as the 1930s, while the facultative association between *Wolbachia* and weevils dates from the late 1990s [17–20]. The more intricate effects of SPE on weevil physiology, such as improved methionine metabolism, vitamin provision, energy metabolism and flight take-off were soon recognized upon full inactivation/suppression of the endosymbiont (i.e., using aposymbiotic weevils) [17,28–30]. The recognition

(A) 1st Generation

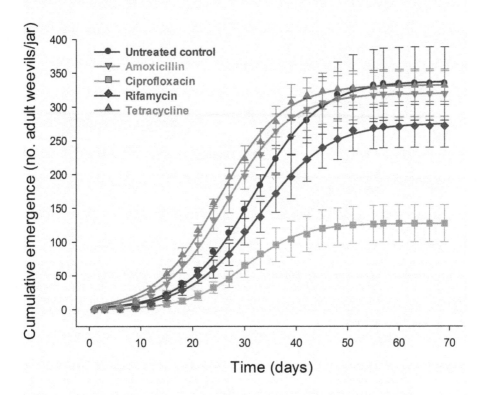

(B) 2nd Generation

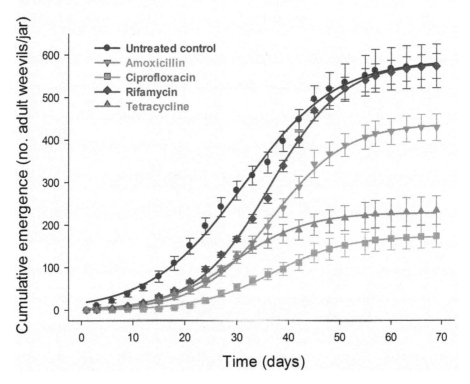

Figure 6. Cumulative emergence of F_1 (A) and F_2 progenies (B) of maize weevils (*Sitophilus zeamais*) exposed to different endosymbiont-reducing treatments. The symbols and vertical bars represent the means and standard errors of four replicates and the equation parameters are exhibited in Table 4.

$$\chi^2 = 9.37 \text{ , } df = 9, \text{ } P = 0.40$$

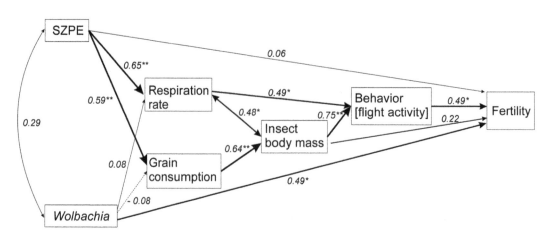

Figure 7. Path analysis diagram for the influence of the endosymbiont load of SZPE and *Wolbachia* on the respiration rate, grain consumption, body mass, behavior (flight activity, and fertility of the maize weevil (*Sitophilus zeamais*). The result of χ^2 goodness-of-fit for the path model is indicated. One-headed arrows indicate causal relationships (regression), while doubled-headed arrows indicate correlation between the variables. Significance levels are represented by asterisks (*P<0.05, **P<0.01), and the thickness of each line is proportional to the strength of the relationship. Solid arrows indicate positive relationships, while dashed arrows indicate negative relationships. Direct, indirect and total values for path coefficients are fully presented in Table 5.

of the role of *Wolbachia* associated with grain weevils has been circumscribed to cytoplasmic incompatibility [18,19], again using aposymbiotic weevils. Here we hypothesized that endosymbiont load and co-occurrence may interfere with weevil respiration rate, grain consumption, body mass, behavior, and reproduction.

Thermal treatment is the strategy usually employed to obtain aposymbiotic weevils, but tetracycline is also frequently used to suppress *Wolbachia* populations [17–19,28–30]. Indeed the thermal treatment is very effective at fully inactivating not only SZPE but also *Wolbachia* in maize weevils. However, the thermally treated weevils obtained in our studies were unable to reproduce and were used only for parental determinations of respiration rate, body mass, and behavior. In contrast, the provision of antibiotics to maize weevils via ingested water was also effective at providing different endosymbiont loads of both SZPE and *Wolbachia*, allowing more comprehensive assessments up to the F_2 progeny of treated individuals and demographic estimates and assessment of grain consumption. Therefore, the antibiotic-treated progeny was used to test our hypothesized relationship between endosymbiont load and co-occurrence and behavioral and physiological traits potentially affecting reproductive output.

Ciprofloxacin was particularly effective in suppressing SZPE, while tetracycline was fairly effective in suppressing *Wolbachia*, and thermal treatment simultaneously completely inactivated both SZPE and *Wolbachia* from their maize weevil hosts. The full simultaneous inactivation of both SZPE and *Wolbachia* significantly affected insect behavior and respiration rate, resembling the effect of the antibiotic ciprofloxacin that affected mainly SZPE, suggesting the pivotal involvement of this endosymbiont on weevil respiration and behavior, particularly flight and overall insect activity. These findings support earlier evidence of the intricate and important role of SPE in energy metabolism and flight take-off in grain weevils [17,18,29,30]. The remaining antibiotics provided varying levels of suppression of both endosymbionts, allowing the correlations and regressions combined in our path diagram of effects.

Wolbachia load in the maize weevil was only a negligible direct contributor affecting respiration rate and grain consumption and indirectly affecting weevil body mass and behavior. However, *Wolbachia* load significantly affected weevil reproduction. Cytoplasmic incompatibility is frequently reported in arthropods [18,19,50], but our finding suggest that the effect of *Wolbachia* in weevils may go beyond that. *Wolbachia* also seems to potentiate the physiological and behavioral effects of SZPE in maize weevils, both directly (for respiration rate and grain consumption) and indirectly (for body mass and behavior), based on the direct and indirect effects evidenced in our path diagram. Furthermore, the complete suppression of *Wolbachia* and SZPE prevented maize weevil reproduction, although unfertilized eggs were laid by the thermally treated female weevils, suggesting a potentiation effect of the latter, with the former favoring reproductive output. Nonetheless, the thermal stress imposed on the insect may also have contributed to preventing their reproduction, considering that the progeny production was assessed in the thermally treated insects, unlike in the antibiotic-treated weevils, where the progeny was the target of the assessments.

SZPE load was of primary importance for the maize weevil, favoring higher respiration rate and grain consumption, which corresponded to improved gain in body mass in weevils with higher loads of this symbiont. The high body mass also exhibited a significant effect on insect behavior, particularly flight activity, aided by respiration rate. Earlier studies on the physiological role of SPE presence indicated involvement in nutrient provision and energy metabolism [17,18,29,30]. Our results support this role and further indicate that such physiological effects are translated into gain in body mass and higher activity, particularly flight activity.

Although our path analysis did not provide evidence for increased overall progeny production in weevils with endosymbiont loads, the *Wolbachia* load positively affected fertility. Furthermore, daily progeny production was delayed with the reduction in endosymbiont load, particularly the drastic suppression of the SZPE load obtained with ciprofloxacin. This delayed progeny production had a negative effect on the weevil population

Table 5. Direct (DE), indirect (IE), and total (TE) effects in the path diagram of Fig. 6 for the model on the influence of endosymbiont load and co-occurrence on the respiration rate, grain consumption, body mass, behavior (flight activity), and reproduction of the maize weevil *Sitophilus zeamais*.

Variable	Respiration rate (nmol CO$_2$/insect/h)			Grain consumption (g)			Insect body mass (mg)			Behavior [flight activity] (no. taking-off for flight)			Fertility (total no. produced)		
	DE	IE	TE	DE	IE	TE	DE	IE	TE	DE	IE	TE	DE	IE	TE
SZPE (copies/µL DNA)	5.42×10^{-4}	-	5.42×10^{-4}	3.00×10^{-4}	-	3.00×10^{-4}	-	8.71×10^{6}	8.71×10^{6}	-	3.82×10^{-5}	3.82×10^{-5}	-0.002	0.001	-0.001
Wolbachia (copies/µL DNA)	-2.60×10^{-3}	-	2.60×10^{-3}	-0.01	-	-0.01	-	-2.60×10^{-4}	-2.60×10^{-4}	-	-9.30×10^{-4}	-9.30×10^{-4}	0.26	-0.03	0.23
Respiration rate (nmol CO$_2$/insect/h)	-	-	-	-	-	-	-	-	-	0.02	-	0.02	-	0.42	0.42
Grain consumption (g)	-	-	-	-	-	-	0.03	-	0.03	-	0.09	0.09	-	3.19	3.19
Insect body mass (mg)	-	-	-	-	-	-	-	-	-	3.10	-	3.10	49.35	63.37	112.42
Behavior [flight activity] (no. taking-off for flight)	-	-	-	-	-	-	-	-	-	-	-	-	20.42	-	20.42
R^2	0.43			0.41			0.40			0.59			0.31		
P	0.002*			0.01*			0.002*			<0.001*			0.20		

Asterisks indicate significant differences at $P<0.05$.

growth, indicating an important reproductive role of SZPE in the maize weevil. Further evidence of *Wolbachia* and SZPE suppression leading to reproductive impairment is also provided by the inability of thermally treated maize weevils to reproduce (i.e., weevils with full inactivation of both SZPE and *Wolbachia*).

Our results with varying endosymbiont loads and co-occurrence of SZPE and *Wolbachia* in the maize weevil reinforce the notion of the relative independence of the symbionts, which are able to coexist, although the primary effects of the SZPE load in the host seem amplified by the *Wolbachia* load. The γ-Proteobacteria SPE, of which SZPE is a representative, is located in specific and differentiated cells (bacteriocytes) in bacteria-bearing tissue (bacteriome) found only in female germ cells and larval and ovarian bacteriomes [17–19]. This characteristic distribution of SPE in weevils likely maintains these endosymbionts in relative isolation, minimizing potential interactions with co-occurring symbionts such as *Wolbachia*. In contrast, *Wolbachia*, which is a α-Proteobacteria with facultative association in grain weevils, is disseminated throughout the body cells and at noticeably high densities in male and female germ cells, where it induces reproductive abnormalities [18,19,21,22].

The co-occurrence of SZPE and *Wolbachia* in a key pest species of stored cereal grains, such as the maize weevil, has potential practical importance. An obvious possibility is the design of alternative management methods for the control of this pest species, such as sterile insect techniques (or incompatible insect techniques) and/or insertion of fitness reduction factors aiming at pest suppression or replacement [21,22,50]. These endosymbionts may prove important in strain divergence, with implications for grain loss and weevil control because endosymbiont load and co-occurrence affect grain consumption, consequently affecting grain loss and leading to higher economic losses. In addition, both endosymbiont load and co-occurrence affect insect activity, interfering with their dispersal and colonization, producing added potential consequences for pest control, which is variable between populations and even among individuals in a population [16,44,45]. Other unforeseeable consequences may also derive from variable endosymbiont loads and co-occurrence in arthropod pest species in general, and grain weevils in particular, which is likely to draw further attention in the future.

Supporting Information

Figure S1   Standard curve of *Wolbachia* 16S gene in the presence of different concentrations (log) of the plasmid.

Figure S2   Standard curve of SZPE 16S gene in the presence of different concentrations (log) of the plasmid.

Data S1   Threshold cycle (Ct) values for *Wolbachia* 16S gene from the F$_1$ progenies of adult maize weevils (*Sitophilus zeamais*) exposed to different endosymbiont-suppression treatments. Number of copies based on standard curve (y), number of copies corrected by the one-point calibration method (OPC) and number of copies per microliter of DNA.

Data S2   Threshold cycle (Ct) values for *Wolbachia* 16S gene from the F$_2$ progenies of adult maize weevils
(*Sitophilus zeamais*) **exposed to different endosymbiont-suppression treatments.** Number of copies based on standard curve (y), number of copies corrected by the one-point calibration method (OPC) and number of copies per microliter of DNA.

Data S3   Threshold cycle (Ct) values for gene SZPE 16S gene from the F$_1$ progenies of adult maize weevils (*Sitophilus zeamais*) exposed to different endosymbiont-suppression treatments. Number of copies based on standard curve (y), number of copies corrected by the one-point calibration method (OPC) and number of copies per microliter of DNA.

Data S4   Threshold cycle (Ct) values for gene SZPE 16S gene from the F$_2$ progenies of adult maize weevils (*Sitophilus zeamais*) exposed to different endosymbiont-suppression treatments. Number of copies based on standard curve (y), number of copies corrected by the one-point calibration method (OPC) and number of copies per microliter of DNA.

Data S5   Raw data of behavioral traits of F$_1$ and F$_2$ progenies of adult maize weevils (*Sitophilus zeamais*) exposed to different endosymbiont-suppression treatments.

Data S6   Raw data of respiration rate, body mass, grain consumption, and fertility of F$_1$ and F$_2$ progenies of adult maize weevils (*Sitophilus zeamais*) exposed to different endosymbiont-suppression treatments.

Data S7   Raw daily emergence data of 1st generation insects.

Data S8   Raw daily emergence data of 2nd generation insects.

Data S9   Raw cumulative emergence data of 1st generations insects.

Data S10   Raw cumulative emergence data of 2nd generation insects.

Movie S1   Video showing short recordings of each behavioral test.

Acknowledgments

We thank M. Rodrigueiro for guidance provided in dealing with *Wolbachia*, and L. Braga and L. Pantoja for providing technical assistance.

Author Contributions

Conceived and designed the experiments: GAC MMH ASC LOO RNCG. Performed the experiments: GAC JLV MMH. Analyzed the data: GAC MMH RNCG. Contributed reagents/materials/analysis tools: AOBR LOO RNCG. Contributed to the writing of the manuscript: GAC RNCG.

References

1. Moran NA (2006) Symbiosis (A primer). Curr Biol 16: R866–R871.
2. Douglas AE (1998) Nutritional interactions in insect-microbial symbioses: Aphids and their symbiotic bacteria *Buchnera*. Annu Rev Entomol 43, 17–37.
3. Dillon RJ, Dillon VM (2004) The gut bacteria of insects: Nonpathogenic interactions. Annu Rev Entomol 49, 71–92.

4. Nardon P, Lefèvre C, Delobel B, Charles H, Heddi A (2002) Occurrence of endosymbiosis in Dryopthoridae weevils: Cytological insights into bacterial symbiotic structures. Symbiosis 33, 227–241.

5. Chiel E, Zchori-Fein E, Inbar M, Gottlieb Y, Adachi-Hagimori T, et al. (2009) Almost there: Transmission routes of bacterial symbionts between trophic levels. PLoS ONE 4(3): e4767.

6. Visôtto LE, Oliveira MGA, Ribon AOB, Mares Guia TR, Guedes RNC (2009) Characterization and identification of proteolytic bacteria from the gut of the velvetbean caterpillar (Lepidoptera: Noctuidae). Environ Entomol 38: 1078–1085.

7. Visôtto LE, Oliveira MGA, Guedes RNC, Ribon AOB, Good-God PIV (2009) Contribution of gut bacteria to digestion and development of the velvetbean caterpillar, Anticarsia gemmatalis. J Insect Physiol 55: 185–191.

8. Pilon FM, Visôtto LE, Guedes RNC, Oliveira MGA (2013) Proteolytic activity of gut bacteria isolated from the velvet bean caterpillar Anticarsia gemmatalis. J Comp Physiol B 183: 735–747.

9. Wernegreen JJ (2002) Genome evolution in bacterial endosymbionts of insects. Nature Rev Gen 3: 850–861.

10. Zug R, Hammerstein P (2012) Still a host of hosts for Wolbachia: Analysis of recent data suggests that 40% of terrestrial arthropod species are infected. PLoS ONE 7(6): e38544.

11. Xue X, Li S-J, Ahmed MZ, De Barro PJ, Ren S-X, et al. (2012) Inactivation of Wolbachia reveals its biological roles in whitefly host. PLoS ONE 7(10): e48148.

12. Montlor CB, Maxmen A, Purcell AH (2002) Facultative endosymbionts benefit pea aphids Acyrthosiphon pisum under heat stress. Ecol Entomol 27: 189–195.

13. Russell JA, Moran NA (2006) Costs and benefits of symbiont infection in aphids: Variation among symbionts and across temperatures. Proc Biol Sci 273: 603–610.

14. Olivier KM, Russell JA, Moran NA, Hunter MS (2003) Facultative bacterial symbionts in aphids confer resistance to parasitic wasps. Proc Natl Acad Sci USA 100: 1803–1807.

15. Olivier KM, Moran NA, Hunter MS (2005) Variation in resistance to parasitism in aphids is due to symbionts not host genotype. Proc Natl Acad Sci USA 102: 12795–12800.

16. Morales JA, Cardoso DG, Della Lucia TMC, Guedes RNC (2013) Weevil x insecticide: Does "personality" matter? PLoS ONE 8(6): e67283.

17. Wicker C (1983) Differential vitamin and choline requirements of symbiotic and aposymbiotic S. oryzae (Coleoptera: Curculionidae). Comp Biochem Physiol 76: 177–182.

18. Heddi A, Grenier AM, Khatchadourian C, Nardon C, Charles H, et al. (1999) Four intracellular genomes direct weevil biology: Nuclear, mitochondrial, principal endosymbiont, and Wolbachia. Proc Natl Acad Sci USA 96: 6814–6819.

19. Heddi A, Charles H, Khatchadourian C (2001) Intracellular bacterial symbiosis in the genus Sitophilus: The "Biological Individual" concept revisited. Res Microbiol 152: 431–437.

20. Carvalho GA, Corrêa AS, Oliveira LO de, Guedes RNC (2014) Evidence of horizontal transmission of primary and secondary endosymbionts between maize and rice weevils (Sitophilus zeamais and Sitophilus oryzae) and the parasitoid Theocolax elegans. J Stored Prod Res 59: 61–65.

21. Bourtzis K (2008) Wolbachia-based technologies for insect pest population control. Adv Exp Med Biol 627: 104–113.

22. Saridaki A, Boutzis K (2009) Wolbachia: More than just a bug in insect genitals. Curr Opin Microbiol 13: 67–72.

23. Mansour K (1930) Memoirs: Preliminary studies on the bacterial cell-mass (accessory cell-mass) of Calandra oryzae (Linn): The rice weevil. Q J Microsc Sci 73: 421–436.

24. Nardon P, Grenier AM (1988) Genetical and biochemical interactions between the host and its endocytobiotes in the weevils Sitophilus (Coleoptera, Curculionidae) and other related species. In: Scannerini S, Smith D, Bonfante-Fasolo P, Gianinazzi-Peatson V, editors. Cell to Cell Signals in Plant, Animal and Microbial Symbiosis. Berlin: Springer. 255–270.

25. Araújo RA, Guedes RNC, Oliveira MGA, Ferreira GH (2008) Enhanced activity of carbohydrate- and lipid-metabolizing enzymes in insecticide-resistance populations of the maize weevil, Sitophilus zeamais. Bull Entomol Res 98: 417–424.

26. Lopes KVG, Silva LB, Reis AP, Oliveira MGA, Guedes RNC (2010) Modified α-amylase activity among insecticide-resistant and –susceptible strains of the maize weevil, Sitophilus zeamais. J Insect Physiol 56: 1050–1057.

27. Toju H, Tanabe AS, Notsu Y, Sota T, Fukatsu T (2013) Diversification of endosymbiosis: Replacements, co-speciation and promiscuity of bacteriocyte symbionts in weevils. ISME J 7: 1378–1390.

28. Gasnier-Fauchet F, Nardon P (1987) Comparison of sarcosine and methionine sulfoxide levels in symbiotic and aposymbiotic larvae of two sibling species, Sitophilus oryzae L. and S. zeamais Mots. (Coleoptera: Curculionidae). Insect Biochem 17: 17–20.

29. Heddi A, Fefèvre F, Nardon P (1993) Effect of endocytobiotic bacteria on mitochondrial enzymatic activities in the weevil Sitophilus oryzae (Coleoptera: Curculionidae). Insect Biochem Mol Biol 23: 403–411.

30. Grenier AM, Nardon C, Nardon P (1994) The role of symbionts in flight activity of Sitophilus weevils. Entomol Exp Appl 70: 201–208.

31. Vigneron A, Charif D, Vincent-Monégat C, Vallier A, Gavory F, et al. (2012) Host gene response to endosymbiont and pathogen in the cereal weevil Sitophilus oryzae. BMC Microbiol (Suppl 1): S14.

32. Heddi A, Charles H, Khatchadourian C, Bonnot G, Nardon P (1998) Molecular characterization of the principal symbiotic bacteria of the weevil Sitophilus oryzae: A peculiar G + C content of an endocytobiotic DNA. J Mol Evol 47: 52–61.

33. Rio RVM, Lefèvre C, Heddi A, Aksoy S (2003) Comparative genomics of insect-symbiontic bacteria: Influence of host environment on microbial genome composition. Appl Environ Microbiol 69: 6825–6832.

34. Anselme C, Vallier A, Balmand S, Fauvarque MO, Heddi A (2006) Host PGRP gene expression and bacterial release in endosymbiosis of the weevil Sitophilus zeamais. Appl Environ Microbiol 72: 6766–6772.

35. Gil R, Belda E, Gosalbes MJ, Delaye L, Vallier A, et al. (2008) Massive presence of insertion sequences in the genome of SOPE, the primary endosymbiont of the rice weevil Sitophilus oryzae. Int Microbiol 11: 41–48.

36. Guedes RNC, Lima JOG, Santos JP, Cruz CD (1995) Resistance to DDT and pyrethroids in Brazilian populations of Sitophilus zeamais Motsch. (Coleoptera: Curculionidae). J Stored Prod Res 31, 145–150.

37. Ribeiro BM, Guedes RNC, Oliveira EE, Santos JP (2003) Insecticide resistance and synergism in Brazilian populations of Sitophilus zeamais (Coleoptera: Curculionidae). J Stored Prod Res 39: 21–31.

38. Clark TL, Meinke LJ, Foster JE (2001) Molecular phylogeny of Diabrotica beetles (Coleoptera: Chrysomelidae) inferred from analysis of combined mitochondrial and nuclear DNA sequences. Insect Mol Biol 10: 303–314.

39. Untergasser A, Cutcutache I, Koressaar T, Ye J, Faircloth BC, et al. (2012) Primer3 – new capabilities and interfaces. Nucleic Acids Res 40: e115.

40. Le Clec'h W, Chevalier FD, Genty L, Bertaux J, Bouchon D, et al. (2013) Cannibalism and predation as paths for horizontal passage of Wolbachia between terrestrial isopods. PLoS ONE 8(4): e60232.

41. Branlatschk R, Bodenhausen N, Zeyer J, Burgmann H (2012) Simple absolute quantification method correcting for quantitative PCR efficiency variations for microbial community samples. Appl Environ Microbiol 78: 4481–4489.

42. Guedes NMP, Braga LS, Rosi-Denadai CA, Guedes RNC (2015) Desiccation resistance and water balance in populations of the maize weevil Sitophilus zeamais. J Stored Prod Res (accepted).

43. Tomé HVV, Barbosa WF, Martins GF, Guedes RNC (2015) Spinosad in the native stingless bee Melipona quadrifasciata: Questionable environmental safety of a bioinsecticide. Chemosphere (accepted).

44. Guedes RNC, Oliveira EE, Guedes NMP, Ribeiro B, Serrão JE (2006) Cost and mitigation of insecticide resistance in the maize weevil, Sitophilus zeamais. Physiol Entomol 31: 30–38.

45. Corrêa AS, Tomé HVV, Braga LS, Martins GF, Oliveira LO de, et al. (2014) Are mitochondrial lineages, mitochondrial lysis and respiration rate associated with phosphine susceptibility in the maize weevil Sitophilus zeamais? Ann Appl Biol 165: 137–146.

46. Trematerra P, Fontana F, Mancini M (1996) Analysis of developmental rates of Sitophilus oryzae (L.) in five cereals of the genus Triticum. J Stored Prod Res 32: 315–322.

47. Fragoso DB, Guedes RNC, Peternelli LA (2005) Developmental rates and population growth of insecticide-resistance and susceptible populations of Sitophilus zeamais. J Stored Prod Res 41: 271–281.

48. SAS Institute (2008) SAS/STAT User's Guide. SAS Institute, Cary, NC, USA.

49. Mitchell RJ (1993) Path analysis: Pollination. In: Scheiner SM, Gurevitch J, editors. Design and Analysis of Ecological Experiments. New York: Chapman & Hall. 211–231.

50. Werren JH, Baldo L, Clark ME (2008) Wolbachia: Master manipulators of invertebrate biology. Nature Rev Microbiol 6: 741–751.

13

A New Approach to Quantify Semiochemical Effects on Insects Based on Energy Landscapes

Rory P. Wilson, Rebecca Richards, Angharad Hartnell, Andrew J. King, Justyna Piasecka, Yogendra K. Gaihre, Tariq Butt*

Biosciences, College of Science, Swansea University, Singleton Park, Swansea, Wales, United Kingdom

Abstract

Introduction: Our ability to document insect preference for semiochemicals is pivotal in pest control as these agents can improve monitoring and be deployed within integrated pest management programmes for more efficacious control of pest species. However, methods used to date have drawbacks that limit their utility. We present and test a new concept for determining insect motivation to move towards, or away from, semiochemicals by noting direction and speed of movement as animals work against a defined energy landscape (environmentally dependent variation in the cost of transport) requiring different powers to negotiate. We conducted trials with the pine weevils *Hylobius abietis* and peach-potato aphids *Myzus persicae* exposed to various attractants and repellents and placed so that they either moved up defined slopes against gravity or had to travel over variously rough surfaces.

Results: Linear Mixed Models demonstrated clear reductions in travel speed by insects moving along increasingly energetically taxing energy landscapes but also that responses varied according to different semiochemicals, thus highlighting the value of energy landscapes as a new concept to help measure insect motivation to access or avoid different attractants or repellents across individuals.

Conclusions: New sensitive, detailed indicators of insect motivation derived from this approach should prove important in pest control across the world.

Editor: Daniel Doucet, Natural Resources Canada, Canada

Funding: TB and JP thank the Welsh Government for an A4B CIRP grant. AJK was supported by a NERC Postdoctoral Fellowship (NE.H016600.2). The funders had no role in study design, data collection and analysis, decision to publish, or preparation of the manuscript.

Competing Interests: The authors have declared that no competing interests exist.

* Email: t.butt@swansea.ac.uk

Introduction

Many insect species are implicated in a suite of problems ranging from acting as vectors of human and animal diseases [1,2] to reducing crop yield [3,4] that have significant socio-economic consequences worldwide [5]. The control of these pests is a complex issue that, in its infancy, generally relied on using large quantities of pesticides but which rapidly manifest a series of unwanted associated consequences [6,7,8]. Current attempts to minimize the use of inappropriate chemicals increasingly seek to define methods to attract insects so that pesticides and other mechanisms of control [9] can be applied to restricted areas and thus reduce potential ecosystem damage [10]. The most sophisticated approach of this type purports the use of a "push-pull" strategy which may, for example, use a combination of repellents and attractants to rarefy and concentrate insect densities in prescribed areas [11,12,13]. Pivotal in this, however, is the capacity to define the value of chemicals as attractants or repellents. There are two primary methods used for this; the first method places insects in a Y-tube olfactometer and exposes them to a chemical from one arm and a control from another [14]. The proportion of animals that move into each arm indicates the value of the substance as an attractant. The response per individual is,

therefore, effectively binary (but see [15]). The second complementary method based on an electroantennogram (EAG), examines the voltage generated from the insect's antennae as a result of exposure to odours, and measures the neurological degree of excitation in relation to exposure to chemicals [16,17,18]. This method requires appreciable expertise, equipment and time, does not help define the behavioural role of the chemical, may not necessarily allow for predictions with respect to insect behaviour under more natural conditions [19] and may not indicate whether an observed response will actually lead to movement in the field [20].

Studies on human preference typically use questionnaires whereby subjects are asked to quantify the degree to which they like or dislike something [21]. This has profound consequences for marketing because specific strategies can be directed at the variously reacting groups [22]. In non-human studies, motivational tests – as a measure of what animals want – have been developed and refined for some decades [23] and behavioural biologists interested in motivation from a welfare perspective have adopted a "consumer demand approach". This tests an individual's strength of preferences for a variety of potential resources, and classically involves animals working to push a weight-loaded door (with varying cost) in order to have access to the resources.

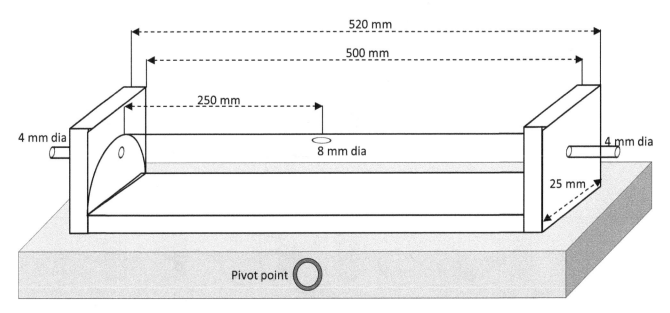

Figure 1. Schematic diagram of the incline energy landscape olfactometer showing the major features and dimensions. Note that the apparatus rests on a board (coloured grey) and that the angle of this board can be varied by rotation about a pivot point.

This push-door paradigm was first used in hens by Duncan and Kite [24], and since then, adapted push-door paradigms have been used in a variety of species and contexts; most recently to test for cichlid fishes motivation for social partners and food [25]. Development of appropriate methodology to measure 'what animals want' has therefore enabled us to quantify motivation vertebrate preference for target resources, and provided a framework in which we can begin to assess animal emotion accurately [26], an important goal in animal welfare science [27].

We sought to create a system whereby insect motivation to move toward an attractant or away from a repellent could be precisely quantified so that all the advantages of similar approaches used in human studies and other vertebrates could

be brought to bear to insect control. We capitalized on the recently proposed concept of 'energy landscapes', that a landscape can be defined in terms of the cost of travel (the energy used per unit distance) for an animal to move over it [28,29]. We used two insect species as a proof of concept, the pine weevil, *Hylobius abietis*, and the peach potato aphid, *Myzus persicae*. In addition, for the pine weevil, we first tested it's preference for different semiochemicals using a traditional Y-tube olfactometer, so that we could evaluate the benefit of our new approach. We created landscapes that could be changed to elicit variable movement costs before letting insects move over the surface, documenting their speed of movement as a function of both the attractant/repellent used and the difficulty of moving over the landscape. The speed of the study animals is a

Table 1. Test chemicals used in the Y-tube and energy landscape olfactometry studies.

Test Chemical	Solvent	Purity
α-pinene	Ethanol	98%
(−)-β-pinene	Ethanol	99%
3-carene	Ethanol	90%
Ethanol	n/a	95.0+%
Methanol	n/a	99.8+%
Hexane	n/a	99.8+%
α-terpineol	Hexane	≥96
(+)-citronella	Hexane	90%
Geraniol	Hexane	98%
Linalool	Hexane	96%
(+)-carvone	Hexane	96%
2-heptanone	Water	
Garlic Metabolic solution (GMS)	Water	100%
Seaweed extract	Hexane	low
Rape Seed Oil	Hexane	100%

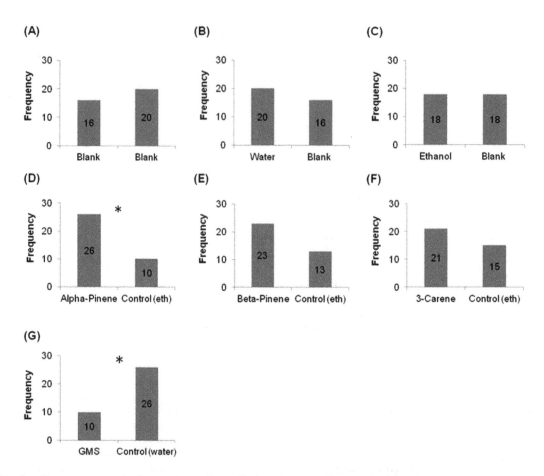

Figure 2. Y-tube olfactometer results for numbers of weevils choosing to walk towards either the semiochemical or the control for; (A) control [for examining arm preferences], (B) water [for examining differences between our control in A and water], (C) ethanol [for examining differences between our control in A and ethanol], (D) α-pinene, (E) β-pinene, (F) 3-carene and (G) garlic metabolic solution [GMS], all against the control for their respective solvents. Significant differences are indicated by (*).

graded response to the prevailing conditions and thus should provide a measure equivalent to those stemming from human questionnaire preference studies, with all the benefits that this brings in designing insect control measures.

Materials and Methods

The Y-Tube Olfactometer

Experiments were conducted using a 90° 24 mm internal diameter, Y-glass tube olfactometer, model numbers OLFM-YT-2425F, OLFM-2425M and OLFM-IN-2425M from Analytical Research Systems, Inc., USA. The external diameter of the tube was 32 mm and arm lengths were 150 mm (long arm) and 85 mm (short arms). The Y-tube olfactometer was placed in a blacked out box to remove visual distractions.

The Energy Landscape Olfactometer (ELO)

We constructed two energy landscapes Both landscapes were housed within a semi-circular high density Perspex tubing with closed ends (50 cm long×2.5 cm width×1.25 cm height) fixed to a flat 10 mm thick wooden base that was painted black to avoid visual stimuli [30]. At each flat end of the tubing 4 mm diameter tubes were inserted to allow for the introduction of an air flow with a chemical mix to be pushed along the tube at a defined rate. The semi-circular tubing had a central dorsal hole through which

insects could be introduced and was marked at regular intervals so that insect speed could be noted (Fig. 1).

The ELO could be manipulated in two ways. The first involved tilting the landscape so that study insects approaching, or moving away from, the impinging chemical had to move up or down a known gradient, thus performing a known amount of work within a specified time to give a work rate metric. This was achieved by shifting the wooden base with pegs at known angles. The second manipulation was a change to the base floor, which was covered with Whatman filter paper. This paper could be variously roughened by sandpapers so that the fibres making up the matrix of the filter paper stood up, emulating trichomes with rough or smooth surfaces. Both systems could be combined. Olfactometers were thoroughly cleaned using appropriate solvents to ensure removal of chemical residues between assays between trials and filter papers renewed.

Test insects and ethics statement

Two distinct insect species were considered as test examples for the energy landscapes, the pine weevil, *Hylobius abietis* Linneaus (Coleoptera: Curculionidae), and the peach potato aphid, *Myzus persicae* Sulzer (Hemiptera: Aphididae), representing very different types of animal. Preliminary tests showed that the weevils were virtually insensitive to the simulated trichome density but highly sensitive to movement gradient whereas the reverse was true of the aphids. Work on the two different species thus concentrated on

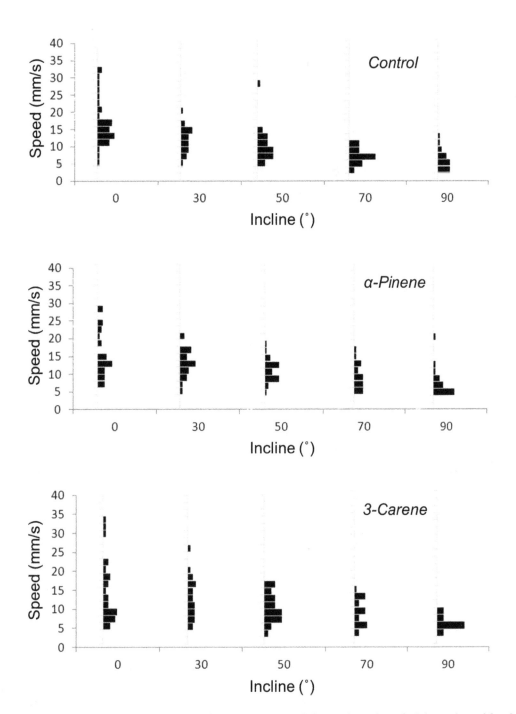

Figure 3. Examples of frequency histograms (expressed as a percentage of the total number of trials conducted for that condition) of the speeds at which pine weevils walked up specified inclines in response to selected semio-chemicals. Each condition consisted of 36 trials for weevils (see text).

using them in the energy landscapes to which they were most sensitive. Fresh insects were used in each assay. No permits were required for the described study, which complied with all relevant regulations.

Semiochemicals and chemical reagents used

We used known attractant/repellent semiochemicals for the weevils and aphids at concentrations appropriate to elicit greatest activity. Test chemicals and solvents (Table 1) were purchased from Sigma Aldrich (UK) with the exception of garlic metabolic

solution (GMS), seaweed extract and rapeseed oil, which were provided by Neem Biotech Ltd (UK).

For the pine weevils, we conducted basic Y-Tube olfactometer preference tests to ensure appropriate controls were used in our ELO experiments. Ethanol was used as a solvent due to the synergising effect it has with attractants such as α-pinene [31]. Ethanol was used as a control. Water was used to dilute the repellent garlic metabolic solution (GMS) and was also used as a control. All testing was undertaken using a 1% solution of either α-pinene, β-pinene, 3-carene and GMS.

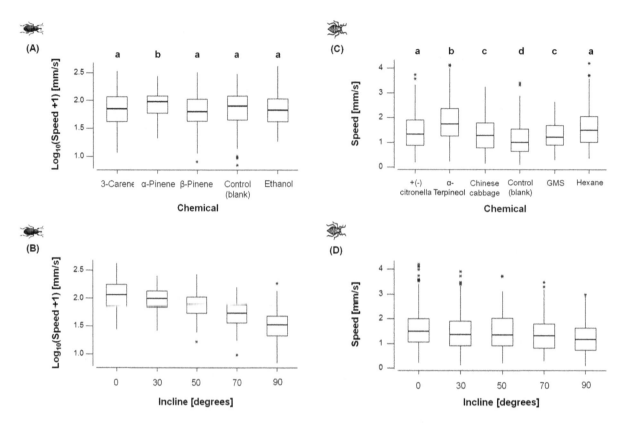

Figure 4. Speed at which aphids and weevils responded to different chemicals and the manner in which this changes with incline, represented by box and whisker plots indicating median (bold line), inter-quartile ranges (box), 95% confidence intervals (whiskers), and outliers (*). (**A**) Speed with which weevils move towards different chemical attractants (the control shown is the blank - Fig. 2a). The type of chemicals to which weevils were exposed significantly altered subjects approach speed (LMM: Wald = 10.86, df = 4, P = 0.028), with subjects moving faster towards α-pinene than all other treatments (significant pairwise differences (P<0.05) across chemical trials are indicated by different letters). (**B**) Speed with which weevils moved at different inclines (LMM: Wald = 75.21, df = 1, P<0.001). (**C**) Speed with which aphids move away from different chemical repellents (the control shown is the blank cf. Fig. 2a), which significantly affected aphid speed (LMM: Wald = 105.09 df = 5, P<0.001). Significant pairwise differences (P<0.05) across chemical trials are indicated by different letters. (**D**) Speed with which aphids move at different inclines(LMM: Wald = 75.21, df = 1, P<0.001). See methods for more details of models.

For aphids, first, the repellency of different solvents (water, methanol, methanol+water, ethanol and hexane) was established. The most repellent solvent, hexane, was then used to dilute and enhance compounds known to have repellent properties. Hexane was used as a control. The exception was 2-heptanone which was dissolved in water in which case water was used as a control. Assays were performed using a 0.1% solution of the test compounds which included α-terpineol, (+)-citronella, geraniol, linalool, (+)-carvone, 2-heptanone, GMS, seaweed extract and rape seed oil. Aphids were attracted to Chinese cabbage, therefore, a Chinese cabbage leaf disc (4 cm diameter) was used as a positive (attractant) control.

The Y-Tube Olfactometer (weevils)

Each arm of the Y-tube olfactometer was connected to an air flow of 0.3 L/min [32] after passing through black carbon filter. Filter paper (4×2 cm) was placed in each of the olfactometer arms with one filter paper being treated with 10 μl of the test solution (1%) and the other with 10 μl of the control solvent (i.e. ethanol or water). The solutions were pipetted onto the centre of the filter paper [33]. The test solutions were replaced every 2 days due to the volatility of the monoterpenes [34]. Twelve pine weevils were tested individually in each trial and trials were repeated three times. Each pine weevil was allowed 2 minutes to acclimatise to the new surroundings and then given 10 minutes to respond and

walk towards either the known attractant or the control. Their response was recorded when the pine weevil had travelled at least 4 cm up the chosen arm and remained there for more than 10 seconds [33]. All experiments were conducted in daylight at 28°C (±2°C) and 60% relative humidity.

Energy Landscape Olfactometer trials

Twelve pine weevils and 10 aphids were tested individually in each trial and trials were repeated three times. Animals were placed through the hole in the centre of the ELO and allowed to choose their travel direction. If the test animals moved toward the control or did not move after a period of 2 minutes, the behaviour was recorded and the animals removed and replaced. If the animals moved in the correct direction, travelling speed (mm/sec) towards the semiochemical at different inclines was recorded. No time was recorded in the 0–5 cm section to allow an 'acclimatisation' segment. The walking speed of individual animals was recorded (in mm/sec) at the 0° incline between 5 cm and 10 cm before the travel speed was then recorded for randomly chosen inclines (from one of four categories; 30°, 50°, 70° and 90°) for next three 5 cm sections. If the animal stopped or turned during this procedure, the behaviour was recorded, the animal removed and the apparatus re-cleaned.

A similar procedure was adopted for the aphids walking within the energy landscape olfactometer except that, instead of only tilting the apparatus to change the energy landscape, the aphids were also obliged to walk on either smooth or roughened filter paper. The latter was prepared by gently passing sandpaper over the surface four times in two alternating perpendicular directions. New filter paper was used for every trial.

Data stemming from this work were deposited in the Swansea University College of Science T-drive.

Statistical analysis

Two-tailed binomial tests (SPSS 19, IBM Corp, 2010) were used to test for weevil preferences among semiochemicals in the Y-tube olfactometer. These tests assume a null hypothesis of no preference, and are commonly applied to Y-tube olfactometer analysis [35,36].

To test the effect of chemical type, and landscape type, upon the speed (mm/s) with which the insects moved towards chemicals attractants and away from repellents, we ran two Linear Mixed Models (LMM) for each species, implemented in MLwiN (v. 2.25, 2011, Bristol University Centre for Multilevel Modelling, Bristol, U.K.). The weevil speed data did not follow a normal distribution and so was $\log_{10}$ transformed. In each model, trial number and date were fitted as random effects to control for potential non-independence across trials and within days. For our weevil dataset we fitted incline (continuous), and chemical attractant (categorical: see Table 1 for chemicals) as fixed effects. For the aphid dataset, we fitted incline (continuous), and chemical repellent (categorical: see Table 1 for chemicals) as fixed effects. We additionally fitted surface type (categorical: smooth, rough), and tested for an interaction between surface type and chemical.

Once we had determined the independent effects of chemical type, and energy landscape (incline, or surface type) in the above models, we ran two further models (one for each species) in which we allowed the effect of 'incline' to vary according to 'chemical type' (i.e. a random intercept, random slope model). This allowed us to see whether the insects would move faster (or slower) up steeper inclines according to the semiochemical present. This is similar to fitting multiple regressions of speed against incline; one for each chemical type, or fitting an interaction between chemical type and incline in our LMM, but is easier to interpret.

Results

Y-Tube Olfactometer

Y-tube olfactometer tests showed weevils had no preference for our blank or control treatments (Figure 2A, 2B, 2C), and revealed an apparent preference for α-pinene, β-pinene, and 3-carene treatments, being apparently repelled by garlic metabolic solution. However, only α-pinene and garlic metabolic solution were statistically significant (Fig. 2D, 2E, 2F, 2G). We therefore investigated the preferences of the weevils to the semiochemical to which they were attracted, namely α-pinene, in our energy landscape olfactometer experiments.

Energy landscape olfactometer

Both insect species reacted to the energy landscapes by walking towards, or away from, the selected semiochemicals at speeds that showed considerable variation (Fig. 3). However, the type of chemicals to which weevils were exposed significantly altered subjects approach speed (Fig. 4a). The incline to which subjects were exposed also had an independent effect upon approach speed (Fig. 4b), and the magnitude of this effect differed with respect to chemical trial with, specifically, speed being maintained during α-pinene trials at steeper inclines than for other chemicals (Fig. 5).

The type of chemicals to which aphids were exposed also significantly altered travel speed (Fig. 4c), although they moved faster away from chemicals on flatter surfaces (Fig. 4d). However, the effect of incline on the speed with which subjects moved away was consistent across chemical repellents, with aphid speed compared to the control condition being fastest away from α terpineol and slowest away from Chinese cabbage (α terpineol> hexane>citrollena>garlic>Chinese cabbage: Fig. 6a). We also found that the aphids moved significantly slower away from chemicals on the rougher surface (Fig. 6b), and this effect differed across chemicals (LMM: Wald = 21.15, df = 5, P<0.001), with subjects moving quickest on the rougher surface when moving away from hexane compared to our control and all other chemicals (pairwise comparisons: P<0.01). Reductions in speeds on rough surfaces for other chemicals were not significantly different from one another (pairwise comparisons: P>0.05 in call cases).

Discussion

The energy landscape olfactometer, using a prescribed energy landscape to assess insect motivation to move towards, or away from, stimuli such as semiochemicals, reveals a number of features in insect response that are not accessible using conventional olfactometers or antennograms. Although the new system effectively mirrors the conventional Y-tube olfactometer in offering one of two travel directions (or the choice to remain stationary), it also goes beyond the individual binary response of the Y-tube olfactometer by revealing motivation manifest in choice of speed. Although increasing the cost associated with different energy landscapes systematically reduced the speed of travel with any given semiochemical (Figs. 3–6), closer inspection revealed two more specific responses displayed to different semiochemicals. In the one case, there was minimal variation in speed of travel for pine weevils moving towards the different semiochemicals on the flat surface (Fig. 5) which would imply that there is apparently either no difference in the attraction of the different semiochemicals or that speed appears to be a poor measure of the attractiveness of semiochemicals. However, weevils responded to increased inclines by changing speed differentially according to semiochemical (Fig. 5). Conversely, aphids did show differences in their speed of movement away from the different semiochemicals at an incline of 0° (Fig. 6) but had decreases in speed with increasing incline that did not vary between semiochemicals with slope (Fig. 6a). Beyond this, however, aphids exhibited variation in speed according to whether the surfaces they travelled over were rough or smooth (Fig. 6b).

In essence, speed indicates motivation because it affects the power costs selected by insects for movement towards, or away from, a defined goal. The more energy animals are prepared to expend per unit time to move towards or away from something, the stronger their motivational state is assumed to be. Full and Tullis [37] show how, in American cockroaches Periplaneta americana Linneaus, the rate of energy expenditure increases linearly with both speed and incline above resting metabolic rate (Fig. 7), a feature that seems common in both vertebrates and invertebrates [38]. We can use the general principles relating to the costs of terrestrial locomotion [38] to set up a framework within which to consider how insect motivational state may be reflected in the speeds and associated power costs that they choose for movement.

Faced with increasingly energetically onerous landscapes, such as increasing hair density or incline, insects may respond with one of two extreme responses or something in-between. They can

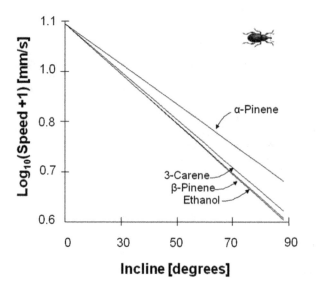

Figure 5. Model results for how incline is predicted to affect weevil speed. In this model, the intercept and slope of the effect of incline were allowed to vary with respect to chemical (i.e. a random slope, random intercept model) and our control condition was used as the reference category. See methods for further details.

either maintain their speed, which will lead to substantial increases in the rate of energy expenditure or they can maintain a constant power output, which will necessitate substantial decreases in speed. The response of our study animals seemed to be somewhere between these two extremes with, importantly, animals changing the way they respond according to the semiochemical used (Fig. 7). The fact that insects clearly do vary their elected power costs of travel according to energy landscape and stimulus, highlights an important difference between conventional Y-tube olfactometers, where incline is always 0°, and energy landscape

olfactometers. This difference is apparent in comparison of the percentage positive responses in a Y-tube olfactometer to different semiochemicals (Fig. 2) with the slope of the regression of speed versus incline (Fig. 5) for the same semiochemicals derived by using energy landscape olfactometers. In pine weevils for example, the correlation coefficient for this is $r^2 = 0.53$ (percentage data arcsin transformed to ensure normalization), which shows some concurrence, but the lack of a tight correlation presumably indicates the extent to which the ELO documents a different response; specifically, with the form of the speed/power change with incline (or hair density) with a given semiochemical, providing important information about motivational state in addition to that given by the conventional approach.

We propose, for our purposes, that an insect's response to a semiochemical may be described hierarchically: The first, most coarse, level is whether the animal moves towards (or away from) the stimulus at all (and here we note that conventional olfactometer approaches do involve a loosely defined element of speed since subjects are required to have moved within a defined period of time). A second level specifically measures that movement speed while a third, and final, level measures how that speed relates to defined, energetically variable movement constraints. Indeed, we could convert our measured speed values with respect to incline and/or trichome density to power costs using respirometry experiments similar to those conducted by Full and Tullis [37,39] on cockroaches. Pragmatically however, the simple measurement of speed should be enough to indicate motivation, especially given that the relationship between speed and metabolic rate is so prescribed [38].

The value in effectively measuring the work rates exhibited by insects in response to semiochemicals is most obvious in 'push-pull' operations [11], designed to attract pests to specific sites where they can be treated with pesticides. Insects operating in the wild will be exposed to natural variation in their own energy landscapes via incline, surface roughness, trichome density, wind strength, surface roughness etc., so a demonstration of metabolically costly

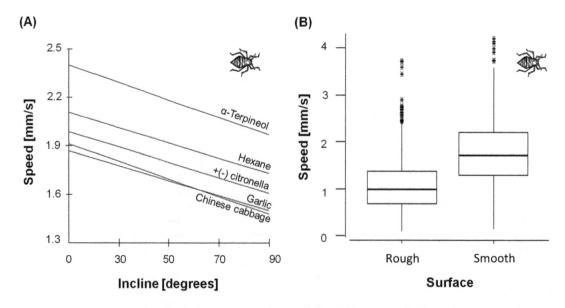

Figure 6. Changes in aphid movement speed as a function of incline and surface roughness. (**A**) The predicted effect of incline upon aphid speed to move away from chemicals, from a model in which the intercept and slope of the effect of incline is allowed to vary with respect to chemical (i.e. a random slope, random intercept model), and in which our control condition is used as the reference category. See methods for further details. (**B**) Box and whisker plot showing the overall effect of surface type upon speed (LMM: Wald = 93.94, df = 1, P<0.001); shown is the median (bold line), inter-quartile ranges (box), 95% confidence intervals (whiskers), and outliers (*).

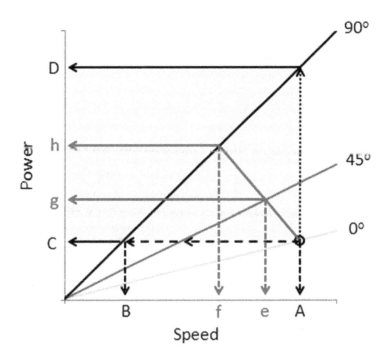

Figure 7. Schematic relationship between speed, incline and power of an insect climbing up 3 slopes of different inclines (after Full & Tullis 1990), showing how the results gained within the energy landscape olfactometer relate to energy expenditure. The grey box shows the proposed operational area of subjects. Walking towards an attractive semio-chemical, insects can either maintain power use at a constant level for the varying inclines, in which case speed is expected to drop with increasing incline (arrows terminating at A and B), or they can maintain speed, in which case power requirements increase with increasing incline (arrows terminating at C and D). In reality (red lines), animals are likely to operate somewhere between these two extremes (see arrows terminating in 'e' and 'f' for speed and 'g' and 'h' for power) with the more motivated subjects tending to maintain speed and incur increased power use with increasing incline.

movement across landscapes is a powerful indicator of semio-chemical value. Beyond consideration of the overall response however, the frequency distribution of speeds used under the varying energy landscapes (Fig. 3) also highlights whether there are differentially motivated groups within the sample taken [40]. Bimodality in travel speed, for example, may indicate a difference in susceptibility to semiochemicals between e.g. sexes or nutritional state [41], something that may prove important in control measures. This issue would have been better defined in our own experimental protocol if we had followed marked individuals through various energy landscapes. Future work can address this, and perhaps clarify how individuals vary over time.

An apparent inconvenience in using inclines as energy landscapes for pine weevils, and a number of other species, is the preference for these insects to walk up inclines anyway (Fig. 3). We effectively dealt with this tendency by having controls, and cognisance of this will be important for trials on any species that displays a similar reaction to inclines. Our controls allow speed as a function of incline to be compared directly to that exhibited during a semiochemical experiment but future work may prefer to simply subtract semio-chemical-induced mean speeds from those of controls although this simpler approach may obscure some patterns. However, experimental protocol with proper controls can deal with this so that the effect can be subtracted from their response. A climbing pattern was not obvious in our aphid experiment. In fact, incline is much less important in modulating travel speed in aphids than in pine weevils. In essence, the work done in climbing should be related to the gain in potential energy (E_p), given by $E_p = mgh$, where m is the mass, g is the gravitational constant and h is the height climbed, so mass-specific work should be the same for both species, with the rate of work given by the

speed and the incline. However, smaller animals have higher mass-specific resting metabolic rates while travel costs scale linearly with mass [39,42], which explains why Lipp et al. [43] found no obvious difference in VO_2 *versus* incline in ants. Indeed, the effect of an elevated resting metabolic rate with respect to travel costs means that anything that slows down rates of travel increases the costs of travel correspondingly and it is for this reason that we designed the hair-density energy landscape as an additional experimental protocol for the aphids. The concept also elicited the expected result (Fig. 6b).

Our work presents two types of energy landscape olfactometers as case studies to assess their utility in insect pest programs. The results are very encouraging, demonstrating that the way insects modulate their speed in relation to the difficulty of traversing variable terrain provides a more complex, and certainly different, response to conventional methods using Y-tube olfactometers or EAGs. Thus, the approach gives both another useful measure of semiochemical attraction or repellency and highlights, and gives metrics for, a more intricate response pattern. Future studies could examine energy landscapes produced in a variety of others ways such as using air currents or a variably constructed 'vegetation'-simulating matrix for flying insects or differential surface textures and substrates for walking or burrowing species, respectively. Such work should help verify the real value of energy landscape olfactometers as an additional tool to help our attempts to control pests with minimum detriment to the environment.

Acknowledgments

We thank Neem Biotech for providing botanicals, Koppert for providing peach potato aphids and UPM Tilhill for providing adult pine weevils.

Author Contributions

Conceived and designed the experiments: RPW TB. Performed the experiments: RR AH JP YKG. Analyzed the data: AJK RR AH. Wrote the paper: RPW TB AJK RR AH JP YKG.

References

1. Githeko AK, Lindsay SW, Confalonieri UE, Patz JA (2000) Climate change and vector-borne diseases: a regional analysis. Bulletin of the World Health Organization 78: 1136–1147.
2. Ansari MA, Pope EC, Carpenter S, Scholte EJ, Butt TM (2011) Entomopathogenic fungus as a biological control for an important vector of livestock disease: the Culicoides biting midge. PLoS One 6: e16108.
3. Gray ME, Sappington TW, Miller NJ, Moeser J, Bohn MO (2009) Adaptation and invasiveness of western corn rootworm: intensifying research on a worsening pest. Annual Review of Entomology 54: 303–321.
4. De Barro PJ, Liu S-S, Boykin LM, Dinsdale AB (2011) *Bemisia tabaci*: a statement of species status. Annual Review of Entomology 56: 1–19.
5. Kesavachandran CN, Fareed M, Pathak MK, Bihari V, Mathur N, et al. (2009) Adverse health effects of pesticides in agrarian populations of developing countries. Reviews of Environmental Contamination and Toxicology 200: 33–52.
6. Alavanja MC, Hoppin JA, Kamel F (2004) Health Effects of Chronic Pesticide Exposure: Cancer and Neurotoxicity 3. Annual Review of Public Health 25: 155–197.
7. Nauen R (2007) Insecticide resistance in disease vectors of public health importance. Pest Management Science 63: 628–633.
8. Rivero A, Vezilier J, Weill M, Read AF, Gandon S (2010) Insecticide control of vector-borne diseases: when is insecticide resistance a problem? PLoS pathogens 6: e1001000.
9. Khan ZR, James DG, Midega CA, Pickett JA (2008) Chemical ecology and conservation biological control. Biological Control 45: 210–224.
10. Margni M, Rossier D, Crettaz P, Jolliet O (2002) Life cycle impact assessment of pesticides on human health and ecosystems. Agriculture, Ecosystems & Environment 93: 379–392.
11. Pickett J, Wadhams L, Woodcock C (1997) Developing sustainable pest control from chemical ecology. Agriculture, Ecosystems & Environment 64: 149–156.
12. Cook SM, Khan ZR, Pickett JA (2006) The use of push-pull strategies in integrated pest management. Annual Review of Entomology 52: 375.
13. Khan Z, Midega C, Pittchar J, Pickett J, Bruce T (2011) Push-pull technology: a conservation agriculture approach for integrated management of insect pests, weeds and soil health in Africa: UK government's Foresight Food and Farming Futures project. International Journal of Agricultural Sustainability 9: 162–170.
14. Blackmer J, Rodriguez-Saona C, Byers J, Shope K, Smith J (2004) Behavioral response of *Lygus hesperus* to conspecifics and headspace volatiles of alfalfa in a Y-tube olfactometer. Journal of Chemical Ecology 30: 1547–1564.
15. Acar E, Medina J, Lee M, Booth G (2001) Olfactory behavior of convergent lady beetles (Coleoptera: Coccinellidae) to alarm pheromone of green peach aphid (Hemiptera: Aphididae). The Canadian Entomologist 133: 389–397.
16. Fraser AM, Mechaber WL, Hildebrand JG (2003) Electroantennographic and behavioral responses of the sphinx moth Manduca sexta to host plant headspace volatiles. Journal of Chemical Ecology 29: 1813–1833.
17. Birkett M, Chamberlain K, Khan Z, Pickett J, Toshova T, et al. (2006) Electrophysiological responses of the lepidopteran stemborers *Chilo partellus* and *Busseola fusca* to volatiles from wild and cultivated host plants. Journal of Chemical Ecology 32: 2475–2487.
18. Solé J, Sans A, Riba M, Guerrero Á (2010) Behavioural and electrophysiological responses of the European corn borer *Ostrinia nubilalis* to host-plant volatiles and related chemicals. Physiological Entomology 35: 354–363.
19. Ballhorn DJ, Kautz S, Heil M (2013) Distance and sex determine host plant choice by herbivorous beetles. PloS one 8: e55602.
20. Ho H-Y, Millar JG (2002) Identification, electroantennogram screening, and field bioassays of volatile chemicals from Lygus hesperus Knight (Heteroptera: Miridae). Zoological Studies Taipei 41: 311–320.
21. Weiner B (1992) Human Motivation: Metaphors, Theories, and Research.: Sage Publications, Thousand Oaks, CA. USA.
22. Ragaert P, Verbeke W, Devlieghere F, Debevere J (2004) Consumer perception and choice of minimally processed vegetables and packaged fruits. Food Quality and Preference 15: 259–270.
23. Dawkins MS (1990) From an animal's point of view: motivation, fitness, and animal welfare. Behavioral and brain sciences 13: 1–9.
24. Duncan I, Kite V (1989) Nest site selection and nest-building behaviour in domestic fowl. Animal Behaviour 37: 215–231.
25. Galhardo L, Almeida O, Oliveira RF (2011) Measuring motivation in a cichlid fish: an adaptation of the push-door paradigm. Applied Animal Behaviour Science 130: 60–70.
26. Boleij A, van Klooster J, Lavrijsen M, Kirchhof S, Arndt SS, et al. (2012) A test to identify judgement in mice. Behavioural Brain Research 233: 45–54.
27. Mendl M, Burman OH, Parker R, Paul ES (2009) Cognitive bias as an indicator of animal emotion and welfare: emerging evidence and underlying mechanisms. Applied Animal Behaviour Science 118: 161–181.
28. Wilson RP, Quintana F, Hobson VJ (2012) Construction of energy landscapes can clarify the movement and distribution of foraging animals. Proceedings of the Royal Society B: Biological Sciences 279: 975–980.
29. Shepard EL, Wilson RP, Rees WG, Grundy E, Lambertucci SA, et al. (2013) Energy landscapes shape animal movement ecology. The American Naturalist 182: 298–312.
30. Koschier EH, de Kogel WJ, Visser J. (2000) Assessing the attractivenss of volatile plant compounds to western flower thrips *Frankliniella occidentalis*. Journal of Chemical Ecology 26: 2643–2655.
31. Nordlander G (1987) A method for trapping *Hylobius abietis* (L.) with a standardized bait and its potential for forecasting seedling damage. Scandinavian Journal of Forest Research 2: 199–213.
32. Jönsson M, Lindkvist A, Anderson P (2005) Behavioural responses in three ichneumonid pollen beetle parasitoids to volatiles emitted from different phenological stages of oilseed rape. Entomologia Experimentalis et Applicata 115: 363–369.
33. Reddy G, Holopainen J, Guerrero A (2002) Olfactory responses of *Plutella xylostella* natural enemies to host pheromone, larval frass, and green leaf cabbage volatiles. Journal of Chemical Ecology 28: 131–143.
34. Peñuelas J, Llusià J (2001) The complexity of factors driving volatile organic compound emissions by plants. Biologia Plantarum 44: 481–487.
35. Legaspi JC, Simmons AM, Legaspi BC (2011) Evaluating Mustard as a Potential Companion Crop for Collards to Control the Silverleaf Whitefly, *Bemisia argentifolii* (Hemiptera: Aleyrodidae): Olfactometer and Outdoor Experiments. Subtropical Plant Science 63: 36–44.
36. Egigu MC, Ibrahim MA, Yahya A, Holopainen JK (2011) *Cordeauxia edulis* and *Rhododendron tomentosum* extracts disturb orientation and feeding behavior of *Hylobius abietis* and *Phyllodecta laticollis*. Entomologia Experimentalis Et Applicata 138: 162–174.
37. Full RJ, Tullis A (1990) Energetics of ascent: insects on inclines. Journal of Experimental Biology 149: 307–317.
38. Watson RR, Rubenson J, Coder L, Hoyt DF, Propert MW, et al. (2011) Gait-specific energetics contributes to economical walking and running in emus and ostriches. Proceedings of the Royal Society B: Biological Sciences 278: 2040–2046.
39. Dickinson MH, Farley CT, Full RJ, Koehl M, Kram R, et al. (2000) How animals move: an integrative view. Science 288: 100–106.
40. Shepherd R (1999) Social determinants of food choice. Proceedings of the Nutrition Society 58: 807–812.
41. Rogowitz GL, Chappell MA (2000) Energy metabolism of eucalyptus-boring beetles at rest and during locomotion: gender makes a difference. Journal of Experimental Biology 203: 1131–1139.
42. Tucker VA (1970) Energetic cost of locomotion in animals. Comparative Biochemistry and Physiology 34: 841–846.
43. Lipp A, Wolf H, Lehmann F-O (2005) Walking on inclines: energetics of locomotion in the ant *Camponotus*. Journal of Experimental Biology 208: 707–719.

14

Insecticide Resistance Status of United States Populations of *Aedes albopictus* and Mechanisms Involved

Sébastien Marcombe[1¤a], Ary Farajollahi[1,2], Sean P. Healy[3¤b], Gary G. Clark[4], Dina M. Fonseca[1]*

1 Center for Vector Biology, Rutgers University, New Brunswick, New Jersey, United States of America, **2** Mercer County Mosquito Control, West Trenton, New Jersey, United States of America, **3** Monmouth County Mosquito Extermination Commission, Eatontown, New Jersey, United States of America, **4** Mosquito and Fly Research Unit, Agriculture Research Service, United States Department of Agriculture, Gainesville, Florida, United States of America

Abstract

Aedes albopictus (Skuse) is an invasive mosquito that has become an important vector of chikungunya and dengue viruses. Immature *Ae. albopictus* thrive in backyard household containers that require treatment with larvicides and when adult populations reach pest levels or disease transmission is ongoing, adulticiding is often required. To assess the feasibility of control of USA populations, we tested the susceptibility of *Ae. albopictus* to chemicals representing the main insecticide classes with different modes of action: organochlorines, organophosphates, carbamates, pyrethroids, insect growth regulators (IGR), naturalytes, and biolarvicides. We characterized a susceptible reference strain of *Ae. albopictus*, ATM95, and tested the susceptibility of eight USA populations to five adulticides and six larvicides. We found that USA populations are broadly susceptible to currently available larvicides and adulticides. Unexpectedly, however, we found significant resistance to dichlorodiphenyltrichloroethane (DDT) in two Florida populations and in a New Jersey population. We also found resistance to malathion, an organophosphate, in Florida and New Jersey and reduced susceptibility to the IGRs pyriproxyfen and methoprene. All populations tested were fully susceptible to pyrethroids. Biochemical assays revealed a significant up-regulation of GSTs in DDT-resistant populations in both larval and adult stages. Also, β-esterases were up-regulated in the populations with suspected resistance to malathion. Of note, we identified a previously unknown amino acid polymorphism (Phe → Leu) in domain III of the VGSC, in a location known to be associated with pyrethroid resistance in another container-inhabiting mosquito, *Aedes aegypti* L. The observed DDT resistance in populations from Florida may indicate multiple introductions of this species into the USA, possibly from tropical populations. In addition, the mechanisms underlying DDT resistance often result in pyrethroid resistance, which would undermine a remaining tool for the control of *Ae. albopictus*. Continued monitoring of the insecticide resistance status of this species is imperative.

Editor: Zach N. Adelman, Virginia Tech, United States of America

Funding: This work was funded by a cooperative Agreement between the United States Department of Agriculture (USDA) and Rutgers University (USDA-ARS-58-6615-8-105) entitled "Area-wide Pest Management Program for the Asian Tiger Mosquito in New Jersey." The funders had no role in study design, data collection and analysis, decision to publish, or preparation of the manuscript.

Competing Interests: The authors have declared that no competing interests exist.

* Email: dinafons@rci.rutgers.edu

¤a Current address: Pasteur Institute, Vientiane, Laos
¤b Current address: Department of Entomology, Louisiana State University Agricultural Center, Baton Rouge, Louisiana, United States of America

Introduction

Aedes (Stegomyia) albopictus (Skuse), the Asian tiger mosquito, is an aggressive human- and day-biting species native to Asia that has recently expanded to at least 28 countries outside its native range, and now occurs in all inhabitable continents [1]. Detailed theoretical analyses indicate that the spread of *Ae. albopictus* may well continue into many more regions of the world [1 3]. Although this species is often considered mostly an urban nuisance, it was the principal dengue vector in Hawaii and other areas were *Aedes aegypti* L. populations have been controlled [4] and in the summer of 2013, an autochthonous case of dengue in Suffolk County, New York has been attributed to thriving populations of *Ae. albopictus* [5]. Furthermore, since recent mutations in the chikungunya virus (CHIKV) increased the vector competence of *Ae. albopictus* for the viral agent [6,7], chikungunya has become epidemic in Africa and the Indian Ocean Basin [8]. Although chikungunya fever has not spread broadly in the temperate zone, an epidemic in northern Italy in 2007 sickened over 200 people [9] and small numbers of locally transmitted CHIKV cases were identified in southern France in 2010 [10], both of which were driven by local populations of *Ae. albopictus*. The European expansion of CHIKV would not have been possible without the prior invasion of that continent by *Ae. albopictus* [11].

Aedes albopictus is a container-inhabiting mosquito strongly associated with human habitats (especially outside its native range) and capable of ovipositing diapause-destined eggs that survive even in cold northern latitudes in parts of its native (*e.g.*, northern Japan, China) and introduced (*e.g.*, Europe and northeastern USA) ranges [12]. The first line of control against *Ae. albopictus* is often source reduction [13], but when containers cannot be removed or emptied, larvicides are used [13]. If adults become a serious nuisance, or disease outbreaks are ongoing or imminent, insecticides targeting the adults are applied [14].

Unfortunately, the development and spread of insecticide resistance represents a serious threat as it can lead to a reduction of the efficacy of larvicide or adulticide-based control programs, as demonstrated in the control of the main dengue vector *Ae. aegypti* [15,16]. In contrast to *Ae. aegypti*, there have been only a few reports of insecticide resistance in *Ae. albopictus* worldwide [16,17]. Several studies implemented in the 1960s and summarized by Mouchet et al. [18] showed that several populations of *Ae. albopictus* from Southeast Asia and India were resistant to some of the insecticides used at the time for vector control (*i.e.*, DDT, dieldrin and fenthion). A recent review by Ranson et al. [16] updated by Vontas et al. [19] summarized the levels of insecticide resistance in *Ae. albopictus* worldwide. It is apparent that resistance to the main families of insecticides currently or historically used for vector control across the world (*i.e.*, DDT, organophosphates and pyrethroids) has been found in *Ae. albopictus* [20–25]. In the USA, to our knowledge, only four studies have reported on insecticide resistance in *Ae. albopictus*: one population in Florida was resistant to the organophosphate malathion [26], populations in Texas and Illinois were also resistant to malathion [25,27], and resistance to a pyrethroid (deltamethrin) was found in a population from Alabama [28].

Insecticide resistance can be associated with mutations in the sequence of the target protein that induce insensitivity to the insecticide (target-site resistance), and/or to the up-regulation of detoxification enzymes (metabolic-based resistance). The main target site resistance mechanisms known in mosquitoes involve 1) amino acid substitutions in the voltage gated sodium channel that cause a resistance phenotype to pyrethroid (DDT) insecticides known as knockdown resistance (Kdr, [29] and 2) mutations in the acetylcholine esterase sequence that lead to insensitivity of this enzyme to organophosphates [30]. Metabolic-based resistance involves the bio-transformation of the insecticide molecule by enzymes and is now considered a key resistance mechanism in insects [31,32]. Three large enzyme families, the cytochrome P450 monooxygenases (P450s), glutathione S-transferases (GSTs), and carboxy/cholinesterases (CCEs) have been implicated in the metabolism of insecticides [32–34]. So far, compared to other mosquito species of importance such as *Anopheles* spp., *Culex* spp., and *Ae. aegypti*, very little is known about the molecular or biochemical basis of resistance in *Ae. albopictus* and, in particular, to our knowledge, no studies have specifically examined the underlying mechanisms of resistance in USA *Ae. albopictus*.

The objective of the present study was to determine the insecticide resistance status of *Ae. albopictus* across the full latitudinal range of the species in the USA. Specifically, we examined populations from New Jersey, Pennsylvania, and Florida (Table 1). We chose eleven chemicals that represent the main classes of insecticides historically or currently used for mosquito control (Table 2), including some that have only recently been adopted. We compared the levels of resistance of field-collected specimens to a susceptible strain of *Ae. albopictus* that we characterized for this purpose (reference strain ATM95). In addition, we used biochemical and molecular assays to identify putative resistance mechanisms in *Ae. albopictus* such as target-site mutations and up-regulation of detoxifying enzymes.

Materials and Methods

Ethics statement

No specific permits were required for collection of field specimens, which were performed in urban and suburban backyards in the US states of New Jersey, Pennsylvania, and Florida with homeowners assent by professional county mosquito control personnel. These studies did not involve endangered or protected species. In the laboratory, mosquito colonies were blood fed on quail, *Colinus virginianus*, under the guidelines of the Rutgers University Animal Use Protocol# 86–129 that was approved by the Rutgers IACUC.

Mosquito strains and collection

We characterized a reference laboratory strain (ATM95) and tested eight field populations of *Ae. albopictus* (Table 1). *Aedes albopictus* was first detected in New Jersey (NJ) on August 1, 1995 in a standard NJ light trap collection in Keyport [35]. Surveillance at a marina 300 m from the trap site yielded *Ae. albopictus* larvae from one discarded bucket and 2 tires and a colony started from this population, now named ATM95, has been continuously reared in the laboratory at the Center for Vector Biology at Rutgers University in New Brunswick, NJ without exposure to insecticides. Preliminary bioassays on the ATM95 strain showed that this strain could be considered susceptible in comparison to previous results from the literature. The field caught *Ae. albopictus* samples were collected as larvae, pupae, or eggs (ovitraps) in one site in Bergen county, NJ (NJBer, N 40°47′33′′, W 74°1′32′′), two replicate sites (less than 5 km apart) in Mercer county, NJ (NJMer1, NJMer2, N 40°13′1′′ W 74°44′35′′), two sites in Monmouth county, NJ (NJMon1 and NJMon2, N 40°26′36′′ W 74°13′5′′), one site in York county, Pennsylvania (PA, N 39°57′46′′ W 76°43′41′′) and two sites in St. Johns county, Florida (FL1 and FL2, N 29°53′39′′ W 81°18′48′′) during the 2011 active mosquito season (Figure 1). All stages were reared to adults in the laboratory on a diet of powdered cat food. After emergence of female *Ae. albopictus* they were provided restrained quails (*Colinus virginianus*) as sources of blood for egg development following the Rutgers University Animal Use protocol# 86–129. Larvae and adults obtained from the F_1 progeny were used for bioassays and biochemical and molecular studies.

Bioassays

We chose to test the susceptibility of *Ae. albopictus* to a range of insecticides representative of those historically and currently used for mosquito control in the USA from all main families of insecticides with different modes of action (Table 2).

Larval bioassays. Larval bioassays were carried out using the water-dispersible granule formulation (VectoBac WDG, Valent BioSciences, Libertyville, IL, USA) of *Bacillus thuringiensis* var. *israelensis* (*Bti*) (37.4% ai, 3000 ITU/mg). The remaining insecticides were tested by diluting the active ingredients (ai) purchased from Sigma-Aldrich (Seelze, Germany) in ethanol to required levels according to WHO guidelines [36]. We tested temephos (97.3% active ingredient [ai]), propoxur (99.8%), spinosad (97.6%), methoprene (95.6%), and pyriproxyfen (99.1%). All bioassays were performed using late third and early fourth-instars of *Ae. albopictus*.

To determine the activity range of the larvicides in *Ae. albopictus*, larvae of the susceptible laboratory strain, ATM95, were exposed to 3 replicates of a wide range of test concentrations. For each bioassay, 25 larvae of each population were transferred to plastic cups containing 99 mL of distilled water with 1 mL of the insecticide at the desired concentration. The appropriate volume

Table 1. Detailed description with geographic and socio-economic information of the sources of mosquito populations.

State	County	Municipality	Mosquito population name abbreviations	Coordinates	Altitude	Human density inhabitants/ Km^2
New Jersey	Bergen	Elmwood Park	NJBer	40°54′N74°70′W	14 m	2,829
	Mercer	Trenton	NJMer1	40°13′N74°45′W	15 m	4,286
		Ewing	NJMer2	40°15′N74°47′W	38 m	906
	Monmouth	Middletown	NJMon1	40°24′N74°04′W	30 m	626
		Belmar	NJMon2	40°10′N74°01′W	4 m	2,140
Pennsylvania	York	York	PA	39°57′N76°43′W	121 m	3,061
Florida	St John's	St Augustine south	FL1	29°50′N81°18′W	7 m	1,118
		St Augustine Beach	FL2	29°53′N81°18′W	0 m	936

Population name abbreviations are used throughout the text.

of dilution from the stock solution was added to the water in the cups to obtain the desired target dosage, starting with the lowest concentration. Four cups per concentration (100 larvae) and 4 to 8 concentrations in the activity range of the insecticide (between 10% and 95% mortality) were used to determine LC_{50} and LC_{90} values (LC: lethal concentration). Control treatments were made with 99 mL of distilled water and 1 mL of ethanol. Larval mortality was recorded after 24 h exposure except for pyriproxyfen and methoprene for which mortality was recorded every 24 h until emergence due to the delayed action of these insect growth regulators. In this case, larvae were provided with food at a concentration of 100 mg/L every day. For each bioassay, temperature was maintained at 27°C in an incubator with a 16L:8D photoperiod.

Adult bioassays. Adult bioassays also followed WHO protocols [37], with 3 to 5 day old females of each F_1 progeny used for tarsal contact tests with insecticide-treated filter paper and compared with the susceptible ATM95 strain. We started with technical grade (Pestanal Sigma-Aldrich, Seelze, Germany) deltamethrin (99.7% ai, type II pyrethroid), prallethrin (96.2%, type I), phenothrin (94.4%, type I), malathion (97.2%), and DDT (99.7%). Insecticide was applied to filter paper by dripping evenly onto the paper 2 mL of technical grade chemical dissolved in acetone and silicone oil to the appropriate concentration [37]. Concentrations were expressed in w/w percentage of the active

ingredient in silicone oil. Filter papers were dried for 24 h before the test. The resistance status of *Ae. albopictus* populations from each locality was determined by using WHO discriminating dosages (DD; double concentration of LC_{99}) of deltamethrin (0.05%), malathion (0.8%), and DDT (4%). Preliminary bioassays conducted on the ATM95 strain displayed that the discriminating dosages for prallethrin and phenothrin were 1% and 1.5%, respectively. Those two pyrethroids are used in combination in the newly available Duet dual-action adulticide formulation (Clarke Mosquito Control, Roselle, Illinois, USA) for adult mosquito control. For each strain, five batches of 20 non-blood fed females (2–5 days old; n = 100) were exposed to the insecticides in WHO test kits for 60 min to estimate the knock down effect (KDT_{50} and KDT_{90}) of the insecticides. The number of knocked down mosquitoes in the tubes was counted every 2 minutes. The adults were then transferred into holding tubes, were provided with sugar solution (10%), and kept at 27°C with a relative humidity of 80%. Mortality was recorded 24 h later. Mosquitoes exposed for 1 h to paper impregnated with the carrier (silicone oil) mixed with acetone were used as controls. Tests were replicated twice when the number of available mosquitoes was suitable. Following WHO criteria a population is considered resistant if the mortality after 24 h is under 90%, resistance is suspected with mortality between 90 and 98% and a population is susceptible with mortality over 98%.

Table 2. Name, class, and mode of action of all insecticides tested in this study.

Status	Insecticide	Family	Mode of action
Larvicide	*Bti*	Biolarvicide	Cell membrane destruction
	Spinosad	Naturalyte	Nicotinic acetylcholine receptor
	Temephos	Organophosphate	Acetylcholinesterase inhibitor
	Propoxur	Carbamate	
	Methoprene	Insect Growth Regulator	Juvenile hormone mimics
	Pyriproxyfen		
Adulticide	Malathion	Organophosphate	Acetylcholinesterase inhibitor
	DDT	Organochlorine	Sodium channel modulator
	Deltamethrin	Pyrethroid	
	Prallethrin		
	Phenothrin		

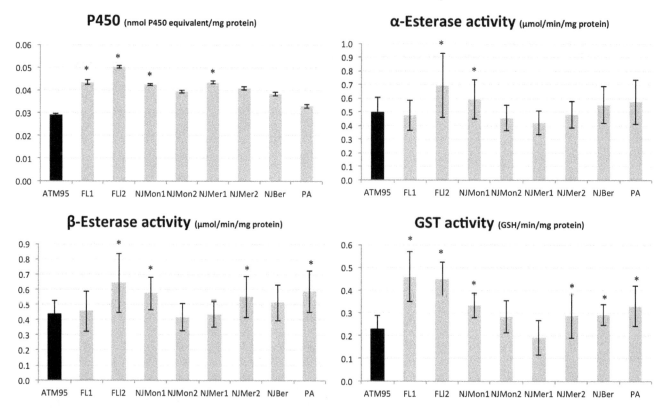

Figure 1. Global amount or activity of detoxification enzymes in *Aedes albopictus* larvae from field populations and the laboratory strain (ATM95): cytochrome P450 monooxygenases (P450s), Esterase (α and β-CCEs), and Glutathione-S transferases (GSTs). Sample sizes are 47 specimens/population (15 for P450, n = 3). Confidence intervals are one standard deviation of the mean. An asterisk (*) denotes significantly up-regulated values compared to the susceptible reference strain ATM95, Tukey-Kramer test.

Larval and adult knock down times (KDT) were analyzed with the log-probit method of Finney [38] using the Sakuma Probit software [39]. Data from all replicates were pooled for analysis. Lethal concentrations (LC$_{50}$ and LC$_{95}$ for larvae) and knock-down time (KDT$_{50}$ and KDT$_{95}$ for adults) were calculated together with their 95% confidence intervals. Adult mortality after 24 h exposure was also recorded for each population. Compared to the susceptible ATM95 strain field populations were considered as having some resistance to a given insecticide when their LC$_{50/95}$ or KDT$_{50/95}$ ratios (resistance ratio: RR$_{50/95}$) had confidence limits that excluded the value 1. We considered resistance to be moderate to strong when RR$_{50/95}$ values rose above 2.

Biochemical assays

The levels of P450 monooxygenases (P450s), and the activities of carboxy/cholinesterases (CCEs) and glutathione S-transferases (GSTs) were assayed from single 3 days-old F$_1$ females (n = 47) following microplate methods described by Hemingway [32] and Brogdon [40] on an Epoch spectrophotometer (BioTek, Vermont, USA). Total protein quantification of mosquito homogenates was performed using Bradford reagent with bovine serum albumin as the standard protein [41] to normalize enzyme activity levels by protein content. For P450 assays, the OD values were measured at 620 nm after 30 min incubation of individual mosquito homogenate with 200 μL of 2 mM 3, 3', 5, 5'-tetramethylbenzidine dihydrochloride (TMBZ) and 25 μL of 3% hydrogen peroxide and the quantity was determined from cytochrome-c standard curve. Nonspecific α- and β-CCEs activities were assayed by 10 min incubation of mosquito homogenate in each well with 100 μL of

3 mM napthyl acetate (either α- or β-) at room temperature and the OD values were measured at 540 nm. The activity was determined from α- or β-naphtol standard curves. Glutathione-S-transferases activity was measured in the reaction containing 2 mM reduced glutathione and 1 mM 1-chloro-2,4-dinitrobenzene (CDNB). The reaction rates were measured at 340 nm after 20 min, and the activity was expressed in nmoles GSH conjugated/min/mg protein.

Statistical comparisons of detoxification enzyme levels between ATM95 and the field populations were assessed with Tukey-Kramer tests in JMP8.0.1 (SAS Institute, Cary, North Carolina, USA) using a P value threshold of 0.05. Tukey-Kramer HSD (honestly significant difference) test is a highly conservative test that accounts for multiple comparisons [42].

Kdr genotyping

We extracted DNA from 14 adult *Ae. albopictus* collected in Florida (FL1 and FL2) using DNAeasy tissue kits (Qiagen, Valencia, California, USA). We chose 6 survivors and 6 dead specimens following DDT exposure and amplified portions of domains II, III, and IV of the voltage-gated sodium channel (VGSC), a known target of DDT and pyrethroid insecticides, using primers from Kasai et al. [43]. Specifically we amplified and sequenced domain II with aegSCF20 and aegSCR21, domain II with aegSCF7 and aegSCR8, and domain IV with albSCF6 and albSCR8. Our PCR was composed of 1× PCR buffer, 2.5 mM of MgCl$_2$ (2.0 mM for Domain III), 200 μM of each dNTP, 0.2 mg/ml of BSA, 0.2 μM of each primer, and 1 unit of *TaqGold* (Applied Biosystems, Foster City, California, USA). The PCR cycle started with a 10 min denaturation (and *TaqGold* activation) at 96°C followed by 40 cycles of 30 s at 96°C, 30 s at 55°C (Domain II and IV) or 53°C (Domain III) and 45 s at 72°C, and a final extension of 10 min at 72°C. The PCR products were cleaned with ExoSAP-IT (USB, Cleveland, Ohio, USA) and cycle sequenced for analyses on an ABI 3100 automated sequencer (Applied Biosystems). Sequences were cleaned and checked with Sequencher 5.0 (Gene Codes, Ann Harbor, Michigan, USA).

Enzymatic phenotyping of Ache1

The phenotypes of the acetylcholine esterase AChE1, encoded by the ace-1 gene, were examined in each population (n = 24) using the previously described TDP test [44] adapted for *Ae. albopictus* with both dichlorvos and propoxur concentrations of 1.10^{-2} M. The TDP test identifies all possible phenotypes containing the G119S, F290V and wild-type (susceptible) alleles.

Results

Larval and adult bioassays

Larval bioassays resulted in low resistant ratios (RRs) indicating that none of the eight USA populations of *Ae. albopictus* were resistant to the larvicides tested (Table 3). However, one of the populations from Florida, FL2, showed significant resistance to both methoprene and pyriproxyfen (IGRs) with RRs of 3.72 and 2.36 fold, respectively. Further, all the populations had values of RRs for propoxur that excluded 1, ranging from 1.47 (NJMon1) to 2.8 fold (FL1 and FL2); the latter indicating significant resistance to this carbamate in Florida populations. The insecticidal activities of the larvicides used against the ATM95 strain (Table 3) can be ranked as follows: pyriproxyfen > methoprene > temephos > *Bti* > spinosad > propoxur with LC_{50} of 9.4E-6, 1.4E-4, 5.4E-3, 0.07, 0.1 and 1.02 mg/L, respectively.

The knockdown times (KDT) for *Ae. albopictus* exposed to DDT indicated that most KDT_{50} values from field populations were higher (non overlapping 95% CIs) than those of the reference strain, ATM95, except for NJMer1 and NJBer that showed lower KDT_{50} (Table 4). The two populations from Florida, FL1 and FL2, showed the highest RRs (1.61 and 1.88 respectively) for DDT. For deltamethrin the RRs ranged from 1.13 (NJMer2) to 1.74 (NJMon2) indicating that all the populations were susceptible. Likewise, for phenothrin the KDTs were lower than those of the susceptible strain and for prallethrin the RR_{50} did not exceed 1.18 (FL1). Of note, the two populations from Florida (FL1 and FL2) showed RRs with values of 2.16 and 2.34, respectively, for malathion. The RR_{50} for malathion for the remaining populations were low but significantly higher than 1 and ranged from 1.15 to 1.67.

Adult mortality after a 24 h exposure to the pyrethroid insecticides (deltamethrin, prallethrin, and phenothrin) at discriminating doses indicated that, like the ATM95 strain, all the field populations tested can be considered susceptible (99–100% mortality; Table 4). However, the two populations from Florida (FL1 and FL2) showed resistance to DDT (75 and 54% mortality, respectively) and a population from New Jersey (NJMon2) also showed resistance to this organochlorine (87% mortality). In addition, resistance to malathion was found in the two populations from Florida (FL1 and FL2) with 86 and 80% mortality, respectively. Finally, the populations from New Jersey (NJMon2, NJMer1, and NJBer) showed suspected resistance to malathion with 95, 96 and 93% mortality, respectively (Table 4).

Table 3. Resistance status of larvae *Aedes albopictus*.

Population	Bti LC₅₀ (95% CI)	RR₅₀	Temephos LC₅₀ (95% CI)	RR₅₀	Propoxur LC₅₀ (95% CI)	RR₅₀	Spinosad LC₅₀ (95% CI)	RR₅₀	Methoprene LC₅₀ (95% CI)	RR₅₀	Pyriproxyfen LC₅₀ (95% CI)	RR₅₀
ATM95	0.07 (0.066-0.071)	1	5.4E-03 (5.1E-03-5.7E-03)	1	1.02 (0.93-1.09)	1	0.10 (0.036-0.106)	1	1.4E-04 (9.9E-05-1.7E-04)	1	9.4E-06 (3.6E-06-2.5E-05)	1
FL1	0.07 (0.01-0.07)	0.99	5.3E-03 (4.9E-03-5.3E-03)	0.99	2.87 (2.67-3.15)	**2.82**	0.15 (0.145-0.162)	1.51	-	-	1.5E-05 (1.2E-05-1.9E-05)	1.57
FL2	0.06 (0.043-0.085)	0.84	5E-03 (5E-03-5.6E-03)		2.83 (2.59-3.29)	**2.77**	0.16 (0.151-0.165)	1.56	5.1E-04 (3.1E-04-1E-03)	**3.72**	2.2E-05 (1.7E-05-2.9E-05)	**2.36**
NJMon1	0.12 (0.108-0.145)	**1.78**	4.7E-03 (4.5E-03-4.9E-03)	0.87	1.50 (1.32-1.79)	**1.47**	0.14 (0.138-0.147)	1.42	1.6E-04 (1.2E-04-2.1E-04)	1.15	4.7E-06 (2.6E-06-6.5E-06)	**0.50**
NJMon2	0.11 (0.10-0.13)	**1.68**	6.1E-03 (5.7E-03-6.6E-03)	1.14	1.62 (1.44-1.9)	**1.59**	0.10 (0.091-112)	1.01	7.4E-05 (3.1E-05-1.2E-4)	**0.54**	3.6E-06 (2.2E-06-8.3E-06)	**0.38**
NJMer1	0.08 (0.06-0.1)	1.16	6.1E-03 (5.6E-03-7E-03)	1.13	1.73 (1.62-1.89)	**1.69**	0.14 (0.109-0.29)	1.38	1.7E-04 (1E-4-2.6E-04)	**1.22**	5.7E-06 (3.3E-06-8.2E-06)	**0.60**
NJMer2	0.08 (0.073-0.089)	1.19	6.3E-03 (5.8E-03-6.8E-03)	1.17	2.09 (1.68-3.26)	**2.05**	0.08 (0.068-0.089)	0.79	9.9E-05 (8.8E-6-3.E-04)	0.71	1.3E-05 (9.5E-06-1.7E-05)	1.37
NJBer	0.05 (0.047-0.057)	0.76	6.9E-03 (6.4E-03-7.4E-03)	1.27	2.13 (1.83-2.82)	**2.09**	0.16 (0.142-0.189)	1.56	4.5E-05 (1.4E-05-8.4E-05)	**0.33**	1.7E-05 (1.2E-05-2.4E-05)	**1.81**
PA	0.08 (0.069-0.085)	1.13	7.6E-03 (6.8E-03-8.8E-03)	1.41	1.94 (1.66-2.41)	**1.90**	0.18 (0.144-0.309)	1.73	1.1E-04 (5.7E-05-2.3E-04)	0.78	1.0E-05 (7.8E-06-1.4E-05)	1.11

ATM95: susceptible reference strain; LC₅₀: Lethal Concentration that kills 50% of the population (mg/L); RR₅₀: Resistant Ratio = LC₅₀ susceptible strain (ATM95)/LC₅₀ field population; CI: Confidence Interval. Significant RRs are shown in bold (P<0.05).

Table 4. Knock down times (min), Resistant ratio, and mortality rates (after 24 h) of *Aedes albopictus* after exposure to insecticides at the diagnostic doses (WHO tube test).

Population	DDT (4%)			Deltamethrin (0.05%)			Phenothrin (1.5%)			Prallethrin (1%)			Malathion (0.8%)		
	KDT$_{50}$ (95% CI)	RR$_{50}$	Mortality (%)	KDT$_{50}$ (95% CI)	RR$_{50}$	Mortality (%)	KDT$_{50}$ (95% CI)	RR$_{50}$	Mortality (%)	KDT$_{50}$ (95% CI)	RR$_{50}$	Mortality (%)	KDT$_{50}$ (95% CI)	RR$_{50}$	Mortality (%)
ATM95	33 (30-35)	1	100	6 (6.3-6.6)	1	100	8.4 (8.2-8.7)	1	100	1.37 (1.29-1.44)	1	100	23 (22.2-23.8)	1	100
FL1	53 (51-56)	**1.61**	72	9.8 (9.6-10.1)	**1.57**	100	7.9 (7.6-8.3)	0.94	100	1.61 (1.46-1.82)	**1.18**	100	49.6 (47.4-52.3)	**2.16**	86
FL2	62 (59-68)	**1.88**	54	10.7 (10.4-10.9)	**1.70**	99	7.7 (7.4-8)	**0.91**	100	1.35 (1.28-1.43)	0.99	99	53.5 (50.5-57.6)	**2.34**	80
NJMon1	42 (41-44)	1.28	100	8.4 (7.9-8.9)	**1.34**	100	6.6 (6.4-6.8)	**0.79**	100	0.81 (0.3-0.88)	**0.59**	99	27.9 (26.9-28.8)	**1.22**	100
NJMon2	42 (40-45)	1.28	87	11.2 (10.9-11.5)	**1.78**	100	7.6 (7.3-7.9)	**0.9**	100	1.37 (1.25-1.5)	1.04	100	36.3 (35.1-37.5)	**1.59**	95
NJMer1	28 (24-30)	0.83	95	10.2 (9.4-11)	**1.62**	100	7.1 (6.9-7.4)	**0.85**	100	1.48 (1.29-1.78)	1.08	100	34.3 (33.5-35.2)	**1.5**	96
NJMer2	47 (45-49)	1.4	100	7.1 (6.8-7.4)	1.13	100	6.6 (5.9-7.2)	**0.78**	100	1.12 (1-1.21)	**0.83**	100	26.4 (25.5-27.3)	**1.15**	99
NJBer	27 (26-27.3)	0.8	99	10.3 (9.9-10.8)	**1.65**	100	8.3 (7.9-8.8)	**0.98**	100	1.13 (0.99-1.25)	**0.83**	100	38.3 (36-40.4)	**1.67**	93

ATM95: susceptible reference strain. KDT$_{50}$: Knock down time where 50% of the mosquitoes are knocked down (min); RR$_{50}$: Resistant Ratio = KDT$_{50}$ ATM95/KDT$_{50}$ field population. Significant RRs are shown in bold.

Detoxification enzyme levels

Comparison of constitutive detoxification enzyme activities between ATM95 and the field strains revealed significant differences in both larval and adult stages (Figure 1 and 2). The P450s levels were significantly higher in larvae from Florida (both FL1 and FL2), NJMon1, and NJMer1 populations. The FL2 and NJMon1 had significantly higher α- and β-ESTs activities and GSTs activities were significantly higher in most populations, particularly in FL1 and FL2, but not in NJMon2 and NJMer1 (Figure 1). In adults, only NJMer2 showed significantly up-regulated P450s, and only NJMer2 had significantly higher α-ESTs activities. The two populations from Florida and NJMer2 had significantly higher β-ESTs activities. Finally, except for NJMer1 and NJBer, all populations had significantly higher GSTs activities (Figure 2) than the susceptible strain.

Kdr genotyping

We obtained clean sequences of exonic regions in domains II (480 bp), III (exon 1 and 2, 347 bp), and IV (280 bp) of the voltage-gated sodium channel. Of note, in approximately half of the specimens in domains II and III we were not able to span the introns due to the presence of insertions or deletions and therefore we could not obtain both forward and reverse exonic sequences. We compensated by sequencing twice in each direction. Although a few silent mutations at codon positions 2 and 3 were seen, no amino-acid changing mutations were detected in the exons of domains II and IV of the mosquitoes tested. However, in one individual, a mutation was found in domain III at position 1534 (base pair positions are numbered according to the amino acid sequence of the most abundant splice variant of the house fly sodium channel, GenBank accession nos. AAB47605 and AAB47604) where a substitution occurred (TTC to CTC), changing the wild type Phenylalanine into a Leucine. The mutation in residue 1,534 that has been associated with pyrethroid resistance in *Ae. aegypti* is F1,534C, resulting in a Cysteine [62].

Enzymatic phenotyping of Ache1

All mosquito test populations from New Jersey, Pennsylvania, and Florida showed similar percentages of AChE inhibition with dichlorvos and propoxur compared to the susceptible ATM95 strain (data not shown), indicating they are all of the susceptible type.

Discussion

The purpose of this study was to evaluate the insecticide resistance status of *Ae. albopictus* populations in several states along the eastern coast of the USA. Insecticides representing the major classes of insecticide (OC, OP, CA, PYR), bio-insecticides (*Bti* and spinosad), and IGRs were used in this study against larvae and adult mosquitoes following WHO protocols. We investigated the possible insecticide resistance mechanisms involved (detoxification enzyme and target site mutations) with biochemical and molecular assays.

For both bioassays and biochemical assays, the eight populations tested were compared to the ATM95 strain, which we first characterized for insecticide susceptibility. The ATM95 strain had similar or higher susceptibilities to the insecticides tested than other *Ae. albopictus* populations used as a reference in previous studies. For example, Ali et al. [26] showed higher LC$_{50}$ for an *Ae. albopictus* strain from Florida maintained for 2 yrs in colony for temephos, *Bti*, methoprene, and pyriproxyfen of 0.01, 0.181, 0.0022, and 0.00011 mg/L respectively, than the ATM95 strain with LC$_{50}$ for the same insecticides of 0.00054, 0.07, 0.00014, and

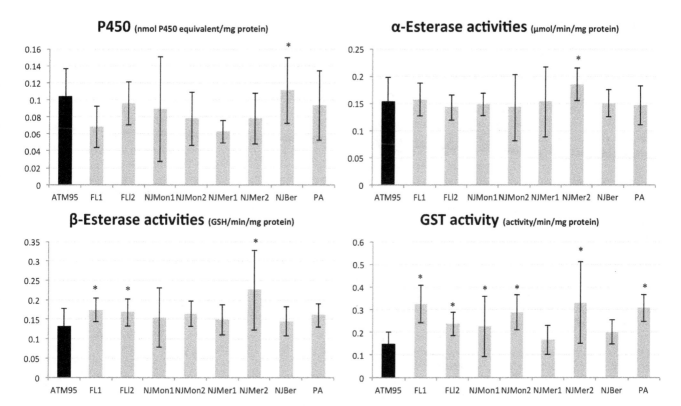

Figure 2. Global amount or activity of detoxification enzymes in adult *Aedes albopictus* from field populations and the laboratory strain (ATM95): cytochrome P450 monooxygenases (P450s), Esterase (α and β-CCEs) and Glutathione-S transferases (GSTs). Sample sizes are 47 specimens/population. Confidence intervals are one standard deviation of the mean. An asterisk (*) denotes significantly up-regulated values compared to the susceptible reference strain ATM95, Tukey-Kramer test.

9.4×10^{-6} mg/L. The susceptible reference strain Ikaken used for the study by Liu et al. [28] presented higher LC_{50} for *Bti*, propoxur, and spinosad (0.1, 3.3, and 0.3 mg/L, respectively) than the LC_{50} of ATM95 (0.07, 1.2, and 0.1 mg/L respectively). Furthermore, the larvae of the ATM95 strain showed higher susceptibility to deltamethrin, permethrin, and malathion than the Ikaken strain or the susceptible strain used by Selvi et al. [45]. In light of these results, we consider the ATM95 as a valid susceptible reference strain for the present study and propose it should be adopted as a reference in future studies of insecticide resistance in temperate *Ae. albopictus*. Reference strains such as the Rockefeller or Bora-Bora used for *Ae. aegypti* studies are essential for the quantification of insecticide resistance across studies [46].

The larval bioassays showed that none of the eight populations examined were strongly resistant to the larvicides tested. Likely because of their specific modes of action, resistance to *Bti*, spinosad, or pyriproxyfen has not been described in mosquitoes, except for a single case of putative resistance to *Bti* in a *Culex pipiens* L. population from New York [47], making these insecticides promising tools for the control of *Ae. albopictus* in the USA. However, we note that spinosad resistance has been reported in several insect pests previously, indicating that it is possible that resistance may occur over time in *Ae. albopictus* if intensive use occurs [48]. Our results showed that temephos was still effective against all the populations tested, although several studies have suggested that temephos resistance selection can develop in *Ae. albopictus* after laboratory selection or prolonged field exposure [49,50]. Indeed, resistant populations have been detected in South-East Asia, South America, and in Europe, where this larvicide is used against *Aedes* species [16,19]. The use of temephos

for control of *Ae. albopictus* larvae in the USA should therefore be carefully evaluated since adult populations from Florida and New Jersey showed resistance or suspected resistance to malathion (OP). Also, the low but significant resistance to propoxur (CA) exhibited by the Florida and New Jersey populations ($RR_{50} > 2$) should be taken into consideration since cross-resistance is known to occur between OPs and CAs.

Methoprene has been used for vector control in Florida for more than 3 decades [51] and even when *Ae. albopictus* is not been the primary control target in this area, populations may have been exposed to this insecticide and developed tolerance over time. One Florida population showed suspected resistance to both methoprene and pyriproxyfen and the adults showed resistance to the adulticide malathion. Previous authors have reported similar findings in mosquitoes exhibiting high resistance to OPs. Specifically, Marcombe et al. [52] and Andrighetti et al. [53] showed that *Ae. aegypti* populations with high resistance to the organophosphate temephos were less susceptible to pyriproxyfen, indicating a possible cross resistance in mosquitoes between these two insecticides families.

The adult bioassays revealed resistance to malathion in Florida and suspected resistance in New Jersey. Resistance to this insecticide, which is used in space spraying treatments was already a concern for the public health authorities in the 1980's [54] when malathion resistance in *Ae. albopictus* was described in Texas only a few years after *Ae. albopictus* became established. Furthermore, other studies report resistance to malathion in populations from Louisiana, Illinois, Alabama, and additional locations in Texas [25,27,28]. Worldwide *Ae. albopictus* resistance to malathion has been extensively reported in Asia, the presumed origin of the USA populations of this species, since the 1960's [55], and it is possible that the introduced populations were already resistant. However, since malathion and other OPs are still being used for mosquito control in the USA, it is also possible that resistance developed locally and is being maintained in this region.

All the populations were susceptible to the three pyrethroids tested at the diagnostic doses. Prallethrin and phenothrin are the components of the Duet formulation that showed promising efficacy in ultra-low volume adulticide applications against *Ae. albopictus* [14]. All the populations were also susceptible to deltamethrin, showing that this insecticide can still be an effective tool for *Ae. albopictus* control. However deltamethrin or pyrethroid resistance has already been detected in China, Japan, and South-East Asia [16,19,22,56] and also more recently in Florida and Alabama, USA [28].

Although we were initially surprised to detect DDT resistance in Florida populations of *Ae. albopictus*, DDT resistance is widespread in *Ae. albopictus* populations worldwide especially in Asia. Since the 1960's very high levels of resistance have been reported from India to the Philippines and from China to Malaysia [18,??]. So as for malathion resistance, it is also likely that the selection for resistance may have occurred in Asia, prior to USA introductions. However, since the use of DDT was terminated in the USA in 1972, before the introduction and establishment of *Ae. albopictus*, the observed levels of resistance in Florida may be explained by a regular exposure of the populations to pyrethroids or other xenobiotics that have the same mode of action as DDT. Alternatively, it is possible that DDT resistance in these populations does not impact fitness and therefore is simply being maintained neutrally or finally, that there have been more recent introductions of DDT resistant *Ae. albopictus* from Asia (Fonseca et al. unpublished data). This last scenario is supported by the study of Kamgang and colleagues [23] that reported DDT resistance in recently introduced populations in Cameroon. The high levels of resistance against DDT found in Florida and the suspected resistance in the populations from New Jersey also underscore the threat of pyrethroid resistance in USA *Ae. albopictus*. Cross resistance mechanisms between DDT and pyrethroids can negatively impact control strategies.

Regarding the various mechanisms of insecticide resistance, we found significant differences in detoxification enzyme activities in several USA resistant *Ae. albopictus* populations suggesting the involvement of metabolic based resistance mechanism. The malathion resistant populations from Florida and New Jersey showed significantly over-expressed β-ESTs and GSTs, which include two detoxification enzyme families known to play a role in organophosphate resistance in mosquitoes [32]. However, because several studies have showed that carboxylesterases do not play a role in resistance to organophosphate in *Ae. albopictus* [45,57], it remains unclear whether one or both of the enzyme families are involved in the resistance at the adult stage. Complementary studies with the use of specific enzyme inhibitors should be implemented to discriminate their roles in malathion resistance in the USA *Ae. albopictus*.

Larvae from Florida populations showed the highest RR_{50} against propoxur but were not resistant to temephos, confirming the absence of insensitive AChE responsible for the cross-resistance between OP and carbamates in mosquitoes. Of note, insensitive AChE was recently detected in *Ae. albopictus* populations in Malaysia [20], underscoring the importance of regular monitoring of this mechanism in the USA. All the populations tested showed a reduced susceptibility against propoxur and all had a significantly increased amount of P450s. It is therefore possible that P450s may be involved in carbamate resistance in *Ae. albopictus* as in other mosquito species [58].

One population from Florida showed significant resistance against the two IGRs, methoprene and pyriproxyfen. The same population also presented over-expressed P450s, ESTs, and GSTs. The P450s are primarily involved in pyrethroid (DDT) resistance and may also be involved in IGR resistance in insects [59]. Indeed, recently the product of the *Ae. aegypti* CYP6Z8 detoxification gene, belonging to the P450s family, was shown to metabolize pyriproxyfen [60]. There are many reports demonstrating elevated P450 activity in insecticide resistant mosquitoes, frequently in conjunction with altered activities of other enzymes [32]. The global overexpression of the four detoxification enzyme families in *Ae. albopictus* from Florida may therefore be leading to a reduced susceptibility to IGRs.

In all populations that presented DDT resistance, GSTs were significantly overexpressed in the adults. This is not surprising since GST-overexpression is the major metabolic mechanism inducing DDT resistance [32,61] and the involvement of the DDT dehydrochlorinase, now classified in the GST family, has been demonstrated in DDT resistant *Ae. albopictus* populations in China. The GSTs probably play an important role in DDT resistance in *Ae. albopictus* in the USA and this should be confirmed by the use of synergists in future studies. The other possible mechanism involved in DDT but also in pyrethroid resistance is a target site modification such as the *kdr* mutation [29]. Although none of the populations showed resistance to pyrethroids we identified a previously unknown amino acid polymorphism (F1534L) in domain III of the VGSC, in a location known to be associated with pyrethroid resistance in *Ae. aegypti* [62], in one of the Florida specimens. Kasai et al. [43] found at the same location a mutation leading to a cytosine in *Ae. albopictus* collected from Singapore (F1534C) but besides the fact that the area where the colony originated was treated with permethrin in the 1980s, there was no information about the current resistance status of this population against pyrethroids. This is the first time such a mutation is detected in *Ae. albopictus* and given the increasing use of pyrethroids for vector control in the USA [63,64] it is important to pursue studies on the global distribution of this allele and its involvement in pyrethroid resistance.

In conclusion, our studies have generated a fully characterized susceptible reference population for temperate *Ae. albopictus*, ATM95, which is available upon request from dinafons@ rutgers.edu. We have also uncovered a complex landscape of populations of *Ae. albopictus* in the USA that are broadly susceptible to larvicides and adulticides. Unexpectedly, we found significant resistance to DDT in two Florida populations and in a New Jersey population. We also found resistance to malathion, an organophosphate, in Florida and suspected resistance in New Jersey plus suspected resistance to several insect growth regulators. Several detoxification enzyme families seemed to be involved in resistance as well, but further studies with the use of synergists should be performed to confirm these findings. All populations tested were fully susceptible to pyrethroids, however, we identified a previously unknown amino acid polymorphism (Phe → Leu) in domain III of the VGSC, in a location known to be associated with pyrethroid resistance in *Ae. aegypti*. We developed a rapid diagnostic PCR to detect this mutation (Marcombe and Fonseca unpublished data) but further studies should be conducted to confirm its implication in DDT/pyrethroid resistance and to assess the frequency of this mutation in *Ae. albopictus*.

This study showed standard larvicides and pyrethroids used for mosquito control are still effective against USA populations of *Ae. albopictus*, but it also demonstrates the importance of research on insecticide resistance and the constant need to develop new tools, new insecticides, and innovative strategies to prevent the development of insecticide resistance in these critical vectors of human diseases. Other strategies such as control using genetically modified male mosquitoes [65], or the use of *Wolbachia* to block disease transmission [66] are very promising because they do not use insecticides but the cost-effectiveness of these strategies and their long term success should be evaluated when compared with conventional control methods.

Acknowledgments

We appreciate the assistance of Linda McCuiston, responsible for the mosquito colonies at the Center for Vector Biology, Rutgers University, and vector control personnel from Mercer and Monmouth counties, particularly Isik Unlu and Taryn Crepeau. We also thank Warren Staudinger, Rui-de Xue, Andrew Kyle, and Mike Hutchinson for providing field-collected specimens from Bergen County, New Jersey, St. Johns County, Florida, and York County, Pennsylvania, respectively. This work was funded by Cooperative Agreement USDAARS-58-6615-8-105 between USDA-ARS and Rutgers University (PI: GGC; PI at Rutgers: DMF).

Author Contributions

Conceived and designed the experiments: SM GGC DMF. Performed the experiments: SM DMF. Analyzed the data: SM DMF. Contributed reagents/materials/analysis tools: AF SPH. Wrote the paper: SM DMF AF GGC SPH.

References

1. Benedict MQ, Levine RS, Hawley WA, Lounibos LP (2007) Spread of the tiger: global risk of invasion by the mosquito *Aedes albopictus*. Vector Borne Zoonotic Dis 7: 76–85.
2. Medley KA (2010) Niche shifts during the global invasion of the Asian tiger mosquito, *Aedes albopictus* Skuse (Culicidae), revealed by reciprocal distribution models. Global Ecology and Biogeography 19: 122–133.
3. Rochlin I, Ninivaggi DV, Hutchinson ML, Farajollahi A (2013) Climate change and range expansion of the Asian tiger mosquito (*Aedes albopictus*) in Northeastern USA: implications for public health practitioners. PLoS One 8: e60874.
4. Rezza G (2012) *Aedes albopictus* and the reemergence of Dengue. BMC Public Health 12: 72.
5. Health Commissioner Reports Dengue Virus Case (November 20, 2013). In: Government SC, editor. Suffolk County Press releases. Available: http://www.suffolkcountyny.gov/SuffolkCountyPressReleases/tabid/1418/itemid/1939/amid/2954/health-commissioner-reports-dengue-virus-case.aspx.
6. Ng LF, Ojcius DM (2009) Chikungunya fever - Re-emergence of an old disease. Microbes Infect. 11(14–15):1163–1164.
7. Tsetsarkin KA, Chen R, Sherman MB, Weaver SC (2011) Chikungunya virus: evolution and genetic determinants of emergence. Current Opinion in Virology 1: 310–317.
8. Enserink M (2007) Infectious diseases. Chikungunya: no longer a third world disease. Science 318: 1860–1861.
9. Moro ML, Gagliotti C, Silvi G, Angelini R, Sambri V, et al. (2010) Chikungunya virus in North-Eastern Italy: a seroprevalence survey. Am J Trop Med Hyg 82: 508–511.
10. Grandadam M, Caro V, Plumet S, Thiberge JM, Souares Y, et al. (2011) Chikungunya virus, southeastern France. Emerging Infectious Diseases 17: 910–913.
11. Lo Presti A, Ciccozzi M, Cella E, Lai A, Simonetti FR, et al. (2012) Origin, evolution, and phylogeography of recent epidemic CHIKV strains. Infection, genetics and evolution: journal of molecular epidemiology and evolutionary genetics in infectious diseases.
12. Mogi M, Armbruster P, Fonseca DM (2012) Analyses of the Northern Distributional Limit of *Aedes albopictus* (Diptera: Culicidae) With a Simple Thermal Index. Journal of Medical Entomology 49: 1233–1243.
13. Fonseca DM, Unlu I, Crepeau T, Farajollahi A, Healy S, et al. (2013) Area-wide management of *Aedes albopictus*: II. Gauging the efficacy of traditional integrated pest control measures against urban container mosquitoes. Pest Manag Sci 69(12): 1351–1361.
14. Farajollahi A, Healy SP, Unlu I, Gaugler R, Fonseca DM (2012) Effectiveness of ultra-low volume nighttime applications of an adulticide against diurnal *Aedes albopictus*, a critical vector of dengue and chikungunya viruses. PLoS One 7: e49181.
15. Marcombe S, Carron A, Darriet F, Etienne M, Agnew P, et al. (2009) Reduced efficacy of pyrethroid space sprays for dengue control in an area of Martinique with pyrethroid resistance. The American Journal of Tropical Medicine and Hygiene 80: 745–751.
16. Ranson H, Burhani J, Lumjuan N, Black WC IV (2010) Insecticide resistance in dengue vectors. TropIKAnet 1: ISSN 2078–8606.
17. McAllister JC, Godsey MS, Scott ML (2012) Pyrethroid resistance in *Aedes aegypti* and *Aedes albopictus* from Port-au-Prince, Haiti. Journal of Vector Ecology 37: 325–332.
18. Mouchet J (1972) [Survey of potential vectors of yellow fever in Tanzania]. Bulletin of the World Health Organization 46: 675–684.
19. Vontas J, Kioulos E, Pavlidi N, Morou E, Torre Ad, et al. (2012) Insecticide resistance in the major dengue vectors *Aedes albopictus* and *Aedes aegypti*. Pesticide Biochemistry and Physiology 104: 126–131.
20. Chen L, Zhao T, Pan C, Ross J, Ginevan M, et al. (2013) Absorption and excretion of organophosphorous insecticide biomarkers of malathion in the rat: implications for overestimation bias and exposure misclassification from environmental biomonitoring. Regul Toxicol Pharmacol 65: 287–293.
21. Chuaycharoensuk T, Juntarajumnong W, Boonyuan W, Bangs MJ, Akratanakul P, et al. (2011) Frequency of pyrethroid resistance in *Aedes aegypti* and *Aedes albopictus* (Diptera: Culicidae) in Thailand. Journal of Vector Ecology 36: 204–212.
22. Cui F, Raymond M, Qiao CL (2006) Insecticide resistance in vector mosquitoes in China. Pest Management Science 62: 1013–1022.
23. Kamgang B, Marcombe S, Chandre F, Nchoutpouen E, Nwane P, et al. (2011) Insecticide susceptibility of *Aedes aegypti* and *Aedes albopictus* in Central Africa. Parasites & Vectors 4: 79.
24. Ponlawat A, Scott JG, Harrington LC (2005) Insecticide susceptibility of *Aedes aegypti* and *Aedes albopictus* across Thailand. Journal of Medical Entomology 42: 821–825.
25. Wesson DM (1990) Susceptibility to organophosphate insecticides in larval *Aedes albopictus*. Journal of the American Mosquito Control Association 6: 258–264.
26. Ali A, Nayar JK, Xue RD (1995) Comparative toxicity of selected larvicides and insect growth regulators to a Florida laboratory population of *Aedes albopictus*. Journal of the American Mosquito Control Association 11: 72–76.
27. Khoo BK, Sutherland DJ, Sprenger D, Dickerson D, Nguyen H (1988) Susceptibility status of *Aedes albopictus* to three topically applied adulticides. Journal of the American Mosquito Control Association 4: 310–313.
28. Liu H, Cupp EW, Guo A, Liu N (2004) Insecticide resistance in Alabama and Florida mosquito strains of *Aedes albopictus*. Journal of Medical Entomology 41: 946–952.
29. Brengues C, Hawkes NJ, Chandre F, McCarroll L, Duchon S, et al. (2003) Pyrethroid and DDT cross-resistance in *Aedes aegypti* is correlated with novel mutations in the voltage-gated sodium channel gene. Medical and Veterinary Entomology 17: 87–94.
30. Raymond M, Berticat C, Weill M, Pasteur N, Chevillon C (2001) Insecticide resistance in the mosquito *Culex pipiens*: what have we learned about adaptation? Genetica 112–113: 287–296.
31. Hemingway J, Field L, Vontas J (2002) An overview of insecticide resistance. Science 298: 96–97.
32. Hemingway J, Hawkes NJ, McCarroll L, Ranson H (2004) The molecular basis of insecticide resistance in mosquitoes. Insect Biochemistry and Molecular Biology 34: 653–665.
33. Hemingway J, Karunaratne SH (1998) Mosquito carboxylesterases: a review of the molecular biology and biochemistry of a major insecticide resistance mechanism. Medical and Veterinary Entomology 12: 1–12.
34. Ranson H, Hemingway J (2005) Mosquito glutathione transferases. Methods in Enzymology 401: 226–241.
35. Crans WJ, Chomsky MS, Guthrie D, Acquaviva A (1996) First record of *Aedes albopictus* from New Jersey. Journal of the American Mosquito Control Association 12: 307–309.
36. WHO (2005) Guidelines for laboratory and field testing of mosquito larvicides. In: WHO/CDS/WHOPES/GCDPP/13, editor. Geneva, Switzerland: World Health Organization.
37. WHO (2006) Guidelines for testing mosquito adulticides for indoor residual spraying and treatment of mosquito nets. In: WHO/CDS/NTD/WHOPES/GCDPP/3, editor. Geneva, Switzerland: World Health Organization.
38. Finney DJ (1971) Probit Analysis. Cambridge, UK: Cambridge University Press.
39. Sakuma M (1998) Probit analysis of preference data. Applied Entomology 33: 339–347.
40. Brogdon WG, McAllister JC, Vulule J (1997) Heme peroxidase activity measured in single mosquitoes identifies individuals expressing an elevated oxidase for insecticide resistance. Journal of the American Mosquito Control Association 13: 233–237.
41. Bradford MM (1976) Rapid and sensitive method for the quantitation of microgram quantities of protein utilizing the principle of protein-dye binding. Anal Biochem 72: 248–254.
42. Hayter AJ (1984) A proof of the conjecture that the Tukey-Kramer multiple comparisons procedure is conservative. The Annals of Statistics 12: 1–401.
43. Kasai S, Ng LC, Lam-Phua SG, Tang CS, Itokawa K, et al. (2011) First detection of a putative knockdown resistance gene in major mosquito vector, *Aedes albopictus*. Japanese Journal of Infectious Diseases 64: 217–221.

44. Alout H, Labbe P, Berthomieu A, Pasteur N, Weill M (2009) Multiple duplications of the rare ace-1 mutation F290V in *Culex pipiens* natural populations. Insect Biochem Mol Biol 39: 884–891.

45. Selvi S, Edah MA, Nazni WA, Lee HL, Tyagi BK, et al. (2010) Insecticide susceptibility and resistance development in malathion selected *Aedes albopictus* (Skuse). Tropical biomedicine 27: 534–550.

46. Kuno G (2010) Early history of laboratory breeding of *Aedes aegypti* (Diptera: Culicidae) focusing on the origins and use of selected strains. J Med Entomol 47: 957–971.

47. Paul A, Harrington LC, Zhang L, Scott JG (2005) Insecticide resistance in *Culex pipiens* from New York. J Am Mosq Control Assoc 21: 305–309.

48. Sparks TC, Dripps JE, Watson GB, Paroonagian D (2012) Resistance and cross-resistance to the spinosyns – A review and analysis. Pesticide Biochemistry and Physiology 2012: 1–10.

49. Hamdan H, Sofian-Azirun M, Nazni WA, Lee HL (2005) Insecticide resistance development in *Culex quinquefasciatus* (Say), *Aedes aegypti* (L.) and *Aedes albopictus* (Skuse) larvae against malathion, permethrin and temephos. Tropical Bomedicine 22: 45–52.

50. Romi R, Toma L, Severini F, Di Luca M (2003) Susceptibility of Italian populations of *Aedes albopictus* to temephos and to other insecticides. Journal of the American Mosquito Control Association 19: 419–423.

51. Nayar JK, Ali A, Zaim M (2002) Effectiveness and residual activity comparison of granular formulations of insect growth regulators pyriproxyfen and s-methoprene against Florida mosquitoes in laboratory and outdoor conditions. J Am Mosq Control Assoc 18: 196–201.

52. Marcombe S, Darriet F, Agnew P, Etienne M, Yp-Tcha MM, et al. (2011) Field efficacy of new larvicide products for control of multi-resistant *Aedes aegypti* populations in Martinique (French West Indies). Am J Trop Med Hyg 84: 118–126.

53. Andrighetti MTM, Cerone F, Rigueti M, Galvani KC, Macoris MdLdG (2008) Effect of pyriproxyfen in *Aedes aegypti* populations with different levels of susceptibility to the organophosphate temephos. Dengue Bulletin: 186–198.

54. Robert LL, Olson JK (1989) Susceptibility of female *Aedes albopictus* from Texas to commonly used insecticides. Journal of the American Mosquito Control Association 5: 251–253.

55. Hawley WA, Reiter P, Copeland RS, Pumpuni CB, Craig GB Jr (1987) *Aedes albopictus* in North America: probable introduction in used tires from northern Asia. Science 236: 1114–1116.

56. Kawada H, Maekawa Y, Abe M, Ohashi K, Ohba SY, et al. (2010) Spatial distribution and pyrethroid susceptibility of mosquito larvae collected from catch basins in parks in Nagasaki city, Nagasaki, Japan. Japanese Journal of Infectious Diseases 63: 19–24.

57. Chen CD, Nazni WA, Lee HL, Sofian-Azirun M (2005) Weekly variation on susceptibility status of *Aedes* mosquitoes against temephos in Selangor, Malaysia. Tropical Biomedicine 22: 195–206.

58. Coleman M, Hemingway J (2007) Insecticide resistance monitoring and evaluation in disease transmitting mosquitoes. Journal of Pesticide Science 32: 69–76.

59. Brogdon WG, McAllister JC (1998) Insecticide resistance and vector control. Emerg Infect Dis 4: 605–613.

60. Chandor-Proust A, Bibby J, Regent-Kloeckner M, Roux J, Guittard-Crilat E, et al. (2013) The central role of mosquito cytochrome P450 CYP6Zs in insecticide detoxification revealed by functional expression and structural modelling. Biochem J 455: 75–85.

61. Neng W, Yan X, Fuming H, Dazong C (1992) Susceptibility of *Aedes albopictus* from China to insecticides, and mechanism of DDT resistance. Journal of the American Mosquito Control Association 8: 394–397.

62. Harris AF, Rajatileka S, Ranson H (2010) Pyrethroid resistance in *Aedes aegypti* from Grand Cayman. Am J Trop Med Hyg 83: 277–284.

63. Davis RS, Peterson RKD, Macedo PA (2007) An ecological risk assessment for insecticides used in adult mosquito management. Integrated Environmental Assessment and Management 3: 373–382.

64. Peterson RKD, Macedo PA, Davis RS (2006) A Human-Health Risk Assessment for West Nile Virus and Insecticides Used in Mosquito Management. Environ Health Perspect 114: 366–372.

65. Harris AF (2011) Field performance of engineered male mosquitoes. Nature Biotechnology 29: 1034–1037.

66. Hoffman AA (2011) Successful establishment of *Wolbachia* in *Aedes* populations to suppress dengue transmission. Nature 476: 454–457.

15

Area-Wide Ground Applications of *Bacillus thuringiensis* var. *israelensis* for the Control of *Aedes albopictus* in Residential Neighborhoods: From Optimization to Operation

Gregory M. Williams[1,2]*, Ary Faraji[2,3], Isik Unlu[2,4], Sean P. Healy[2,5], Muhammad Farooq[6], Randy Gaugler[2], George Hamilton[2], Dina M. Fonseca[2]

1 Hudson Regional Health Commission, Secaucus, New Jersey, United States of America, 2 Center for Vector Biology, Rutgers University, New Brunswick, New Jersey, United States of America, 3 Salt Lake City Mosquito Abatement District, Salt Lake City, Utah, United States of America, 4 Mercer County Mosquito Control, West Trenton, New Jersey, United States of America, 5 Department of Pathobiological Sciences, Louisiana State University School of Veterinary Medicine, Baton Rouge, Louisiana, United States of America, 6 Navy Entomology Center of Excellence, Jacksonville, Florida, United States of America

Abstract

The increasing range of *Aedes albopictus*, the Asian tiger mosquito, in the USA and the threat of chikungunya and dengue outbreaks vectored by this species have necessitated novel approaches to control this peridomestic mosquito. Conventional methods such as adulticiding provide temporary relief, but fail to manage this pest on a sustained basis. We explored the use of cold aerosol foggers and misting machines for area-wide applications of *Bacillus thuringiensis* var. *israelensis* (VectoBac WDG) as a larvicide targeting *Aedes albopictus*. During 2010–2013 we performed initially open field trials and then 19 operational area-wide applications in urban and suburban residential areas in northeastern USA to test three truck-mounted sprayers at two application rates. Area-wide applications of WDG in open field conditions at 400 and 800 g/ha killed on average 87% of tested larvae. Once techniques were optimized in residential areas, applications with a Buffalo Turbine Mist Sprayer at a rate of 800 g/ha, the best combination, consistently provided over 90% mortality. Importantly, there was no significant decrease in efficacy with distance from the spray line even in blocks of row homes with trees and bushes in the backyards. Under laboratory conditions *Bti* deposition in bioassay cups during the operational trials resulted in over 6 weeks of residual control. Our results demonstrate that area-wide truck mounted applications of WDG can effectively suppress *Ae. albopictus* larvae and should be used in integrated mosquito management approaches to control this nuisance pest and disease vector.

Editor: Nigel Beebe, University of Queensland & CSIRO Biosecurity Flagship, Australia

Funding: This work was funded by Cooperative Agreement USDA-ARS-58-6615-8-105 between the US Department of Agriculture, Agricultural Research Service (USDA-ARS) and Rutgers University (PI at Rutgers: DMF) and Regional IPM Grant 2008-34103-18978 between USDA-ARS and Rutgers University (PI at Rutgers: GH). The funders had no role in study design, data collection and analysis, decision to publish, or preparation of the manuscript.

Competing Interests: The authors have declared that no competing interests exist.

* Email: gwilliams@hudsonregionalhealth.org

Introduction

The range of *Aedes albopictus*, the Asian tiger mosquito, is expected to expand in highly urbanized temperate regions in North America and Europe in response to climate change [1–3]. An aggressive human biter, this mosquito is often the primary pest species eliciting complaints from the public in areas where it occurs [4] and it also presents a significant health risk [1,5,6]. Unfortunately, conventional mosquito abatement methods do not control *Ae. albopictus* effectively. Local mosquito control agencies across the US mount aggressive control campaigns against salt marsh, floodwater, and many other rural-based mosquito pest species [2]. Since the onset of the WNV epidemic mosquito control agencies in urban and suburban areas have targeted *Culex* spp. in residential storm drains and other large stagnant water sources [2]. These programs rarely impact *Ae. albopictus*, which

thrive in small pockets of water in artificial containers such as buckets, toys and trash primarily within private yards [7]. Therefore, efforts tend to be limited to public education, placing the responsibility for control on unqualified citizens [8,9]. Having individual homeowners bear considerable responsibility for source reduction has hampered control efforts primarily because residents are not usually able to coordinate their efforts with enough of their neighbors to eliminate the larval sources that contribute biting *Ae. albopictus* to a neighborhood.

For any significant suppression, this pest must be managed on an area-wide basis [9]. Intensive source reduction efforts which involve teams of professionals emptying, removing or treating all sources of standing water in thousands of private yards can be effective. However, they are extremely labor intensive leading to unsustainably high costs [9–11] and often also temporary since new containers are continuously being added by uninformed homeowners [7,12]. Area-wide efforts require relatively swift and repeatable applications of insecticides at a large scale and low volume usually from a ground vehicle or aircraft.

Although *Ae. albopictus* is a day-biting mosquito, adulticide applications at night when thermal inversions optimize likelihood of insecticide deposition, non-target species are mostly inactive and residents are less likely to be outside, can be effective [13]. Unfortunately, results are temporary [13] because such methods kill adults without addressing the large pool of immature stages that continuously emerge into adults.

Area-wide application of mosquito larvicides involves applying liquid or emulsified larvicides with a low-volume sprayer so the droplets drift onto backyards. The droplets of larvicide then settle into the myriad of containers of water where *Ae. albopictus* can be found, delivering a lethal dose. Using this approach hundreds of residential yards can be treated in a single application during one night, an action that can be repeated as needed, as opposed to intensive source reduction approaches that would take minimally weeks to treat the same number of yards [9].

Bacillus thuringiensis subspecies *israelensis* (*Bti*) is an excellent candidate for a low-volume area-wide strategy. The efficacy of *Bti* has been demonstrated for many mosquito species in a variety of habitats. Today, formulations of *Bti* are one of the primary products used for larval mosquito control in the U.S. and other countries [14]. After ingestion and activation in the mosquito larvae, which have a highly basic gut pH, proteinaceous toxins are released which cause lysis of the epithelial cells of the larval midgut, killing the larvae. *Bti* contains four different larvicidal proteins, each acting in slightly different ways making the development of resistance to *Bti* difficult [15]. The specificity of these proteins to organisms with catalytic gut pH results in very few non-target impacts. With the exception of several families of Nematocera (many of them also pest species such as blackflies and biting midges), *Bti* has no known effects on other insects, invertebrates, or vertebrates [16].

Although *Bti* has often been applied as a solid directly to larval sources, recent studies have demonstrated that *Bti* liquid can be applied as ULV or low volume (LV) sprays to control mosquitoes. Several such studies have successfully used the VectoBac 12AS formulation (Valent BioSciences Corporation, Libertyville, IL, USA) [17–19], however 12AS can cause persistent spotting on automotive paint [20,21] and is therefore unsuitable for use in residential areas in the US.

Until now, studies that used the VectoBac WDG formulation (Valent BioSciences Corporation, Libertyville, IL, USA) focused on backpack mist sprayers that are insufficient for area-wide control efforts because they hold small volumes of product and require the operator to walk into backyards during the application [17,22,23]. These manual applications are problematic because it takes too much time to treat a neighborhood and access to yards is often restricted by fences, dogs, or uncooperative homeowners. Most studies were conducted in Southeast Asia where differences in the strains of *Ae. albopictus*, housing construction and condition, and accepted pest management practices limit the value of those data in the US. The present study describes the development of an area-wide larviciding strategy using VectoBac WDG applied with truck-mounted equipment. We evaluated several sprayers and application rates under both staged open-field as well as operational trials in urban and suburban settings, and used larval bioassays to examine the efficacy of this method over distance (from the road into the backyards) and over time (residual effects). Our ultimate goal is to develop a novel strategy for the control of *Ae. albopictus* larvae to be used in an integrated mosquito

management program in hopes of reducing the nuisance and disease risk associated with this species.

Materials and Methods

Ethics Statement

No specific permits were required for the described field studies. All pesticide applications were made by county mosquito control agencies under the authority of Title 26:9 of the New Jersey Administrative Code. These studies did not involve endangered or protected species.

Pesticide

VectoBac WDG is a water-dispersible granule formulation containing 3,000 International Toxic Unit/mg of *Bti* (strain AM 65-52). Containing only *Bti* and food-grade inert ingredients, it has been approved for use in organic crops and sensitive habitats around the world. We chose it for this study because of the favorable environmental profile, safety for non-target organisms, and fast results; all mortality usually occurs within 12–72 hrs. The product is mixed by weight with tap water. The US product label allows for application rates of 123–988 g/ha for ground spraying equipment.

Equipment

We tested three sprayers: a Cougar ULV cold aerosol generator (Clarke, Roselle, IL), a CSM2 Mist Sprayer (Buffalo Turbine, Springville, NY), and an Ag-Mister LV-8 low volume sprayer (Curtis Dyna-Fog, Westfield, IN). We chose the Cougar to examine the feasibility of using conventional ULV equipment with WDG. The standard pump was replaced with an optional QB3CSC 1/2 inch pump (Fluid Metering, Inc., Syosset, NY) to increase flow rate. However, despite being rated for a flow rate of over 2 L/min, we were unable to achieve more than 1 L/min. That flow rate was too low to reach our lowest application rate of 400 g/ha and the Cougar was not evaluated further.

The Buffalo Turbine Mist Sprayer (CSM2) has a 189 L stainless spray tank. A mechanical twin piston pump sends fluid to a spray head contained within an air chute. A gas-powered wind turbine propels the fluid from the spray head. The standard spray head consists of four flat-fan spray nozzles. For this study, we replaced the standard head with an optional Monsoon gyratory atomizing spray head (Buffalo Turbine, Springville, NY) for increased flow rate and more precise control over the droplet spectra. With the atomizing head, fan blades cause the head to rotate in relation to the wind speed. Fluid released from the center of the head is forced through stainless mesh screens; larger mesh producing larger droplets. We evaluated 0.33, 0.51, and 0.91 mm screens. The smallest screen was too fragile to be used alone and was supported with the 0.51 mm screen. Pump pressure was set to 1,034 kPa and a ball valve on the pump outlet was adjusted to a flow rate of 8.3 L/min. Wind speed was controlled with the engine throttle and set at maximum (177 km/h) for all tests for greatest swath width.

The LV-8 is a low volume sprayer designed for orchard spraying. An electric pump pushes liquid to eight nozzles arranged in two vertical rows of four. A gas-powered roots type blower generates air pressure to atomize the pesticide. A ball valve adjusts the flow rate on the fluid supply line and was set to 8.3 L/min. Air pressure was adjusted by changing engine speed. Pump pressure was maintained at 68.9 kPa. As equipped from the factory, the nozzles point out to the sides of the spray truck and may not release droplets high enough to make it over buildings and other obstacles. Given the narrow streets in many urban neighborhoods, the LV-8 may end up applying material directly to cars parked along the streets. To minimize this occurrence, we cut and welded the booms containing the nozzles to angle at 45 degrees so that the spray would be released over parked cars.

Droplet Characterization

We characterized the droplet spectra at the Navy Entomology Center of Excellence (Jacksonville, FL) with a 2-D Phase Doppler Particle Analyzer system (TSI Inc., Shoreview, MN). The transmitter and receiver were outfitted with 500 and 300 mm lenses, resulting in a measurable droplet size range of 0.6–211 μm respectively. Equipment was positioned 1.2 m from the lens array. We measured the droplet size at the horizontal center of the spray scanning the whole plume vertically in a continuous mode and characterized the droplet spectra under a variety of flow rates and pressures with three replicated measurements recorded for each test.

We set all equipment for a flow rate of 8.3 L/min. The fluid pump of the LV-8 was therefore set at 68.9 kPa for each test and we adjusted the engine speed to change the air pressure through the nozzles, thereby varying the droplet size. With the CSM2 the fluid pump pressure was maintained at 1,034.2 kPa for all tests. Finally, we kept the Buffalo turbine wind speed at a maximum of 177 km/h for greatest swath width and varied droplet size by changing screen sizes on the atomizing spray head. VectoBac WDG was mixed with tap water at a concentration of 59 or 118 g/L with an electric paint mixer in 19 L plastic buckets. Target application rates were 400 and 800 g/ha (representing mid and high label rates on the USA pesticide label) assuming a swath width of 91.4 m and an application speed of 8 km/h.

Open-field Trials

We conducted open-field trials to determine efficacy, swath width and droplet deposition under ideal conditions. The LV-8 was tested on private land over an open grass field at Arnold Groves Ranch (Clermont, FL) in September 2010 with permission from the property owner, Robert Arnold. The CSM2 was tested over a blacktop tarmac at the Trenton-Mercer Airport (Ewing, NJ) in October 2010 with permission from the Airport Manager, Melinda Montgomery. We set up stations in a 3×8 grid out to 76.2 m for the LV-8 and a 3×9 grid out to 91.4 m for the CSM2. The grids were set up parallel to the prevailing wind. Each station contained an empty 500 ml polyethylene container for larval bioassays and one Kromekote C1S card (Mohawk Fine Papers, Inc., Cohoes, NY) on a plastic compact disc case (for ballast) at ground level to determine droplet deposition and characteristics.

Sprayers were mounted on pickup trucks and driven at 8 km/h perpendicular to the wind during applications. We mixed the WDG with tap water at 59 and 118 g/L for the 400 and 800 g/ha rates respectively. A 2% solution FD&C Red 40 granular dye (Glanbia Nutritionals, Carlsbad, CA) was added to the mixture for droplet analysis. We operated the LV-8 at 62 kPa and 8.3 L/min with a volume mean diameter (VMD) of 107 μm from the nozzles. The CSM2 was operated with a 0.51 mm screen at 1,034 kPa and 8.3 L/min with a VMD of 233 μm from the spray head with the CSM2 head positioned at a 60° angle from horizontal. We made control applications with dye and water (CSM2) or water only (LV-8). Applications were made when wind speeds were between 3.2 and 8 km/h and stations were moved as needed to remain parallel to wind direction. We retrieved cards and cups 15 min after the applications allowing time for droplets to settle. Lids were placed on the cups to prevent contamination or loss of product. We replicated each application three times and performed one control application with each machine. Cups were returned to the

laboratory for analysis where we added 400 ml water and 10 (CSM2) or 20 (LV-8) 3rd instar *Ae. albopictus* larvae to each cup. Mortality was recorded at 24, 48 and 72 hours and final mortality rates were corrected for control mortality using the Schneider-Orelli formula [24]. We analyzed the Kromekote cards with the DropVision AG system (Leading Edge Associates, Inc., Waynesville, NC) a system that uses a portable scanner to digitize droplet cards. Stains on the card were counted and measured and the software provided droplet and volume data using user-defined application and mix rates.

Operational Trials

We developed operational field trials in Trenton (Mercer County), New Jersey, USA (40° 12′ N, 74° 44′ W) and Belmar (Monmouth County), NJ USA (40° 13′ N, 74° 44′ W). The two Trenton sites were urban residential neighborhoods: S. Olden (40° 22′ N, 74° 73′ W) was 48.6 ha consisting of 24 residential blocks, each containing a residential street on all four sides, sometimes divided lengthwise by a drivable alley. St. Olden included 1,250 parcels (i.e. house with surrounding yard) most often built as adjoining row homes or duplexes [25]. Parcel sizes were relatively constant at approximately 200 m². The second site, Cummings (40° 21′N, 74° 74′W), was 30 ha with 23 residential blocks, occasionally divided lengthwise by a drivable alley. Cummins included 1,122 parcels many adjoining (i.e. row houses). Almost all adjoining parcels contained a sheltered alcove area between homes, where vegetation and trash proliferated creating mosquito resting areas. Socioeconomic conditions in both St. Olden and Cummins have led to a large number of abandoned homes with neglected yards that accumulate containers such as buckets, cans and other trash that can hold water [26].

The field site in Belmar was a seaside resort community located along the Atlantic coast and consisting mainly of traditional suburban single- family home parcels with yard space on all four sides, interspersed with seasonal rental properties. A small section of the plot contained commercial storefront and several large condominiums and motels. This 121.9 ha site consisted of 59 residential blocks and 2,356 parcels averaging 0.1 ha in size. The socioeconomic conditions of the site can best be described as upper-middleclass. The majority of the homes are well maintained and container mosquito habitat, such as planters and flexible downspout extensions, tend to be cryptic but abundant. Rental properties often had more traditional container habitats such as buckets, toys, and recyclable containers.

We used either a LV-8 or a CSM2 mist sprayer for each application. We set up the equipment as previously described for a flow rate of 8.3 L/min. We mixed VectoBac WDG with tap water at a rate of 59 or 118 g/L and drove the vehicles at an average speed of 8 km/h for a final application rate of 400 or 800 g/ha. Meteorological data was recorded for wind speed, direction, humidity, and temperature at 1 m and 10 m heights for monitoring thermal inversion. A Vantage Pro2 wireless weather station (Davis Instruments, Hayward, CA) was set up within the treatment site 14 h prior to application and maintained until 8 h post application. We made applications from April through September 2011, June 2012, and July 2013 between 01:00 and 06:00 hours. All pesticide applications were made by county mosquito control agencies under the authority of Title 26:9 of the New Jersey Administrative Code.

We selected 30 residential parcels within our treatment sites and 10 parcels within a control site in either Mercer or Monmouth for placement of bioassay cups. Within each parcel, dry bioassay cups were placed in front, alongside, and in the backyard of each home. A total of 210 polyethylene 500 ml bioassay cups per replicate were placed in the field 2 hrs prior to application and removed 1 hr post application to allow time for droplets to settle into cups. Bioassay cups were used in lieu of sampling natural larval populations to get consistent and comparable data points. Existing larval habitats were too transient to be of statistical value and sampling of varied containers could not be standardized.

All cups were transported to the laboratory and loaded with 400 ml of distilled water and 40 mg of a 50:50 mix of yeast:lactalbumin (Sigma-Aldrich, St. Louis, MO) added as an emulsion in 1 ml of water as a food source for larvae. Then we added 10–15 late 2nd to early 3rd instar *Ae. albopictus*. Mortality was recorded at 24, 48, and 72 hrs post-application. Mortality rates were corrected for control mortality using the Schneider-Orelli formula [24].

Based on the outcome of the applications, we selected three of the operational trials to determine the residual efficacy of the treatments. For these experiments, we retained any cups which achieved 100% mortality after 72 hours. Dead larvae were removed from the cups and ten early 3rd instar *Ae. albopictus* were added. After 72 hours mortality was recorded, dead larvae were removed, and ten fresh larvae were added. We repeated this process until larval mortality dropped to or below 20%.

Statistical analyses

All mortality data were reported as proportions and arcsin square root transformed to normalize the distributions [27]. After the open field trials we performed least squares repeated measures analyses of variance in JMP (SAS Institute, Cary, NC) to examine the effects of equipment type and application rates on mortality rates. As mentioned, there were three spatial replicates at increasing distances from the application site and each trial was replicated 3 times. To assess the reliance of the results on expectations of homoscedasticity of the data (hard to assess with a feasible number of replicates), we also performed the repeated measures analysis with ranked data.

After the operational field trials, we examined the effects of equipment type, application rate and location of bioassay containers (front, middle, and back) with a least squares analysis of variance after confirming normality of the distributions using a goodness of fit test in JMP (SAS Institute, Cary, NC).

Results

Droplet Analysis

Unlike ULV sprayers that are optimized to produce droplets below 30 μm, the LV-8 is designed to produce larger droplets for agricultural use. Under laboratory conditions, median droplet sizes ranged from 74–141 μm (Table 1). In general, increasing air pressure led to larger droplet sizes with the LV-8. Beyond 34.5 kPa there was a direct relationship between air pressure and droplet size. With the CSM2, VMD's ranged from 214–233 μm.

Open-Field Trials

Weather conditions during the LV-8 tests averaged 23.9°C, 85% RH and 4.8 km/h wind speed. Average conditions during the CSM2 runs were 14.4°C, 55% RH and 3.2 km/h wind speed. The LV-8 resulted in an average mortality of 89.4±6.4 and 76.3±9.2% out to 80 m (maximum distance tested) for the 400 and 800 g/ha rates, respectively. The CSM2 averaged 98.9±0.9% mortality out to 100 m at the 800 g/ha rate and 85.1±5.1% at the 400 g/ha rate (Fig. 1). There was no significant difference in the overall average mortality at each distance from deployment between rates or machines. However, when only data for the CMS2 is examined, the average mortality was greater at the 800 g/ha rate than at the 400 g/ha rate ($P = 0.04$) (Fig. 1). Repeating the analysis with ranked data yielded the same results.

We conducted a droplet analysis to determine droplet size, swath width and deposition rates (Fig. 2). The LV-8 had greater deposition at the 10 m stations than the CSM2. At distances beyond 10 m, there was little difference in the deposition rates regardless of equipment or application rate. Minimal deposition at the farther distances still resulted in significant mortality (Fig. 1). Application rates were determined by the mix ratio alone. The 800 g/ha rate resulted in greater volume deposition for the LV-8 but lower deposition for the CSM2. Generally, larger droplets settled out closer to the spray line and decreased swath width (Fig. 2). There was a clear correlation between VMD and distance for the CSM2 (average $r^2 = 0.9$) while the droplets from the LV-8 appear to be relatively uniform across the measured swath width.

Operational Trials

The efficacy of control in urban and suburban neighborhoods in 2011 were mixed with an average efficacy of 72.0±1.3%. In contrast, by the end of 2012 we were able to consistently achieve high mortalities in urban settings (over 91%), which we further increased in 2013 to 94.5±1% (Fig. 3). However, operational results varied greatly for the LV-8. Of the ten applications made with the LV-8, we were only able to exceed 80% mortality on three occasions. In contrast, the CSM2 averaged 89.4±1.0% mortality over all nine applications and exceeded 90% mortality for five of those applications.

Although we had slightly greater success in urban (Mercer) than in suburban areas (Monmouth) (Fig. 4), we only performed two applications in suburban Monmouth and both were performed with the CSM2 at the 400 g/ha rate. Since these were the only operational trials with the CSM2 at the lower rate we did not include these two suburban trials in the statistical analysis. Despite a slight decline in mortality as distance from the spray line increased, operational mortality rates were similar across all cup locations and the differences were not significant ($F_{2,2} = 1.31$, $P = 0.28$). We did however see significantly greater efficacy at the higher application rate of 800 g/ha ($F_{1,1} = 4.26$, $P = 0.04$), and the CSM2 outperformed the LV-8 ($F_{1,1} = 8.64$, $P = 0.005$). Under laboratory conditions, the residual activity of the applications achieved over 80% mortality for 6 weeks and lasted nearly 10 weeks (Fig. 5).

Discussion

Unlike most pressurized spray equipment where air pressure has an inverse relationship with droplet size, the LV-8 exhibited a direct relationship between air pressure and VMD. It is unclear if this trend was a result of nozzle design or was an artifact of the properties of the liquid WDG, which is an emulsion rather than a solution. However, the largest droplets did occur at the lowest pressure setting indicating that the nozzles were not operating within their optimal range when spraying the viscous mixture of WDG. Also of note, screen size did not seem to have a predictable effect on droplet size for the CSM2. It is plausible that running the head at maximum rpm canceled out any effect from the screens by creating additional wind shear. Reducing turbine speed may result in different droplet sizes but will reduce swath width. We only tested the 59 g/L mix ratio for the particle analyzer tests since existing literature [28] demonstrated that there was little advantage to application rates above 400 g/ha.

Based on equipment design, the droplets from these mist blowers were significantly larger than the droplets created by a ULV machine applying the same material [29]. Similar studies

Table 1. Droplet spectra for VectoBac WDG determined by 2-D phase doppler particle analyzer ± SEM.*

Equipment	Setting	DV$_{0.1}$ (µm)	DV$_{0.5}$ (µm)	DV$_{0.9}$ (µm)
LV-8	34.5 kPa	35.7±2.2	140.7±22.4	276.0±13.0
	41.4 kPa	31.5±0.5	73.5±1.5	202.5±0.5
	42.7 kPa	34.0±0.6	81.0±6.2	217.0±26.0
	63.4 kPa	28	92	250
	65.5 kPa	34.7±0.9	109.7±9.7	262.0±6.4
CSM2	0.33 & 0.51 mm screen	71.3±8.2	214.0±11.5	313.7±6.6
	0.51 mm screen	82.0±3.1	232.7±2.4	324.7±2.4
	0.91 mm screen	64.3±9.2	218.0±10.1	313.3±16.2

*VectoBac WDG mixed with tap water at 59 g/l. Equipment set for final application rate of 400 g/ha.

with back pack mist blowers produced droplets comparable in size to the LV-8 [22]. The insufficient flow rates we experienced with the modified Clarke Cougar suggest that most ULV sprayer will be unsuitable for WDG application without extensive modification.

In open field trials we observed the expected decrease in mortality with increasing distance from the spray line regardless of equipment or rate; however the CSM2 outperformed the LV-8 in both swath width and efficacy. The LV-8 had greater deposition at the 10 m stations than the CSM2 indicating that larger drops settled out earlier likely because the LV-8 is a passive sprayer (i.e. it does not induce a strong directional force to the droplets the way the CSM2 does). The wind speed generated by the turbine of the CSM2 propelled the large droplets further, resulting in a more uniform deposition rate. The performance of the LV-8 was more dependent on wind speed than the CSM2: for example, with the LV-8, the 800 g/ha rate did not perform as well as the 400 g/ha rate probably because of environmental conditions as the wind averaged 4.8–9.6 km/h for the 400 g/ha trials and only 0–3.2 km/h for the 800 g/ha trials. While the CSM2 was less dependent on environmental conditions, the wind from the turbine pushed droplets beyond the closest stations, resulting in lower deposition rates near the spray vehicle. Importantly, we

observed very high toxicity of WDG to mosquito larvae such that the lower deposition rates did not result in lower mortality.

Mortality in the controls was low for the LV-8 but significant mortality was observed in the CSM2 controls. This is likely because the LV-8 is an easier machine to clean and load and had been used for orchard spraying of agricultural pests with agricultural pesticides, not mosquito larvicides. In contrast, the CSM2 had routinely been used in a mosquito control program for several years, exclusively with mosquito larvicides. Further, on the LV-8 the removable spray tank is translucent and the fluid lines are transparent, making it easy to see when the unit has been sufficiently cleansed. The spray tank on the CSM2, is stainless steel and is welded into the machine. The tank is filled through a 51 mm diameter neck and emptied through a 19 mm drain plug. Flushing the tank is difficult and there is no way to see inside the tank to determine when the tank is clean. The fluid lines also have metal fittings that corrode and provide recesses for pesticide residue to collect. Despite repeated efforts to clean the CSM2 tank, residual material may have persisted during the control applications. We have since also conducted experiments to confirm that the food dye used in the controls was not toxic to *Ae. albopictus* at the rates applied [30].

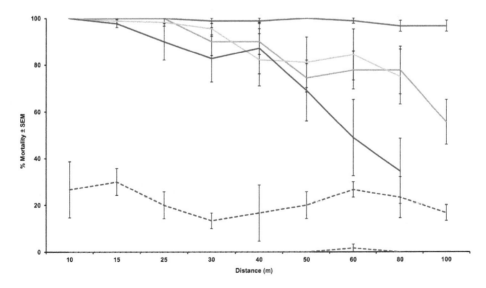

Figure 1. *Aedes albopictus* **72 hr mortality rates for VectoBac WDG applied with a CSM2 and LV-8 at 400 and 800 g/ha in an open-field.** Red = CSM2, blue = LV-8, light colors = 400 g/ha, dark colors = 800 g/ha, dashed line = control.

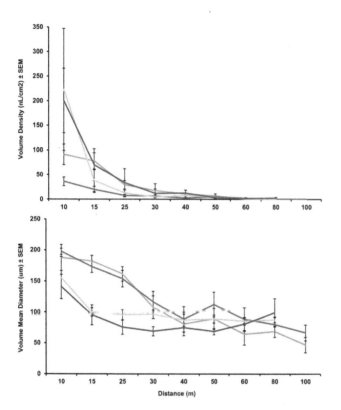

Figure 2. Volume density and droplet sizes of VectoBac WDG applied with a CSM2 and LV-8 at 400 and 800 g/ha in an open-field. Red = CSM2, blue = LV-8, light colors = 400 g/ha, dark colors = 800 g/ha.

The VMD values at the 10 m stations suggest that the droplet data from the spray cards are reliable as they are close to the VMD values from the nozzle as determined by the laser analysis. While it is logical to presume that the droplets from the Buffalo Turbine are too large (VMD = 233 μm) to travel 100 m, the 177 km/h wind created by the turbine was sufficient to push the drops, compensating for the extra mass of the large drops. The smaller drops from the LV-8 (VMD = 107 μm) were still small enough to drift out to 80 m from wind currents alone. We did not find

significant differences in droplet size based on the mix ratio of the product which contradicts previous studies that found that higher mix ratios resulted in larger droplets [22]. This is likely because of the low power of the backpack mist blower used in that study. The more powerful motors of the truck-mounted machines we used appear better equipped to handle the more viscous mix. The problem with many water-based formulations is the evaporation of droplets. Personal communication from Valent technical staff (Peter DeChant, 21 July 2010) indicated that WDG will not evaporate beyond a terminal droplet size of around 80 μm. Our data support this assertion. Regardless of the initial size, beyond 40 m droplet sizes converged to an average VMD of 80.7 μm.

Smaller droplets are lighter than larger droplets and therefore should drift farther yielding greater swath width. Other open-field trials with ULV sprayers produced small droplets of 20–50 μm yet a swath of only 30 m [31]. These sprayers are normally used for mosquito adulticides where small drops are expected to drift up to 90 m without settling. It is possible that many of the droplets created by the ULV sprayer in Lee et al. [31] drifted beyond the sample area. For the purposes of applying mosquito larvicides, larger droplets are preferable so that they settle into containers before drifting past the target area. Based on our results we recommend that equipment should be optimized for larger droplets (≥80 μm).

We expected a positive correlation between mortality and volume density, higher volume density resulting in greater mortality. However, the LD_{50} values for WDG against *Ae. albopictus* are so low, 3 ppb as determined against our laboratory strain, that even a single droplet in the bioassay cups was sufficient to cause 100% mortality. Any deposition over 18 nL/cm^2 resulted in 100% mortality and in many cases rates below 1 nL/cm^2 were still able to provide 100% mortality. While there is no correlation between volume density and mortality, the data demonstrate that equipment can be calibrated without bioassays. As long as droplets are visible on spray cards, mortality should be acceptable at that distance.

The open-field trials tested area-wide larval control from truck-mounted equipment under ideal conditions in the absence of any obstructions. Operational conditions present many obstacles to any type of space spray. Houses, automobiles, and vegetation all serve to potentially impede the drift of pesticide sprays. To evaluate the operational utility of this technique, we conducted operational applications under operational conditions. These data

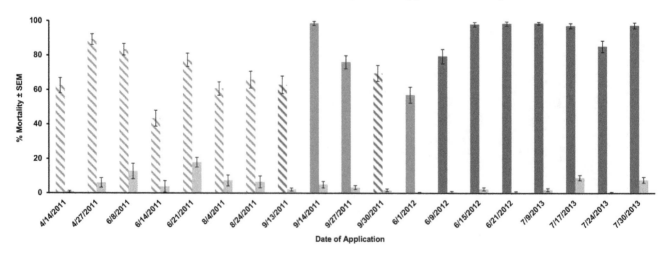

Figure 3. Efficacy summary of area-wide VectoBac WDG applications by date. Red = CSM2, blue = LV-8, green = control, dashed lines = 400 g/ha, solid lines = 800 g/ha.

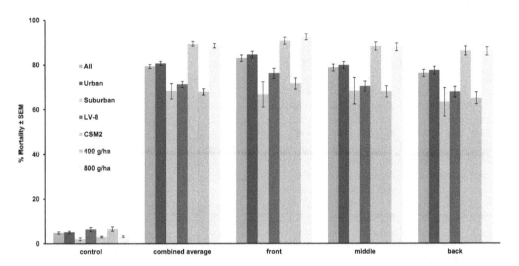

Figure 4. Efficacy by location of VectoBac WDG applied with a CSM2 and LV-8 at 400 and 800 g/ha in residential areas. Orange = average efficacy, dark blue = urban plots, light blue = suburban plots, red = LV-8, pink = CSM2, dark green = 400 g/ha, light green = 800 g/ha.

show a clear improvement in our area-wide technique over the three seasons of the study. Early in the study we focused on a 400 g/ha rate from the LV-8. Equipment problems such as pump failure, solenoid malfunctions, and short circuits of the electrical system led to poor results in several of the early trials. Further, contrary to what we found in the open field trials, bioassay results indicated that the 400 g/ha rate was not delivering enough material into distant containers. In late 2011 we increased the rate to 800 g/ha and found that this change yielded improved yet inconsistent results from the LV-8. In 2012 and 2013 all applications were made at the 800 g/ha rate and all but one application was made with the CSM2. This combination of higher application rate and better machine led to high mortality rates in all trials.

Within each parcel, the consistent results of the operational trials demonstrate that this technique can be used effectively in any type of residential environment with different types of equipment. The WDG spray effectively traveled around obstructions and

settled in containers. In total, larval mortality was observed in 90% of the treatment cups despite efforts to place cups in inaccessible areas such as under decks and in dense vegetation. Our results are limited to bioassay cups but we expect similar results in other habitats such as roof gutters although these areas were not monitored. Additionally, the observed persistence of WDG indicates that rainfall following WDG applications would likely wash *Bti* to non-target areas such as catch basins and sump pumps leading to increased control.

As a result of the thorough coverage, there was no significant difference in mortality based on the location of the cups. Bioassay cups placed in the backyards of homes did just as well as cups placed in front of homes nearest the spray line. This is similar to Jacups et al. [22] using a backpack mist blower over various distances. However our truck mounted strategy has the advantage of being more consistent and keeps the operators inside the truck cab away from the *Bti* spray.

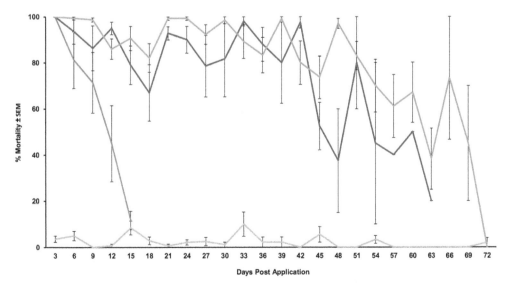

Figure 5. Residual efficacy of VectoBac WDG applied with a CSM2 and LV-8 at 400 and 800 g/ha in residential areas. Orange and red = CSM2 at 800 g/ha, blue = LV-8 at 400 g/ha, green = control.

The greater efficacy we observed in the urban areas was not a result of the smaller parcel sizes or easier access in those plots, as there was no difference in efficacy across distances even in the suburban parcels. Instead, it is more likely due to the fact that the suburban trials were only run at the 400 g/ha rate at a time when we thought based on the results of the open-field trials, and the results of Jacups et al. [22] in a bushland setting, that the higher application rate was unnecessary. Obstacles present during operational trials likely resulted in lower deposition rates which would explain the increased mortality at the higher application rate. We were unable to measure deposition during operational trials because we could not apply dyed material in a residential setting. High environmental humidity precluded the use of water sensitive paper and ultraviolet dye did not mix well with the WDG solution. However, it is logical to conclude that deposition rates would decrease given the physical obstacles present under operational conditions. With fewer droplets reaching the cups, the amount of Bti in each droplet becomes increasingly critical especially at the farthest distances. Significantly, our application rate was below the maximum of 988 g/ha specified on the product label. Higher application rates could provide even greater mortality, although as discussed mixing and spraying at high rates has limitations.

The CSM2 performed significantly better than the LV-8 overall most likely because the LV-8 is a passive machine relying on wind to carry the spray cloud. Wind conditions were less than optimal for most applications, rarely exceeding 4.8 km/h, putting the LV-8 at a disadvantage. By contrast the turbine of the CSM2 generates its own wind and is therefore less reliant on environmental conditions. This is further evidenced by the consistency of the CSM2 across all distances. It is possible, however, that under more favorable wind conditions, the efficacy of the LV-8 would be higher.

Although not considered a residual product, Bti has been shown to give extended control in containers [32–35]. Our applications achieved over 80% mortality for 6 weeks and lasted nearly 10 weeks. It is important to note that the residual trials were conducted in the laboratory. While environmental conditions could reduce the residual activity of Bti, trials conducted under semi-field conditions have provided residual control for up to 11 weeks [22]. In that study, WDG was applied closer to the containers with a backpack mist blower leading to droplet densities up to 300 drops/cm². In our open-field trials, droplet densities averaged 10 drops/cm² and never exceeded 81 drops/cm² and Jacups et al. [22] found that higher application rates resulted in longer residual control. Given that there were significant differences in mortality between machines and rates, it is likely that the reduced residual efficacy of the LV-8 application at 400 g/ha was due to fewer Bti particles reaching the containers.

Based on our open-field and operational trials, both sprayers are suitable for area wide larval control operations. The CSM2 offers the advantages of higher potential flow rates, adjustable nozzle angle, and increased swath width due to the 177 km/h wind drift created by the turbine. However, capacity is limited to 189 L (about 25 min of spray time), and the hurricane force wind from the turbine may present problems in a residential areas rattling windows, blowing over potted plants, scattering debris, etc. The LV-8 has the advantage of an optional 568 L spray tank with an onboard mixer, which increases the potential run time and makes loading the WDG easier. However, as discussed the LV-8 is a passive sprayer that relies on the wind to carry the pesticide to the target areas. Wind speeds often averaged less than 4.8 km/h during our operational trials, a standard condition in very urban areas at night, leading to inconsistent efficacy with LV8-

applications. With the reconfigured nozzles, we found that WDG deposition to parked cars was minimal and similar to what was left by the CSM2. Observations by the team indicate that applications left a fine dust on automobiles not easily distinguished from dirt or pollen. The WDG rinsed easily from vehicles and caused no damage to automotive paint. Although citizens were in regular contact with the mosquito control programs, we received no negative comments from the public regarding visible residues from any of the 19 area-wide applications over 3 years.

The present study indicates that area-wide applications of larvicides with truck-mounted equipment have the clear potential to reduce Ae. albopictus populations in urban and suburban residential areas by killing immature mosquitoes in open containers. Both larvicide doses effected high larval mortality that could be sustained for over one week and high dose applications exhibited up to 90 days of residual control. Street-level applications were able to drift past obstructions and settle into containers located up to 100 m away with no loss of efficacy at greater operational distances. In conclusion, we provide evidence that a nighttime LV application of VectoBac WDG can be efficacious in killing larvae of Ae. albopictus in urban and suburban neighborhoods.

It is important to note, however, that effective larval control does not necessarily translate into a reduction in the adult mosquito population or human landing rates. In fact, a reduction in larval population can actually result in an increase in adult emergence for some species [36], but this does not appear to be the case with Ae. albopictus [37]. We deployed a high-density of BG Sentinel (Biogents AG, Regensburg, Germany) mosquito traps throughout each operational research plot for the duration of the study and have measures of adult populations of Ae. albopictus. These data were not presented here because we were performing multiple control strategies simultaneously, as needed operationally to safeguard the well being of the residents. Those data are being developed into a separate publication. In brief, we found that when area-wide applications of WDG were coordinated with a degree-day model and traditional adulticide spray missions, there was a synergistic effect resulting in a significant decrease in the adult Ae. albopictus populations.

This study was conducted with readily available equipment adapted for larvicide applications. Purpose-built equipment could provide even better efficacy and easier operational use. New formulations of Bti developed for this technique that mix with water more easily or come pre-mixed would greatly reduce the labor involved in preparing and loading the product. Future studies are needed to determine the optimal timing of applications so as to have maximum impact on adult mosquito populations. Based on our findings, we recommend that nighttime applications of LV larvicides in areas with large Ae. albopictus populations be considered as part of an integrated approach for public health protection.

Acknowledgments

We acknowledge Peter DeChant and Jim Andrews from Valent BioSciences for providing product and technical assistance. We thank Linda McCuiston and Rafael Valentin for laboratory assistance, and multiple students and postdocs from the Center for Vector Biology, Rutgers University, and personnel from Mercer, Monmouth, and Hudson Counties for laboratory and field work. In addition, we thank LCDR Jeff Stancil for access to NECE facilities and staff, Robert Arnold, the owner of Arnold Groves Ranch, and Melinda Montgomery, the Mercer Airport Manager, for permission to develop open field trials on their lands.

Author Contributions

Conceived and designed the experiments: GMW AF SPH MF RG GH DMF. Performed the experiments: GMW AF SPH MF DMF. Analyzed the data: GMW AF IU SPH MF DMF. Contributed reagents/materials/analysis tools: GMW AF SPH MF DMF. Contributed to the writing of the manuscript: GMW AF IU SPH RG DMF.

References

1. Benedict MQ, Levine RS, Hawley WA, Lounibos LP (2007) Spread of the tiger: global risk of invasion by the mosquito Aedes albopictus. Vector Borne Zoonotic Dis 7: 76–85.
2. Rochlin I, Ninivaggi DV, Hutchinson ML, Farajollahi A (2013) Climate change and range expansion of the Asian tiger mosquito (Aedes albopictus) in northeastern USA: implications for public health practitioners. PLOS ONE 8: e60874.
3. Unlu I, Farajollahi A (2012) Vectors without borders: Imminent arrival, establishment and public health implications of the Asian bush (Aedes japonicus) and the Asian tiger (Aedes albopictus) mosquitoes in Turkey. Hacettepe J Biol Chem, 40: 23–36.
4. Farajollahi A, Nelder MP (2009) Changes in Aedes albopictus (Diptera: Culicidae) populations in New Jersey and implications for arbovirus transmission. J Med Entomol 46: 1220–1224.
5. Armstrong P, Anderson JK, Farajollahi A, Healy SP, Unlu I, et al. (2013) Isolations of Cache Valley virus from Aedes albopictus (Diptera: Culicidae) in New Jersey and evaluation of its role as a regional arbovirus vector. J Med Entomol 50: 1310–1314.
6. Gratz NG (2004) Critical review of the vector status of Aedes albopictus. Med Vet Entomol 18: 215–27.
7. Unlu I, Farajollahi A, Strickman D, Fonseca DM (2013) Crouching tiger hidden trouble: urban sources of Aedes albopictus (Diptera: Culicidae) refractory to source reduction. PLoS One 8: e77999.
8. Bartlett-Healy K, Unlu I, Obenauer P, Hughes T, Healy S, et al. (2012) Larval mosquito habitat utilization and community dynamics of Aedes albopictus and Aedes japonicus (Diptera: Culicidae). J Med Entomol 49: 813–824.
9. Fonseca DM, Unlu I, Crepeau T, Farajollahi A, Healy SP, et al. (2013) Area-wide management of Aedes albopictus: II. Gauging the efficacy of traditional integrated pest control measures against urban container mosquitoes. Pest Manag Sci 69: 1351–1361.
10. Halasa YA, Shepard DS, Fonseca DM, Farajollahi A, Healy S, et al. (2014) Quantifying the impact of mosquitoes on quality of life and enjoyment of yard and porch activities in New Jersey PLOS ONE 9: e89221. doi:10.1371/journal.pone.0089221.
11. Zhou YB, Zhao TY, Leng PE (2009) Evaluation on the control efficacy of source reduction to Aedes albopictus in Shanghai, China. Chin J Vector Biol Control 20: 3–6.
12. Bartlett-Healy K, Hamilton G, Healy SP, Crepeau T, Unlu I, et al. (2011) Source reduction behavior as an independent measurement of the impact of a public health education campaign in an integrated vector management program for the Asian tiger mosquito. Int J Environ Res Public Health 8: 1358–1367.
13. Farajollahi A, Healy SP, Unlu I, Gaugler R, Fonseca DM (2013) Effectiveness of ultra-low volume nighttime applications of an adulticide against diurnal Aedes albopictus, a critical vector of dengue and chikungunya viruses PLoS One 7: 11. e49181 doi:10.1371/journal.pone.0049181.
14. Lacey LA (2007) Bacillus thuringiensis serovariety israelensis and Bacillus sphaericus for mosquito control. J Am Mosq Control Assoc 23: 133–163.
15. Regis L, Nielson-LeRoux C (2000) Management of resistance to bacterial vector control. In: Charles J-F, Delécluse A, Nielsen-LeRoux C, editors. Entomo-pathogenic Bacteria: From laboratory to field application. Dordecht: Kluwer Academic Publishers. 419–438.
16. Lacey LA, Merritt RW (2003) The safety of bacterial microbial agents used for black fly and mosquito control in aquatic environments. In: Hokkanen HMT, Hajek AE, editors. Environmental Impacts of Microbial Insecticides: Need and Methods for Risk Assessment. Dordecht: Kluwer Academic Publishers. 151–168.
17. Lee HL, Chen CD, Mohd Masri S, Chiang YF, Chooi KH, et al. (2007) Impact of larviciding with a Bacillus thuringiensis israelensis formulation of VectoBac WDG, on dengue mosquito vector population in a dengue endemic site in Selangor State, Malaysia. SE Asian J Trop Med Pub Health 39: 601–609.
18. Seleena P, Lee HL (1998) Field trials to determine the effectiveness of Bacillus thuringiensis subsp. israelensis application using an ultra low volume generator for the control of Aedes mosquitoes. Israel J Entomol 32: 25–31.
19. Yap HH, Chong ASC, Adanan CR, Chong NL, Rohaizat B, et al. (1997) Performance of ULV formulations (Pesguard 102/VectoBac 12AS) against three mosquito species. J Am Mosq Control Assoc 13: 384–388.
20. Valent BioSciences Corporation (2003) Technical Use Bulletin for VectoBac 12AS Mosquito and Black Fly Larvicide. Valent Biosciences Corporation Public Health Products website. Available: http://publichealth.valentbiosciences.com/docs/public-health-resources/vectobac-12as-technical-use-bulletin_mosquito-and-black-fly. Accessed 25 April 2014.
21. Valent BioSciences Corporation (2012) VectoBac 12AS specimen label. Valent Biosciences Corporation Public Health Products website. Available: http://publichealth.valentbiosciences.com/docs/public-health-resources/vectobac-sup-12as-specimen-label. Accessed 25 April 2014.
22. Jacups SP, Rapley LP, Johnson PH, Benjamin S, Ritchie SA (2013) Bacillus thuringiensis var. israelensis misting for control of Aedes in cryptic ground containers in North Queensland, Australia. Am J Trop Med Hyg 88: 490–496.
23. Sun D, Williges E, Unlu S, Healy S, Williams GM, et al. (2014) Taming a tiger in the city: Comparison of motorized backpack mist applications and source reduction against the Asian tiger mosquito, Aedes albopictus. J Am Mosq Control Assoc 30: 99–105.
24. Punter W (1981) Manual for field trials in plant protection. Second edition. Basle: Agricultural Division, Ciba-Geigy Limited.
25. Unlu I, Farajollahi A, Healy SP, Crepeau T, Bartlett-Healy K, et al. (2011) Area-wide management of Aedes albopictus: choice of study sites based on geospatial charateristics, socioeconomic factors and mosquito populations. Pest Manag Sci 8: 965–974.
26. Unlu I, Farajollahi A (2012) To catch a tiger in a concrete jungle: operational challenges for trapping Aedes albopictus in an urban environment. J Am Mosq Control Assoc 28: 334–337.
27. Sokal RR, Rohlf FJ (1995) Biometry: the principles and practice of statistics in biological research, 3rd ed. Freeman: New York. 887.
28. World Health Organization (2004) Report of the seventh WHOPES working group meeting: Review of VectoBac WG.
29. Lam PH, Boon CS, Yng NY, Benjamin S (2010) Aedes albopictus control with spray application of Bacillus thuringiensis israelensis strain AM 65–52. SE Asian J Trop Med Pub Health 41: 1071–1081.
30. Dow GW, Leisnham P, Williams GM, Kirchoff N, Jin S, et al. (2014) Effects of a red marker dye on Aedes and Culex mosquitoes: Implications for operational mosquito control. J Am Mosq Control Assoc. In press.
31. Lee HL, Gregorio ER Jr, Khadri MS, Seleena P (1996) Ultralow volume applications of Bacillus thuringiensis ssp. israelensis for the control of mosquitoes. J Am Mosq Control Assoc 12: 651–655.
32. Batra CP, Mittal PK, Adak T (2000) Control of Aedes aegypti breeding in desert coolers and tires by use of Bacillus thuringiensis var. israelensis formulation. J Am Mosq Control Assoc 16: 321–323.
33. Mulla MS, Tharavara U, Tawatsin A, Chompoosri J (2004) Procedures for the evaluation of field efficacy of slow-release formulations of larvicides against Aedes aegypti in water-storage containers. J Am Mosq Control Assoc 20: 64–73.
34. Vilarinhos PTR, Monerat R (2004) Larvicidal persistence of formulations of Bacillus thuringiensis var. israelensis to control larval Aedes aegypti. J Am Mosq Control Assoc 20: 311–314.
35. Farajollahi A, Williams GM, Condon G, Kesavaraju B, Unlu I, et al. (2013) Assesment of a direct application of two Bacillus thuringiensis israelensis formulations for immediate and residual control of Aedes albopictus. J Am Mosq Control Assoc 29: 385–388.
36. Agnew P, Haussy C, Michalakis Y (2000) Effects of density and larval competition on selected life history traits of Culex pipiens quinquefasciatus (Diptera: Culicidae). J Med Entomol 37: 732–735.
37. Reiskind MH, Lounibos LP (2009) Effects of intraspecific larval competition on adult longevity in the mosquitoes Aedes aegypti and Aedes albopictus. Med Vet Entomol 23: 62–68.

16

Invasive Plants and Enemy Release: Evolution of Trait Means and Trait Correlations in *Ulex europaeus*

Benjamin Hornoy[1], Michèle Tarayre[1], Maxime Hervé[2], Luc Gigord[3], Anne Atlan[1]*

1 Ecobio, Centre National de la Recherche Scientifique, Université de Rennes 1, Rennes, France, **2** BIO3P, Institut National de la Recherche Agronomique - Agrocampus Ouest, Université de Rennes 1, Rennes, France, **3** Conservatoire Botanique National de Mascarin, Saint-Leu, La Réunion, France

Abstract

Several hypotheses that attempt to explain invasive processes are based on the fact that plants have been introduced without their natural enemies. Among them, the EICA (Evolution of Increased Competitive Ability) hypothesis is the most influential. It states that, due to enemy release, exotic plants evolve a shift in resource allocation from defence to reproduction or growth. In the native range of the invasive species *Ulex europaeus*, traits involved in reproduction and growth have been shown to be highly variable and genetically correlated. Thus, in order to explore the joint evolution of life history traits and susceptibility to seed predation in this species, we investigated changes in both trait means and trait correlations. To do so, we compared plants from native and invaded regions grown in a common garden. According to the expectations of the EICA hypothesis, we observed an increase in seedling height. However, there was little change in other trait means. By contrast, correlations exhibited a clear pattern: the correlations between life history traits and infestation rate by seed predators were always weaker in the invaded range than in the native range. In *U. europaeus*, the role of enemy release in shaping life history traits thus appeared to imply trait correlations rather than trait means. In the invaded regions studied, the correlations involving infestation rates and key life history traits such as flowering phenology, growth and pod density were reduced, enabling more independent evolution of these key traits and potentially facilitating local adaptation to a wide range of environments. These results led us to hypothesise that a relaxation of genetic correlations may be implied in the expansion of invasive species.

Editor: Mari Moora, University of Tartu, Estonia

Funding: This work was supported by an ATIP (Action Thématique et Incitative sur Programme) grant of the CNRS (Centre National de la Recherche Scientifique). The funders had no role in study design, data collection and analysis, decision to publish, or preparation of the manuscript.

Competing Interests: The authors have declared that no competing interests exist.

* E-mail: anne.atlan@univ-rennes1.fr

Introduction

Biological invasions are a major threat to global biodiversity, and thus many studies aimed at identifying species that are potential invaders and habitats that are prone to invasion. If some characteristics of invaded environments have been identified, such as a high level of disturbance or low species diversity [1,2], no general biological properties of invasive species were identified [3,4]. One explanation is that the spreading phase of an invasion process is habitat-specific so that invasive success is associated with different traits in different environments [1,5]. Another explanation is that it is not the nature of life history traits but their capacity to adapt to a novel environment that is relevant. Indeed, the evolutionary potential of introduced populations is often considered a key factor in their invasiveness [6,7]. This is supported by the observation of rapid evolutionary changes in many invasive species (e.g. in allelopathic chemicals, reproductive phenology or vegetative growth) in response to new biotic interactions or abiotic conditions [8,9,10].

Biotic interactions are always different in the native and introduced ranges, partly because exotic plants have usually been introduced without their pathogens, parasites or herbivores [2]. This release from natural enemies has been proposed as one of the most important ecological factors contributing to the invasiveness of numerous species [11,12,13]. Arguments on this Enemy Release

Hypothesis (ERH) have been reviewed by Keane and Crawley [14] and Liu and Stiling [15]. Based on this release from natural enemies, an evolutionary mechanism was proposed by Blossey and Nötzold [16]: the Evolution of Increased Competitive Ability (EICA) hypothesis states that, because a plant has limited resources to invest in defence against enemies, growth and reproduction, an exotic species in an environment devoid of natural enemies will evolve to invest less in defence and more in other fitness components. The resulting increase in vegetative growth and/or reproductive effort would result in a better competitive ability of the species. This hypothesis makes clear and testable predictions: genotypes from the invaded regions should (i) be less well defended against natural enemies than genotypes from the native regions, and (ii) grow faster and/or produce more seeds.

The EICA hypothesis has been tested on many invasive plant species. It seems to play an important role in species such as *Silene latifolia* [11] and *Sapium sebiferum* [12], but in many others, evidence for increased susceptibility to natural enemies or increased growth and reproduction were not found [17,18]. A shift in resources from defence to growth and/or reproduction is expected only if these defences are costly and directed against specialist enemies [5,19,20]. Moreover, the traits usually measured to test the EICA hypothesis (growth, reproduction, resistance to natural enemies) are often genetically correlated with one another or with other life history traits [21,22]. This may prevent the evolution of traits predicted by the EICA hypothesis because the rate of trait evolution is affected by the respective intensity and direction of genetic correlations and selection [23]. More generally, genetic correlations can constrain the local adaptation of ecologically important traits such as vegetative growth and flowering phenology, and hence prevent further range expansion of introduced species [24]. In studies that aim to compare plants from native and invaded regions, it might thus be important to compare not only the trait means, but also the correlations among traits. Yet, due to the release from natural enemies in the introduced regions of the species, we might expect a relaxation of the strength of genetic correlations between defence against these enemies and life history traits, potentially leaving these life history traits more free to evolve during invasion.

The common gorse (*Ulex europaeus*, Fabaceae) is a suitable model for this type of comparison. It is one of the 30 most invasive plant species in the world according to IUCN [25]. In its native range, the Atlantic face of western Europe, the climate is oceanic and temperate and gorse populations are always found at the sea level (below 300 meters). The species is considered as invasive in many parts of the world at different latitudes, including New Zealand, Australia, South and North America (Chile, Colombia, California, Oregon...) and tropical islands (Hawaii, Reunion), in altitudes that vary from zero to 4000 meters ([25] and pers). obs. Introduced populations had thus to adapt to a wide range of climates. In Europe, gorse is associated with several pathogens and herbivores, of which the specific seed-eating herbivores, the weevil *Exapion ulicis* and the moth *Cydia succedana*, are the most harmful, since they can infest 90% of a plant's pods [26]. In the invaded range, *U. europaeus* was initially introduced without any natural enemies. Some seed predators were later introduced for biological control [27], but in all the invaded areas, gorse experienced a release of selective pressures induced by seed predators for over a century. Previous studies have shown that in the native range, seed predation occurs in spring and is reduced by a genetically-based polymorphism of flowering strategy: long flowering individuals that flower in both winter and spring partly escape seed predation in time, while short flowering individuals that flower in spring reduce seed predation through mass bloom and predator satiation [28]. This polymorphism of strategy is present within all gorse populations studied and maintains a high genetic diversity for reproductive phenology (implied in escape in time), and for pod density and plant size (implied in predator satiation) [29]. As a consequence, the coexistence of these two strategies induces strong genetic correlations between these traits and infestation rates [29]. Therefore, one can expect that in the absence of selective pressure induced by seed predators, not only the trait means, but also the strength of genetic correlations among traits, may have evolved in the invaded regions.

While gorse is a good model to explore the evolution of trait means and trait correlations, it is however not suitable to produce large experimental samples: this big perennial shrub flowers at the age of three, and reaches four meters high and two meters wide. This clearly limits the number of plants that can be cultivated in an experimental garden. Since most studies were performed in annual herbaceous plants, the study of such perennial shrubs is however needed to enlarge our understanding of the invasion process. Furthermore, we knew from previous studies [28,29] that population differences and trait correlations were so strong that they can generate significant effects even with a low number of individuals.

The aim of the present study was to explore the potential evolutionary changes that occurred in gorse invasive populations in the absence of natural enemies. We focused on seed predators, but we also considered predators that attack vegetative parts, when present. Our goals were (i) to test whether plants from invaded regions have an increased growth, reproduction, and susceptibility to seed predators compared with gorse from native regions (EICA hypothesis), (ii) to compare other life history traits related to environmental adaptation and strongly associated with susceptibility to seed predators in the native range (growth pattern and reproductive phenology), and (iii) to explore the evolution of the correlations among these traits in native and introduced regions. For this purpose we used a common garden in which gorse plants from two native regions and two invaded regions were randomly grown, and measured their trait means and trait correlations. The results were interpreted in regard of previous knowledge on gorse and theoretical expectations.

Materials and Methods

Study species

The common gorse *Ulex europaeus* is a spiny hexaploid shrub. It lives up to 30 years, and its adult height varies from 1 to 4 m. Plants begin to flower during their third year. Flowers are hermaphrodite and contain 10 stamens and an ovary with up to 12 ovules, enclosed in a carina. They are pollinated by large insects such as honeybees or bumblebees [30,31]. In Europe, the peak of flowering is in spring, but there is a genetically based polymorphism of flowering phenology, with the co-existence of long-flowering individuals that flower from autumn to spring and short-flowering individuals that only flower in spring [28,29]. Seed dispersal is primarily by ejection from the pod. Gorse seeds are very long-lasting, and may germinate over a period of up to 30 years [32].

In Europe, the most common and harmful herbivores attacking adult gorses (that are thus used for biological control), are two specific seed-eaters: the weevil *Exapion ulicis* (Coleoptera: Apionidae) [33,34], and to a lesser extent the moth *Cydia succedana* (Lepidoptera: Tortricidae) [33]. *Exapion ulicis* is univoltine and lays its eggs in spring inside young pods. Larvae develop at the expense of seeds, and adults emerge about two months later at pod dehiscence. The larvae of *E. ulicis* can be attacked by a parasitoid wasp, *Pteromalus sequester* (Hymenoptera: Pteromalidae). Vegetative parts of adult gorses are attacked by other predators or parasites, of which the most common in the study area are the aphid *Aphis ulicis* (Homoptera: Aphididae), the spider mite *Tetranychus lintearius* (Acari: Tetranychidae), and the rust fungus *Uromyces genistae-tinctoriae* (Pucciniaceae). Young seedlings are devoid of thorns and are attacked by several generalist herbivores (e.g. rabbits and slugs).

In the invaded range, *U. europaeus* was initially introduced without its natural enemies, and is not attacked by local seed predators. Some seed predators from the native range were however introduced later for biological control. In New Zealand, *U. europaeus* was present before 1835, while *E. ulicis* was introduced in 1931 and *C. succedana* in 1992 [35]. In Reunion Island, there is no biological control program, and gorse plants still have no natural enemies.

Experimental design

We compared plants from two native regions (Brittany, France and Scotland, UK) with plants from two invaded regions (Reunion Island, Indian Ocean and New Zealand, Pacific Ocean). In these regions, gorse was introduced by Europeans: Reunion Island was initially colonized by the French (and Brittany is the region of France where most gorse populations are located) while New Zealand was initially colonized by the Scots (and Scotland is a region that exhibits large gorse populations).

Three populations were sampled in each region (Table 1). Seeds were collected on an individual basis between 1999 and 2005, and stored at 4°C. No specific permits were required for seed collection. *Ulex europaeus* is not endangered, neither protected in none of the regions sampled: it is invasive in Reunion and New Zealand, and is an abundant native in Scotland and Brittany. The collection of its seeds does thus not require any specific authorization in any of these regions. None of the locations were privately owned. Two were situated in protected Areas and the corresponding authority was informed of our research program and did support our field work. These are The Conservatoire du Littoral in Brittany (contact: Mr. Denis Bredin) and the National Forest Office in Reunion (contact: Mr. Julien Triolo).

In October 2006, the seeds were scarified and allowed to germinate on Petri dishes. One seed per mother plant was randomly chosen and the resulting seedlings were grown in a greenhouse (N = 265, with a mean of 22±4 individuals per population) in cell trays and then in two-litre pots. These were filled with a mix of sand (ca 15%), perlite (ca 15%), potting soil (ca 25%) and soil collected behind gorses in a natural population (ca 45%), thus containing nitrogen-fixing bacteria gorse needs to develop. Seedlings were grown for one year in the greenhouse under horticultural lighting, at a temperature ranging from 10 to 20°C. In November 2007, seedlings were transplanted in a common garden. Due to the fact that gorse plants need three years to flower, and may grow quickly high and wide, we had to keep a minimum spacing of 1.20 m between each plant and restrict the number of seedlings transplanted. Ten seedlings per population were randomly chosen and randomly transplanted in a common garden (N = 120). The common garden is located on the Campus of Rennes (Brittany, France), an area with several gorse populations nearby and prone to natural infestation by seed predators and pathogens. Some of the individuals were left out of the analyses of reproductive traits due to non-flowering, heavy rust infestation or chlorosis, which left 8 to 10 plants per population.

Plant height and growth pattern

Vegetative growth of the one-year-old seedlings was measured in the glasshouse in June 2007. We estimated plant height, as well as the number and length of lateral shoots. The total shoot length of each individual was estimated by adding the length of lateral shoots to the length of the main axis.

Vegetative growth at two years was measured in October 2008. We measured the height and the basal area of the individuals in the common garden. To estimate basal area, we measured the plant's largest width and the width perpendicular to it, and used the formula of an ellipse.

Vegetative growth at three years was measured in October 2009. We measured plant height and the mean length of five randomly chosen shoots per plant. It was not possible to measure their width anymore, since they had grown large enough to touch each other.

Flowering and fruiting phenologies

We monitored reproductive phenology, reproductive output, and pod infestation every two weeks from October 2008 to July 2009. The date of flowering onset corresponded with the first occurrence of flowers together with large flower buds. The date of fruiting onset corresponded with the first occurrence of ripe pods together with browning pods.

Reproductive output

The mean number of pods per centimetre (pod density) was estimated in spring 2009 between the end of flowering and pod dehiscence. For most individuals, this occurred in late May. For each plant, we randomly chose five shoots and counted the pods on the last 30 apical centimetres of each shoot. If the chosen shoot was shorter than 30 cm, we measured its length and counted all the pods it contained. If the chosen shoot was branched on the last 30 cm, we only retained the longest branch. We then divided the total number of pods by the total shoot length measured to get the estimation of the number of pods per centimetre.

The number of seeds per pod and seed mass were estimated when all plants had produced ripe pods, *i.e.* in late June in 2009. The number of seeds per pod was estimated on ten uninfested ripe pods per individual. Seed mass was estimated on ten seeds, from at least three different ripe pods. Flat, rotten or chewed seeds were not taken into account.

Infestation by seed predators

The infestation rates were estimated on ripe pods. At each visit between October 2008 and July 2009, we opened 30 ripe pods,

Table 1. Main characteristics of the gorse populations sampled.

Region	Location	ID	Latitude	Longitude	Elevation (m)
Native range					
Brittany	Cap de la Chèvre	BCC	48.1°N	04.5°W	0
	Château de Vaux	BCV	48.0°N	01.6°W	50
	Kergusul	BKE	48.0°N	03.2°W	200
Scotland	Banchory	SBA	57.1°N	02.5°W	100
	Crail	SCR	56.1°N	02.6°W	0
	Stirling	SST	56.0°N	03.9°W	300
Invaded range					
Reunion	Luc Boyer	RLB	21.1°S	55.6°E	1200
	Piton Maido	RMA	21.1°S	55.4°E	2200
	Piton de Brèdes	RPB	21.2°S	55.6°E	1500
New Zealand	Auckland	ZAU	37.3°S	175.1°E	0
	Christchurch	ZCH	43.6°S	172.5°E	50
	Wellington	ZWE	41.3°S	174.9°E	100

when available, and noted the number of seeds and insects. A pod was considered as infested by weevils if it contained weevils at any developmental stage (from larva to adult) or weevil parasitoids. A pod was considered as infested by a moth if it contained a moth larva, or when its presence was revealed by droppings or holes. Infestation rates were estimated when at least 10 ripe pods were available.

To compare the infestation rate of pods that were all exposed to the same abundance and activity of seed-eaters, we compared pod infestation rate at the same time for all individuals, i.e. in late June in 2009, when the highest proportion of individuals was fruiting synchronously.

The presence of predators and parasites that attack vegetative parts (the aphid *Aphis ulicis*, the spider mite *Tetranychus lintearius*, and the rust fungus *Uromyces genistae-tinctoriae*) was also recorded.

Statistical analyses

Trait-by-trait analyses were performed using the GLM procedure of SAS software [36]. The significance of each effect was determined using type III F-statistics. We first used a three-level model, in which populations were nested within regions and regions were nested within ranges (native range and invaded range). Regions were tested as a fixed effect and populations as a random effect. Whatever the variable, the range effect was never

significant. We then used a two-level model, in which populations were nested within regions. This model had a lower AIC and better fits the data. Plant height in 2007, 2008 and 2009 was tested with a repeated statement, to investigate the interaction between years and populations within regions, and between years and regions. As for the other analysis, there was no range effect and the best model only included populations nested within regions. Pod infestation rates were submitted to arcsine transformation before analysis.

The correlations between traits were estimated with the CORR procedure of SAS, using Spearman's rank-order correlation coefficient, since the data were not normally distributed. The pairwise relationships between traits were investigated using partial correlations, in which untested life history traits were used as covariates. To test for differences in correlation strength between the invaded range and the native range, we wrote a nested analysis of variance model using R software [37]. Correlation coefficients were first transformed using Fisher's z transformation. Then, between-regions (within ranges) and between-ranges mean squares were estimated using a weighted analysis of variance on correlation coefficients, where weights were the number of couples n on which each correlation coefficient was estimated. Residual mean squares were computed manually using the formula for the variance of z, that depends only on n. Since we

Table 2. Results of nested ANOVA for 18 traits of *Ulex europaeus* plants grown in a common garden.

			Populations		Regions	
	N	Mean±SD[a]	d.f.	F	d.f.	F
Plant height						
June 2007 (cm)	265	42.7±13.5	8	2.29*	3	10.97**
October 2008 (cm)	120	126.0±30.3	8	1.05	3	6.44*
October 2009 (cm)	107	180.9±34.4	8	1.03	3	4.92*
Vegetative growth						
Shoot number 2007	265	4.12±3.21	8	2.56*	3	0.57
Total shoot length 2007 (cm)	265	100.1±52.2	8	3.03**	3	0.54
Basal area 2008 (m²)	120	1.01±0.46	8	0.55	3	10.07**
Shoot length 2009 (cm)	112	60.7±13.6	8	2.18*	3	2.67
Reproductive phenology[b]						
Flowering onset (days)	106	173.8±55.3	8	4.93****	3	1.38
Flowering duration (days)	106	90.8±52.0	8	3.68***	3	1.34
Fruiting onset (days)	106	268.2±44.4	8	4.78****	3	0.59
Fruiting duration (days)	105	36.8±44.2	8	4.68****	3	0.68
Reproductive output[c]						
Pod density (cm⁻¹)	106	2.31±1.45	8	2.80**	3	1.77
Seeds per pod	103	3.12±0.82	8	1.74#	3	0.33
Seed mass (mg)	105	6.48±0.89	8	1.97#	3	1.24
Pod infestation rates[d]						
Weevils (%)	103	21.6±23.4	8	4.48****	3	0.55
Moths (%)	103	2.53±3.67	8	0.99	3	3.67#
Total (%)	103	25.3±25.7	8	4.54****	3	0.69
Hymenoptera (%)	103	4.17±7.72	8	2.62*	3	0.92

[a]Standard deviation of the whole sample.
[b]Monitored in 2008–2009, with Sept 1st 2008 taken as the first day of the reproductive season.
[c]Pod density was measured in May 2009; seeds per pod and seed mass were measured in late June 2009.
[d]Measured in late June 2009.
#P<0.10, * P<0.05, ** P<0.01, *** P<0.001, **** P<0.0001.

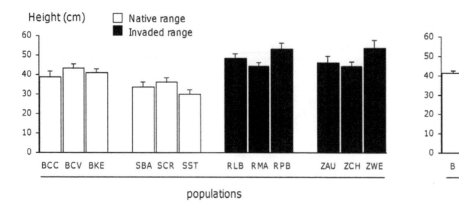

Figure 1. Height of one-year-old *U. europaeus* plants grown in a glasshouse. Population and regional means are given with 1 SE. N = 265.

performed partial correlations, we removed one degree of freedom for each covariate used. Finally, the significance of the effects were tested using hierarchical F-ratio tests, in which the range effect was tested against variation between regions (within ranges).

Results

Growth

In 2007 (one-year-old plants), plant height showed significant population and region effects, the latter being stronger (Table 2). All six populations of the invaded regions were taller than the six populations of the native regions (Figure 1), and plants from the invaded regions were on average 30% taller than plants from the native regions. The number of lateral shoots and the total shoot length showed a significant population effect, but no region effect (Table 2). In 2008 (two-year-old plants), plant height still showed a significant region effect but no more population effect (Table 2). The differences between regions were less pronounced: plants from the invaded regions were on average 15% taller than plants from the native regions (Figure 2A). For basal area, there was no population effect but the region effect was significant (Table 2). Mean basal area ± SE was 1.26 ± 0.09 m^2 for Brittany, 0.82 ± 0.07 m^2 for Scotland, 1.05 ± 0.09 m^2 for Reunion and 0.92 ± 0.07 m^2 for New Zealand. In 2009 (three-year-old plants), like in 2008, plant height showed a significant region effect but no population effect (Table 2). The differences were still less pronounced: plants from the invaded regions were on average 10% taller than plants from native regions (data not shown). The mean shoot length showed a significant population effect, but no region effect (Table 2). Finally, when plant height at the three years was analysed in a repeated statement, the interaction between years and regions was significant (N = 107, F = 3.20, P<0.01), confirming that the differences among regions depended on the year. There was no interaction between years and populations (N = 107, F = 0.95, P = 0.51).

Reproduction

Flowering and fruiting phenologies. The earliest plants began to flower in October and the latest plants in April. Whatever their date of flowering onset, all individuals flowered until spring, inducing a strong negative correlation between flowering onset and flowering duration (N = 106, $R_{Spearman} = -0.88$, P<10^{-4}). Flowering and fruiting onset and duration showed significant population effects, but no region effects (Table 2). The variation among populations was indeed very high and larger than the variation among regions (Figure 2B).

Pod and seed production. The region effect was never significant for pod density, the number of seeds per pod and seed mass (Table 2). The population effect was significant only for pod density, which exhibited a wide variability between populations (see Figure 2C). The number of seeds per pod and seed mass showed a low variability and a marginally significant population effect (Table 2).

Infestation rates

Infestation of pods. The first weevil-infested pods were observed in late May, and their proportion increased continuously until the end of the fruiting season (early July). Weevil infestation rate in late June reached 22% and showed a very high population effect, but no region effect (Table 2). No region effect was detected in the other dates of observation (data not shown). Pod infestation rate by the parasitoid wasp *Pteromalus sequester* was dependent on the infestation rate by weevils, thus showing a similar pattern (Table 2). Pod infestation rate by the moth *Cydia succedana* was much lower than the infestation rate by weevils (2.5% in late June) and did not exhibit any population or region effect (Table 2). Total pod infestation rate in late June (by both weevils and moths) was mainly due to the infestation rate by weevils, and consequently showed a highly significant population effect and no region effect (Table 2, Figure 2D). These latter effects stayed the same when we used pod density and plant height, traits known to directly affect pod infestation rate, as covariates in the ANOVA ($F_{reg} = 0.69$, P = 0.59; $F_{pop} = 4.40$, P = 0.0002;).

Infestation of vegetative parts. In September 2008, some plants were naturally infested by the rust fungus *Uromyces genistae-tinctoriae*, and in May 2009, some plants were infested by aphids. In both cases, we recorded, for each plant, if it was infested or not, and in both cases, as for seed predators, we did not find any difference between the two invaded regions and the two native regions (data not shown). No attack by the spider mite *Tetranychus lintearius* was observed during the time of the experiment.

Population differences

Populations within regions exhibited strong and significant differences for most traits studied, including growth, flowering phenology, reproductive output and infestation rates (Table 2). Despite the low number of populations, mean flowering onset was correlated with latitude (N = 12, $R_{Spearman} = +0.59$, P<0.05).

Correlations between traits

We tested the correlations between traits in the 2008–2009 reproductive season. We retained four main traits: plant height

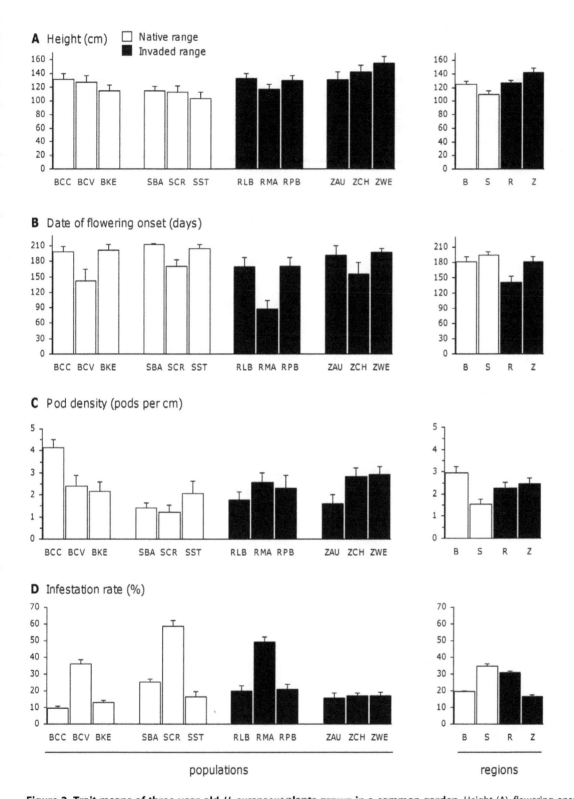

Figure 2. Trait means of three-year-old *U. europaeus* plants grown in a common garden. Height (A), flowering onset (B), pod density (C) and infestation rate (D). Population and regional means are given with 1 SE. Measures were done as in Table 3. N = 103 to 106 (see Table 2).

(for growth), flowering onset (for reproductive phenology), pod density in late May (for reproductive output) and total infestation rate in late June (for seed predation). When all the individuals were pooled, most of the correlations among these traits were significant (Table 3). Further comparisons were thus

performed with partial correlations in which untested life history traits were used as covariables. For example, when estimating the correlation between pod infestation rate and the date of flowering onset, pod density and plant height were used as covariates.

Table 3. Spearman's rank order correlation coefficients between four main traits of *Ulex europaeus* plants grown in the common garden.

	height[a]	pod density[b]	Infestation rate[c]
flowering onset[d]	−0.168[#]	−0.231*	−0.442***
height		0.235*	−0.218*
pod density			−0.277**

N = 103 to 106 (see Table 2).
[a]measured in Oct. 2008.
[b]measured in May 2009.
[c]total infestation rate (weevil + moth) in late June 2009.
[d]monitored between Oct. 2008 and July 2009.
[#]P<0.10, * P<0.05, ** P<0.01, *** P<0.001.

When correlations within regions were compared, the pattern obtained was different for correlations involving infestation rate (Figure 3A) and for correlations not involving infestation rate (Figure 3B). For correlations involving infestation rate, the correlation coefficients were negative and significant in the two native regions. They were also negative but weaker and rarely significant in the two invaded regions. The strength of the correlations was thus higher in the native regions than in the invaded regions (Figure 3A). The probability of observing such a consistent pattern three times can be estimated with a permutation test, where the null hypothesis is that the rank of correlation coefficients is independent of region status, and the risk alpha is the probability to obtain the observed pattern by chance. The probability for the two values of the native regions to be both higher than the two values of the invaded regions is 1/6. The probability of observing the same pattern three times is thus

$(1/6)^3 = 0.0046$. For correlations not involving infestation rate, correlations were low and rarely significant, and no special pattern was observed (Figure 3B). Moreover, the range effect was significant or nearly so for correlations involving infestation rate (flowering onset x infestation rate: $F_{1,2} = 15.04$, $P = 0.06$; pod density x infestation rate: $F_{1,2} = 64.32$, $P = 0.02$ height x infestation rate: $F_{1,2} = 27.86$, $P = 0.03$), while it was not significant for the correlations not involving infestation rate (P>0.20).

Discussion

Trait-by-trait analysis

According to the EICA hypothesis, we expected plants from regions of the invaded range to show an increase in growth and/or reproduction and a decrease in defence against seed predators compared to regions of the native range [16]. While the population or region effects were often highly significant, the range effect was significant for none of the traits studied. Furthermore, for all the traits studied but one, the differences between regions were not related to their range. This implies that whatever the statistical power of the range effect, the variance in these traits depended more on regions and populations of origin than on their native or invasive status.

Plant height is the only trait for which we observed a difference between the two invaded and the two native regions: plants from the two invaded regions grew taller than plants from the two native regions. Such an increased growth of plants from invaded areas was also observed in several studies performed in common gardens [12,38,39] where it was interpreted as a shift in resource allocation from defence to growth. In *U. europaeus*, increased growth was mainly observed on seedlings: the difference among native and invaded regions reached 30% the first year, but was reduced the second and third years. In this first year, gorse individuals were kept in a greenhouse, in an environment devoid of herbivores.

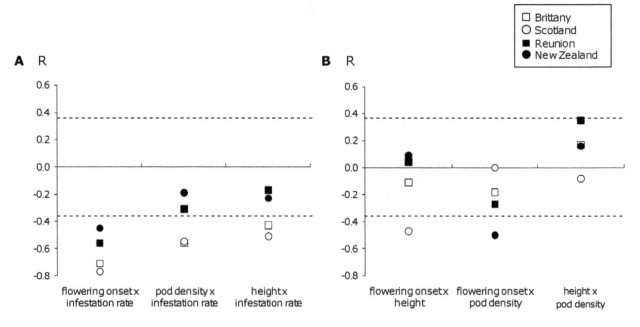

Figure 3. Correlations between traits of three-year-old *U. europaeus* individuals grown in a common garden. (A) correlations between infestation rate and three life history traits, (B) correlations among the three life history traits. Each point represents the Spearman's correlation coefficient for a given region. White symbols represent native regions, black symbols represent invaded regions. Measures were done as in Table 3. For each pairwise correlation, untested life history traits were used as covariables. N = 24 to 28. Dashed lines represent the 0.05 significance threshold (10^{-2} significant threshold is ±0.53, 10^{-3} significant threshold is ±0.66).

Their better growth is thus in agreement with the EICA hypothesis. However, the increased growth of seedlings cannot result from reallocation of resource devoted to defence against seed predators, because individuals do not produce fruits before their third year. As a consequence, the difference in seedling height could rather result from the reallocation of resource devoted to defence against generalist herbivores that attack vegetative parts of gorse seedlings. Alternatively, it may result from maternal effects [40], or from selective pressures that were not studied here, such as competition for sunlight. Indeed, no such consistent difference between invaded and native regions was observed for the other growth parameters.

Reproductive traits did not show any significant difference among regions. This result is in agreement with the absence of difference among native and invaded regions observed in a previous study in a global comparison of seed mass, performed in natural populations of *U. europaeus* [41]. Increased fecundity of plants from the invaded range has sometimes been shown in other species (e.g. *Eschscholzia californica* [42], *Silene latifolia* [11]). However, such an increased fecundity was less often observed than increased vegetative growth [5,17].

Finally, infestation rates were different depending on the population, but plants from the invaded and native regions exhibited similar susceptibility to the two main seed predators, *Exapion ulicis* and *Cydia succedana*. The fact that the same significance pattern remained after using pod density and plant height as covariates reveals that these traits are not the only one involved. Other features such as physical or chemical defence may be implied. Overall, these results contrast with studies on other plant species, which have often shown an increase of susceptibility to natural enemies in plants from the invaded range (e.g. in *Silene latifolia* [11], *Hypericum perforatum* [43], *Triadica sebifera* [44]). This phenomenon is however not the rule, and seems to be of importance only in some invasive species [17,18].

Although we found that invasive gorse genotypes grew faster than native ones in their first year in the greenhouse, our results in the common garden do not support the EICA hypothesis as a major mechanism responsible for the invasiveness of *U. europaeus*. This does not exclude any involvement of EICA in this species, since growing the plants in an enemy-free environment may have revealed a reallocation of resources that could not be detected here. Still, in the classic experimental conditions used, when traits are considered independently, the native and the invaded regions were not very different, although studying more regions in each range would increase the statistical power and allow to know how general this finding is in *U. europaeus*. In any case, in our system, the variability seemed to lie mainly among populations.

Population differences

While significant differences between regions were observed only for a few vegetative traits, populations within regions exhibited strong and significant differences for most traits studied, including growth, flowering phenology, reproductive output and infestation rates. This strong differentiation among populations was already observed in natural populations of the native regions [28], where it has been shown to be genetically determined [29]. In the invaded regions, the differences among populations observed in this study could result from genetic differentiation among source populations, drift or local adaptation.

The EICA hypothesis does not make any prediction on reproductive phenology. However, in the absence of seed predators, flowering and fruiting phenologies are expected to evolve to fit local conditions [5], independently of the constraints exerted by the avoidance of seed predation. In *U. europaeus*,

flowering phenology and plant height depend on latitude and altitude [27,45,46]. Evidence of adaptation to altitude and latitude has been observed for vegetative growth, reproduction, and/or flowering phenology in introduced plant populations of several invasive species [9,10,47,48] and, although more populations are needed to conclude, it is likely that adaptation of these traits to new environmental factors did also occur in *U. europaeus*.

Regarding infestation rates, local adaptation cannot be invoked in regions devoid of seed predators. This is notably the case in Reunion, where large differences between populations for susceptibility to seed predators are observed, even though no seed predators were ever introduced onto the island. However, pod infestation rate is an integrative measure that results from a set of factors influencing predator reproductive success, but also host choice and detection [22]. These include plant size and architecture [49], flowering phenology [50], and chemicals [51]. It is thus likely that selection acting on these traits may interact with direct selection on predator avoidance [52]. This is especially true when the different traits involved are genetically correlated, as in *U. europaeus* [29].

Correlations between traits

Most of the traits studied here were correlated with each others, revealing again the strong relationships among them already observed in Atlan *et al.* [29]. Despite the low sample size, the correlation coefficients reached very high values and significance levels when regions were analysed separately.

For the three correlations involving the infestation rate and a life history trait (flowering onset, pod density, and plant height), the strength of the correlations was lower in the invaded regions than in the native regions. A reduction in the strength of correlations may result from a reduced diversity [53]. However, in *U. europaeus*, the diversity present in the invaded populations is very high and similar to that of the native regions, both for phenotypic traits (this study) and for neutral molecular markers (microsatellites and allozymes studies on 28 populations, Hornoy *et al.*, in prep.). Moreover, the reduction of correlation coefficients in the invaded regions was only observed in correlations involving infestation rate, suggesting that the absence of seed predation was the main causing factor reducing the strength of these correlations.

Differences between native and invaded regions thus appeared to lie in trait correlations rather than in trait means. Indeed, the trait-by-trait analysis did not reveal any clear difference between the native and the invaded ranges, and would have led to underestimating the role of the release from natural enemies in the invasive process. Such trait-by-trait analyses are often misleading when correlational selection is involved [54], which is the case in *U. europaeus* [29].

An interpretation of the differences in trait correlations observed here would need to assume that they have a genetic basis, which cannot be tested with the experimental design used in this study. However, the genetic nature of the correlations observed in the native range was ascertained in Atlan *et al.* [29] on a set of maternal families. In that previous study, correlations between family means were always greater than correlations between individuals, indicating that the small environmental differences inevitably present in a common garden, far from creating the correlations observed, were only creating noise that reduced the values of the correlations. Since the plants studied here belong to the same species, were grown in the same homogeneous conditions, randomized in the same manner, and planted in the same area as gorses studied by Atlan *et al.* [29], it seems reasonable to make the hypothesis that the phenotypic correlations observed correspond to genetic correlations. Such a hypothesis allows proposing a speculative scenario to explain the reduced correlations observed in the invaded range.

Genetic correlations can be generated by pleiotropy and/or by linkage disequilibrium [55]. In contrast to pleiotropy, that generates correlations which hold whatever the spatial scale considered, linkage disequilibrium generates correlations at the region level [56], such as those observed here. In the native range of *U. europaeus*, seed predator avoidance is achieved through a polymorphism of strategy involving phenology and growth [28,29]. Correlational selection acting on these life history traits could thus generate strong linkage disequilibria and maintain a high level of polymorphism [21,57].

During the invasive process, the selective pressure induced by seed predation disappeared, so that recombination and segregation could reduce the linkage in a few generations [58]. When this occurred, the polymorphic traits could have been selected to meet local conditions regardless of their consequences for susceptibility to seed predators, providing *U. europaeus* with new ecological potentialities. Interestingly, in New Zealand, where the weevil *E. ulicis* has been introduced in 1931, correlations between infestation rate and life history traits were still much lower than in native regions. Thus, the reintroduction of a biological agent was not sufficient to recreate the defence strategy observed in the native range, at least on such time scale. It would be interesting to study more invaded regions with or without biological control to see how general this relaxation of genetic correlations is in *U. europaeus*, and how long it takes for them to re-form. In any case, even if this phenomenon occurred only in a subset of invaded regions, it may still play an important role in the invasive potential of the species.

This scenario will remain speculative because the results obtained on gorse are not sufficient to ascertain the evolution of genetic correlations in the invaded range, and because the species is not suitable to generate the large dataset necessary to explore it further. However, by enhancing the importance of trait correlations, these results led us to propose a theoretical hypothesis that may provide new insights into the ecological and evolutionary mechanisms involved in the expansion of invasive species.

Toward a new hypothesis?

Genetic constraints on life history traits have recently been shown to strongly influence the invasion dynamics and the range limits of introduced species [24,59,60]. Indeed, they can reduce the evolutionary potential of a species, despite the existence of genetic variance for traits considered [23,61]. One of the main sources of genetic constraints are the genetic correlations resulting from correlational selection [21,57]. As a consequence, if life history traits of a species are constrained by correlational selection for the avoidance of natural enemies, enemy release in the introduced range can relax these genetic correlations and enable more independent evolution of key life history traits.

We thus suggest that a relaxation of genetic constraints, and in particular a Relaxation of Genetic Correlations (or RGC) may follow enemy release and potentially enhance the adaptive potential of some introduced species. It may contribute to their invasive success by facilitating the optimisation of life history traits in the invaded range, as observed for flowering onset in *Silene latifolia* [11,62], and flowering size in *Carduus nutans* [63]. It may also contribute to explain the niche shift or niche expansion observed in many invasive species, such as *Ulex europaeus* or *Centaurea maculosa* [64].

The RGC hypothesis takes into account the fact that the evolvability of life history traits depends not only on their genetic variability, but also on the genetic links among the traits [23,65]. Although it does not lead to an increase of competitive ability *per se*, it may facilitate local adaptation for traits relevant for the invasive success. The RGC hypothesis may be a non-exclusive alternative to the EICA hypothesis. The proposed mechanism is more general since it can involve other features than those related to growth and reproductive effort. Also, it does not necessarily require a cost of resistance to natural enemies and is not limited to negative correlations (trade-offs) between traits. Finally, it may involve the relaxation of other genetic constraints than those related to natural enemies.

Acknowledgments

The authors thank Louis Parize and Thierry Fontaine for technical assistance, Yvonne Buckley for providing us with some seed samples, Jean-Sébastien Pierre for statistical help and Alan Scaife for English edition. We are grateful to Gabrielle Thiébaut and Arjen Biere for their helpful comments. We also thank the Editor and the anonymous reviewers for their constructive comments on a previous version of this paper.

Author Contributions

Conceived and designed the experiments: MT AA. Performed the experiments: BH MH MT AA. Analyzed the data: BH MH LG AA. Contributed reagents/materials/analysis tools: LG MT AA. Wrote the paper: BH AA. Edited the manuscript: BH MT MH LG AA.

References

1. Alpert P, Bone E, Holzapfel C (2000) Invasiveness, invasibility and the role of environmental stress in the spread of non-native plants. Perspectives in Plant Ecology, Evolution and Systematics 3: 52–66.

2. Mitchell CE, Agrawal AA, Bever JD, Gilbert GS, Hufbauer RA, et al. (2006) Biotic interactions and plant invasions. Ecology Letters 9: 726–740.

3. Kolar CS, Lodge DM (2001) Progress in invasion biology: predicting invaders. Trends in Ecology and Evolution 16: 199–204.

4. Richardson DM, Pysek P (2006) Plant invasions: merging the concepts of species invasiveness and community invasibility. Progress in Physical Geography 30: 409–431.

5. Müller-Schärer H, Steinger T (2004) Predicting evolutionary change in invasive, exotic plants and its consequences for plant-herbivore interactions. In: Ehler LE, Sforza R, Mateille T, eds. Genetics, evolution and biological control. Wallingford, UK: CABI. pp 137–162.

6. Gilchrist GW, Lee CE (2007) All stressed out and nowhere to go: does evolvability limit adaptation in invasive species? Genetica 129: 127–132.

7. Lee CE, Gelembiuk GW (2008) Evolutionary origins of invasive populations. Evolutionary Applications 1: 427–448.

8. Callaway RM, Ridenour WM (2004) Novel weapons: invasive success and the evolution of increased competitive ability. Frontiers in Ecology and Environment 2: 436–443.

9. Montague JL, Barrett SCH, Eckert CG (2008) Re-establishment of clinal variation in flowering time among introduced populations of purple loosestrife (Lythrum salicaria, Lythraceae). Journal of Evolutionary Biology 21: 234–245.

10. Monty A, Mahy G (2009) Clinal differentiation during invasion: Senecio inaequidens (Asteraceae) along altitudinal gradients in Europe. Oecologia 159: 305–315.

11. Wolfe LM, Elzinga JA, Biere A (2004) Increased susceptibility to enemies following introduction in the invasive plant Silene latifolia. Ecology Letters 7: 813–820.

12. Zou J, Rogers WE, Siemann E (2008) Increased competitive ability and herbivory tolerance in the invasive plant Sapium sebiferum. Biological Invasions 10: 291–302.

13. Norghauer JM, Martin AR, Mycroft EE, James A, Thomas SC (2011) Island invasion by a threatened tree species: evidence for natural enemy release of Mahogany (Swietenia macrophylla) on Dominica, Lesser Antilles. PLoS ONE 6(4): e18790.

14. Keane RM, Crawley MJ (2002) Exotic plant invasions and the enemy release hypothesis. Trends in Ecology and Evolution 17: 164–170.

15. Liu H, Stiling P (2006) Testing the enemy release hypothesis: a review and meta-analysis. Biological Invasions 8: 1535–1545.

16. Blossey B, Notzold R (1995) Evolution of increased competitive ability in invasive nonindigenous plants: a hypothesis. Journal of Ecology 83: 887–889.
17. Bossdorf O, Auge H, Lafuma L, Rogers WE, Siemann E, et al. (2005) Phenotypic and genetic differentiation between native and introduced plant populations. Oecologia 144: 1–11.
18. Orians CM, Ward D (2010) Evolution of plant defenses in nonindigenous environments. Annual Review of Entomology 55: 439–459.
19. Joshi J, Vrieling K (2005) The enemy release and EICA hypothesis revisited: incorporating the fundamental difference between specialist and generalist herbivores. Ecology Letters 8: 704–714.
20. Ridenour WM, Vivanco JM, Feng Y, Horiuchi J-I, Callaway RM (2008) No evidence for trade-offs: Centaurea plants from America are better competitors and defenders. Ecological Monographs 78: 369–386.
21. Roff DA (2002) Life history evolution. Sunderland, MA: Sinauer Associates Inc. 527 p.
22. Johnson MTJ, Agrawal AA, Maron JL, Salminen J-P (2009) Heritability, covariation and natural selection on 24 traits of common evening primrose (Oenothera biennis) from a field experiment. Journal of Evolutionary Biology 22: 1295–1307.
23. Agrawal AA, Stinchcombe JR (2009) How much do genetic covariances alter the rate of adaptation? Proceedings of the Royal Society of London B 276: 1183–1191.
24. Colautti RI, Eckert CG, Barrett SCH (2010) Evolutionary constraints on adaptive evolution during range expansion in an invasive plant. Proceedings of the Royal Society Biological Sciences Series B.
25. Lowe S, Browne M, Boudjelas S, De Poorter M (2000) 100 of the World's worst invasive alien species. A selection from the global invasive species database. The Invasive Species Specialist Group (ISSG) a specialist group of the Species Survival Commission (SSC) of the World Conservation Union (IUCN). 12 p.
26. Davies WM (1928) The bionomics of Apion ulicis Forst. (Gorse weevil) with special reference to its role in the control of Ulex europaeus in New Zealand. Annals of Applied Biology 15: 263–286.
27. Hill RL, Gourlay AH, Martin L (1991) Seasonal and geographic variation in the predation of gorse seed, Ulex europaeus L., by seed weevil Apion ulicis Forst. New Zealand Journal of Zoology 18: 37–43.
28. Tarayre M, Bowman G, Schermann-Legionnet A, Barat M, Atlan A (2007) Flowering phenology of Ulex europaeus: ecological consequences of variation within and among populations. Evolutionary Ecology 21: 395–409.
29. Atlan A, Barat M, Legionnet AS, Parize L, Tarayre M (2010) Genetic variation in flowering phenology and avoidance of seed predation in native populations of Ulex europaeus. Journal of Evolutionary Biology 23: 362–371.
30. Herrera J (1999) Fecundity above the species level: ovule number and brood size in the Genistae (Fabaceae: Papilionideae). International Journal of Plant Sciences 160: 887–896.
31. Bowman G, Tarayre M, Atlan A (2008) How is the invasive gorse Ulex europaeus pollinated during winter? A lesson from its native range. Plant Ecology 197: 197–206.
32. Chater EH (1931) A contribution to the study of the natural control of gorse. Bulletin of entomological research 22: 225–235.
33. Barat M, Tarayre M, Atlan A (2007) Plant phenology and seed predation: interactions between gorses and weevils in Brittany (France). Entomologia Experimentalis et Applicata 124: 167–176.
34. Barat M, Tarayre M, Atlan A (2008) Genetic divergence and ecological specialisation of seed weevils (Exapion spp.) on gorses (Ulex spp.). Ecological Entomology 33: 328–336.
35. Hill RL, Gourlay AH (2002) Host-range testing, introduction, and establishment of Cydia succedana (Lepidoptera: Tortricidae) for biological control of gorse, Ulex europaeus L., in New Zealand. Biological Control 25: 173–186.
36. SAS Institute (2005) Version 9.1. SAS institute, Cary, NC.
37. R Development Core Team (2010) R: A language and environment for statistical computing. R Foundation for Statistical Computing. Vienna, Austria.
38. Stastny M, Schaffner U, Elle E (2005) Do vigour of introduced populations and escape from specialist herbivores contribute to invasiveness? Journal of Ecology 93: 27–37.
39. Blumenthal DM, Hufbauer RA (2007) Increased plant size in exotic populations: a common-garden test with 14 invasives species. Ecology 88: 2758–2765.
40. Roach DA, Wulff RD (1987) Maternal effects in plants. Annual Review of Ecology and Systematics 18: 209–235.
41. Buckley YM, Downey P, Fowler SV, Hill RL, Memmott J, et al. (2003) Are invasives bigger? A global study of seed size variation in two invasive shrubs. Ecology 84: 1434–1440.
42. Leger EA, Rice KJ (2003) Invasive California poppies (Eschscholzia californica Cham.) grow larger than native individuals under reduced competition. Ecology Letters 6: 257–264.
43. Maron JL, Vila M, Arnason JT (2004) Loss of enemy resistance among introduced populations of St. John's wort (Hypericum perforatum). Ecology 85: 3243–3253.
44. Huang W, Siemann E, Wheeler GS, Zou J, Carrillo J, et al. (2010) Resource allocation to defence and growth are driven by different responses to generalist and specialist herbivory in an invasive plant. Journal of Ecology 98: 1157–1167.
45. Millener LH (1962) Day-length as related to vegetative development in Ulex europaeus L.: II. Ecotype variation with latitude. New Phytologist 61: 119–127.
46. Markin GP, Yoshioka E (1996) The phenology and growth rates of the weed gorse (Ulex europaeus) in Hawaii. Newsletter of the Hawaiian Botanical Society 35: 45–50.
47. Kollmann J, Banuelos MJ (2004) Latitudinal trends in growth and phenology of the invasive alien plant Impatiens glandulifera (Balsaminaceae). Diversity and Distributions 10: 377–385.
48. Alexander JM, Edwards PJ, Poll M, Parks CG, Dietz H (2009) Establishment of parallel altitudinal clines in traits of native and introduced forbs. Ecology 90: 612–622.
49. Leimu R, Syrjänen K (2002) Effects of population size, seed predation and plant size on male and female reproductive success in Vincetoxicum hirundinaria (Asclepiadaceae). Oikos 98: 229–238.
50. Elzinga JA, Atlan A, Biere A, Gigord L, Weis AE, et al. (2007) Time after time: flowering phenology and biotic interactions. Trends in Ecology and Evolution 22: 432–439.
51. Hilker M, Meiners T (2002) Induction of plant responses to oviposition and feeding by herbivorous arthropods: a comparison. Entomologia Experimentalis et Applicata 104: 181–192.
52. Lahti DC, Johnson NA, Ajie BC, Otto SP, Hendry AP, et al. (2009) Relaxed selection in the wild. 24: 487–496.
53. Roff DA (2000) The evolution of the G matrix: selection or drift? Heredity 84: 135–142.
54. Walsh B, Blows MW (2009) Abundant genetic variation + strong selection = multivariate genetic constraints: a geometric view of adaptation. Annual Review of Ecology, Evolution and Systematics 40: 41–59.
55. Lynch M, Walsh B (1998) Genetics and analysis of quantitative traits. Sunderland, Massachussetts, USA: Sinauer. 980 p.
56. Futuyma DJ (2005) Evolution. Sunderland MA: Sinauer Associates. 603 p.
57. Sinervo B, Svensson E (2002) Correlational selection and the evolution of genomic architecture. Heredity 89: 329–338.
58. Falconer DS, Mackay TFC (1996) Introduction to quantitative genetics. Essex, UK: Longman.
59. Levin DA (2004) The ecological transition in speciation. New Phytologist 161: 91–96.
60. Alexander JM, Edwards PJ (2010) Limits to the niche and range margins of alien species. Oikos 119: 1377–1386.
61. Etterson JR, Shaw RG (2001) Constraint to adaptive evolution in response to global warming. Science 294: 151–154.
62. Biere A, Antonovics J (1996) Sex-specific costs of resistance to the fungal pathogen Ustilago violacea (Microbotryum violaceum) in Silene alba. Evolution 50: 1098–1110.
63. Metcalf CJE, Rees M, Buckley YM, Sheppard AW (2009) Seed predators and the evolutionarily stable flowering strategy in the invasive plant, Carduus nutans. Evolutionary Ecology 23: 893–906.
64. Treier UA, Broennimann O, Normand S, Guisan A, Schaffner U, et al. (2009) Shift in cytotype frequency and niche space in the invasive plant Centaurea maculosa. Ecology 90: 1366–1377.
65. Blows MW, Hoffman AA (2005) A reassessment of genetic limits to evolutionary change. Ecology 86: 1371–1384.

17

Population Genetics of *Ceratitis capitata* in South Africa: Implications for Dispersal and Pest Management

Minette Karsten[1,4], Bettine Jansen van Vuuren[2], Adeline Barnaud[3], John S. Terblanche[4*]

1 Evolutionary Genomics Group, Department of Botany and Zoology, University of Stellenbosch, Matieland, South Africa, 2 Centre for Invasion Biology, Department of Zoology, University of Johannesburg, Auckland Park, South Africa, 3 IRD, Montpellier, France, 4 Centre for Invasion Biology, Department of Conservation Ecology and Entomology, Stellenbosch University, University of Stellenbosch, Matieland, South Africa

Abstract

The invasive Mediterranean fruit fly (medfly), *Ceratitis capitata*, is one of the major agricultural and economical pests globally. Understanding invasion risk and mitigation of medfly in agricultural landscapes requires knowledge of its population structure and dispersal patterns. Here, estimates of dispersal ability are provided in medfly from South Africa at three spatial scales using molecular approaches. Individuals were genotyped at 11 polymorphic microsatellite loci and a subset of individuals were also sequenced for the mitochondrial cytochrome oxidase subunit I gene. Our results show that South African medfly populations are generally characterized by high levels of genetic diversity and limited population differentiation at all spatial scales. This suggests high levels of gene flow among sampling locations. However, natural dispersal in *C. capitata* has been shown to rarely exceed 10 km. Therefore, documented levels of high gene flow in the present study, even between distant populations (>1600 km), are likely the result of human-mediated dispersal or at least some form of long-distance jump dispersal. These findings may have broad applicability to other global fruit production areas and have significant implications for ongoing pest management practices, such as the sterile insect technique.

Editor: Nadia Singh, North Carolina State University, United States of America

Funding: This study was funded by Fruitgro Science (JST), NRF-THRIP (P. Addison) and NRF Scarce Skills. The funders had no role in study design, data collection and analysis, decision to publish, or preparation of the manuscript.

Competing Interests: The authors have declared that no competing interests exist.

* E-mail: jst@sun.ac.za

Introduction

Through globalization and increased economic trade, species are frequently transported outside of their natural ranges [1]. For an introduced species to become established and ultimately invasive in new environments, a number of barriers need to be overcome (see e.g. [2]). The impact of invasive species can be wide-ranging, from direct impacts on natural biodiversity and resources to affecting human well-being and agriculture.

A case in hand concerns the Mediterranean fruit fly (medfly), *Ceratitis capitata* (Weidemann) (Diptera: Tephritidae), one of the economically costly pest species worldwide [3,4]. This species has, through fruit production and associated trade-related transport [5], spread from its native Afrotropical range (following [6]) to several of the main fruit-producing regions across the world [7]. Although somewhat contentious, the geographic extent of the historical (native) range of medfly is now assumed to be Afrotropical [6] but the exact range remains uncertain. The first confirmed presence of medfly in the Western Cape Province, South Africa dates to before the end of the nineteenth century [8,9], and the species is currently widespread throughout South Africa [10]. Several factors may contribute to the successful and wide-spread establishment of *C. capitata* here and elsewhere. Amongst these are the species' polyphagous life-history [6,11], short development time, and high population reproductive potential [12]. *Ceratitis capitata* may also have a broader climate niche compared to its congeners [13,14].

Current management of medfly populations in South Africa predominantly relies on the use of insecticides, including bait application technique (food baits combined with pesticide), bait stations (food bait and pesticide placed in container or trap) and full-cover sprays [15]. Insecticides are problematic not only for human health but also detrimental to the environment. Consequently, more effective and environmentally-friendly techniques are becoming increasingly sought after. Foremost amongst these is the Sterile Insect Technique (SIT) [16] which involves the release of mass-reared sterile males that mate with wild females thereby decreasing population numbers to a threshold from which populations are unable to recover [17]. SIT is currently used in parts of South Africa (Western Cape) with varying success and limited control of medfly [18]. The successful implementation of SIT relies heavily on information regarding the movement of individuals as well as the effective population sizes for different populations and regions [19]. Traditionally, direct methods, including mark-and-recapture studies, were employed to provide this information. However, recent studies have indicated that such methods may significantly under-estimate population size and migration (see e.g. [20,21]).

Accurate information regarding population structure and movement of individuals is crucial in ensuring successful implementation of pest-control strategies. For example, incorrect estimates of the minimum size for area-wide pest management could result in a failed attempt and, given the cost involved, funding may not be available to complete or re-start eradication or control programmes [22]. As such, indirect methods such as gene flow estimated from molecular information are increasingly being used (reviewed in e.g. [23–25]). Furthermore, an understanding of the population structure allows inferences about movement patterns and neighbourhood sizes (see e.g. [20,26–31]). While it is clear that medfly can have population genetic structure on large geographic scales [3,27], it is presently unknown at what spatial scale this pattern breaks down or in fact, whether the spatial patterns is dependent on climate or various environmental factors. To address these questions we use a combination of molecular sequence (mitochondrial cytochrome oxidase subunit I) and microsatellite data to estimate population genetic parameters at different spatial scales. The mutation rate of mitochondrial DNA markers are slower than that of microsatellites and can therefore be used to infer historical rather than contemporary (microsatellites) processes that shape evolutionary processes [32]. The null hypotheses we test is that South African populations are equally connected and that there is no population differentiation. We discuss these results in the context of dispersal and pest management.

Materials and Methods

Sampling Sites and Fly Collection

Our sampling regime aimed to capture genetic diversity at three different spatial scales. *Ceratitis capitata* individuals were collected from eight locations in South Africa (broad scale sampling, N = 198 individuals), 13 locations in the Western Cape (regional scale sampling, N = 385) and 13 locations in the Ceres valley (fine scale sampling, N = 382) (Fig. 1, Table 1). Given the nature of our experimental design which aimed to sample medfly in sites currently occupied, sampling focused on the Western Cape as it is an agricultural production area. To ensure spatial homogeneity of sampling effort in our broad scale analyses (South Africa), four locations chosen at random from the Western Cape were included. To ensure that the selection of locations did not have a significant effect on the spatial structure, analyses were repeated several times with a different set of randomly chosen locations; the results remained the same irrespective of the sampling localities included. We therefore report results from only one of these data sets.

Bucket traps (Chempac, Paarl, South Africa) were set up in fruit orchards and geo-referenced using a hand-held GPS. Traps were dry baited with a three-component attractant Biolure 3C (Chempac) consisting of putrecine, ammonium acetate and trimethylamine (attractiveness <30 m; [15]). Traps were collected every two weeks and flies were transferred to absolute ethanol for storage. Flies were identified and sexed under a stereomicroscope in the laboratory. To ensure that flies included in the study were not part of a SIT release (and, as such, not a true reflection of the wild populations), all specimens were inspected under UV light and SIT flies were discarded. Pupae from the SIT program are covered in fluorescent dye before release that accumulates in the head suture during emergence and are thus easily identified under a UV light [18]. DNA was extracted from whole flies using a DNeasy® tissue kit (QIAGEN Inc.).

Mitochondrial DNA Amplification and Sequencing

For mitochondrial DNA, a 782-bp segment of the cytochrome oxidase subunit I (COI) gene was targeted for ten randomly selected individuals per sampling location where possible. This resulted in a sequence data set of 125 individuals from the Western Cape (Table 2) and 74 individuals from across South Africa (Table 2). The primers C1-J-2183 and TL2-N-3014 [33] were used for amplification following standard procedures for medfly

[34]. Sequencing reactions were performed using BigDye chemistry (Applied Biosystems, Foster City, California, USA) and analyzed on an ABI 3170 automated sequencer (Applied Biosystems) (Genbank accession numbers: JX855840- JX855921).

Mitochondrial DNA Analysis

DNA sequences were aligned and edited in GENEIOUS Pro™ 5.0 software (Biomatters Ltd, New Zealand). We calculated the number of haplotypes (N_h), haplotype diversity (h) and nucleotide diversity (π) [35] (ARLEQUIN v3.5.1.2; [36]). To test for selective neutrality we used Fu's F-statistic [37] (ARLEQUIN v3.5.1.2). A parsimony haplotype network was constructed to investigate the relationship between COI haplotypes with a 95% connection limit (TCS v1.21; [38]).

Overall (i.e. all sampling locations considered together) and population pairwise Φ_{ST} values were calculated using 1000 permutations (ARLEQUIN v3.5.2.1). We tested for isolation by distance (IBD) using a Mantel test [39] with 1000 permutations by testing for a correlation between genetic distance and geographic distance (ARLEQUIN v3.5.2.1). The minimum straight line distance (i.e. as-the-crow-flies) between the GPS coordinates of sampling sites were taken as the geographic distance.

Microsatellite Genotyping

Specimens were genotyped for 12 microsatellite markers obtained from previously published studies [40–42] (see Table S1). Forward primers were 5'-labelled with one of four fluorophores (6-FAM, HEX, VIC or NED) and microsatellite loci were pooled for amplification if there was no signal inhibition during amplification (See Table S1 for further details). Samples were genotyped on an ABI 3130 Automated Sequencer (Applied Biosystems) and alleles were scored using GENEMAPPER v3.7 (Applied Biosystems). A positive control was included to verify that all plates were read consistently.

Microsatellite DNA Analysis

Microsatellite loci were tested for departures from Hardy-Weinberg equilibrium (HWE) and linkage disequilibrium using 10 000 permutations (GENEPOP; [43,44]). Significance levels were adjusted using False Discovery Rates ([45]; QVALUE; [46]). The marker Medflymic88 was found to have limited polymorphism and was subsequently excluded from further analyses (Table S1).

Levels of genetic diversity were assessed by computing basic statistics for 11 microsatellite loci. The average number of alleles (N_A), expected heterozygosity (H_E, expected allele frequencies given under Hardy-Weinberg equilibrium) and observed heterozygosity (H_O, actual heterozygosity measured in a population) were calculated for each microsatellite locus and for each location (GENETIX v4.05.2; [47]; GenAlEx 6.4; [48]). Allelic richness (A_R), which is a measure of genetic diversity independent of sample size, was calculated in FSTAT v2.9.3.2 [49]. The inbreeding coefficient (F_{IS}) was calculated in GENETIX v4.05.2 with 10 000 permutations to assess deviations from the null hypothesis of no inbreeding ($F_{IS} = 0$).

To assess the degree of population differentiation (at broad-, regional- and fine- spatial scales), we used three complementary approaches: F_{ST} and two Bayesian clustering methods, one with and one without prior spatial information. These analyses were performed on the three datasets independently. First, pairwise F_{ST} values (a measure of the genetic variance in a subpopulation compared to the total genetic variance in the entire population) and overall F_{ST} values were calculated in ARLEQUIN v3.5.1.2 using 10 000 permutations [36]. Second, STRUCTURE v2.3.3 [50,51] was run to assign the multilocus genotypes of individuals to

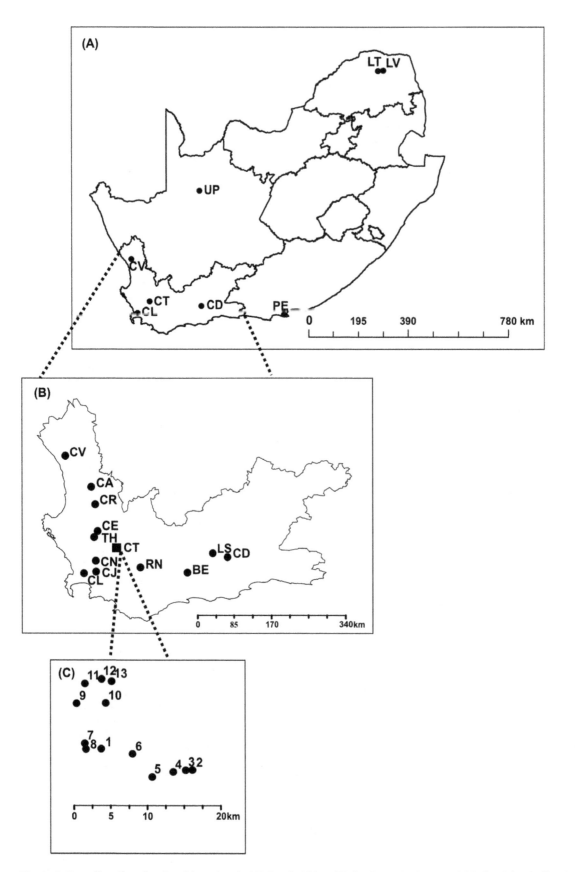

Figure 1. Sampling sites for *Ceratitis capitata* **in (A) South Africa, (B) the Western Cape and (C) the Ceres valley.** Locations include (CV, Lutzville; CA, Clanwilliam; CR, Citrusdal; CE, Porterville; TH, Tulbagh; CT, Ceres; CN, Wellington, CJ, Simondium; CL, Stellenbosch; RN, Robertson; BE,Barrydale; LS, Ladismith; CD, Calitzdorp; PE, Port Elizabeth; UP, Upington; LV, Levubu; LT, Makhado).

Table 1. The locations of *Ceratitis capitata* sampling in the Ceres valley, the Western Cape and South Africa with sample size (N), average number of alleles (N_A), number of private alleles (N_{AP}), allelic richness (A_R, based on a minimum of 3 individuals), expected (H_E) and observed (H_O) heterozygosity ($\pm$ = standard error) and the inbreeding coefficient (F_{IS}).

Location	ID	N	N_A	N_{AP}	A_R	H_E	H_O	F_{IS}
Fine scale (Ceres valley)								
Ceres 1	CT1	30	11.182	4	4.100	0.807±0.102	0.671±0.135	0.185
Ceres 2	CT2	30	11.455	2	4.127	0.812±0.087	0.634±0.184	0.236
Ceres 3	CT3	30	11.182	2	4.099	0.805±0.109	0.662±0.179	0.194
Ceres 4	CT4	30	10.545	1	4.064	0.803±0.100	0.634±0.215	0.227
Ceres 5	CT5	29	10.273	2	3.991	0.793±0.095	0.654±0.192	0.193
Ceres 6	CT6	29	11.182	3	4.147	0.816+0.079	0.660+0.167	0.209
Ceres 7	CT7	25	10.000	0	4.076	0.798±0.097	0.643±0.171	0.215
Ceres 8	CT8	30	10.273	2	4.087	0.808±0.096	0.656±0.127	0.205
Ceres 9	CT9	30	10.545	2	4.127	0.813±0.099	0.665±0.155	0.199
Ceres 10	CT10	29	10.727	3	4.050	0.800±0.087	0.665±0.176	0.187
Ceres 11	CT11	30	10.545	2	4.074	0.808±0.081	0.651±0.168	0.211
Ceres 12	CT12	30	10.364	1	4.117	0.811±0.108	0.617±0.144	0.256
Ceres 13	CT13	30	10.273	3	4.004	0.797±0.094	0.638±0.182	0.217
Total		382	10.657	27	4.082	0.805	0.650	0.210
Regional scale (Western Cape)								
Barrydale	BE	26	9.727	1	4.111	0.812±0.096	0.658±0.215	0.209
Calitzdorp	CD	29	11.455	2	4.210	0.822±0.090	0.632±0.216	0.248
Ceres	CT	30	10.455	2	3.991	0.808±0.096	0.658±0.124	0.203
Citrusdal	CR	30	10.182	0	3.958	0.788±0.119	0.593±0.201	0.264
Clanwilliam	CA	30	10.273	3	3.969	0.793±0.097	0.628±0.168	0.225
Ladismith	LS	30	10.545	3	4.030	0.803±0.089	0.672±0.174	0.18
Lutzville	CV	30	9.909	0	3.950	0.793±0.088	0.697±0.198	0.138
Porterville	CE	30	11.364	2	4.115	0.808±0.101	0.640±0.199	0.226
Robertson	RN	30	10.636	2	4.041	0.802±0.100	0.634±0.180	0.227
Simondium	CJ	30	10.364	3	4.036	0.802±0.096	0.632±0.197	0.228
Stellenbosch	CL	30	10.727	4	4.084	0.810±0.088	0.641±0.189	0.225
Tulbagh	TH	30	10.909	4	4.046	0.802±0.104	0.637±0.185	0.222
Wellington	CN	30	10.818	3	4.103	0.805±0.104	0.675±0.217	0.179
Total		385	10.566	29	4.050	0.803	0.646	0.213
Broad scale (South Africa)								
Calitzdorp	CD	29	11.455	8	4.210	0.822±0.090	0.632±0.216	0.248
Ceres	CT	30	10.455	3	3.991	0.808±0.096	0.658±0.124	0.203
Levubu	LV	30	11.364	15	4.068	0.793±0.120	0.576±0.183	0.29
Makhado	LT	13	8.636	3	4.177	0.794±0.124	0.585±0.227	0.302
Lutzville	CV	30	9.909	4	3.950	0.793±0.088	0.697±0.198	0.138
Port Elizabeth	PE	6	4.727	1	3.443	0.630±0.200	0.530±0.277	0.26
Stellenbosch	CL	30	10.727	7	4.084	0.810±0.088	0.641±0.189	0.225
Upington	UP	30	9.818	4	3.859	0.777±0.102	0.610±0.199	0.233
Total		198	9.636	45	3.973	0.778	0.616	0.237

populations without any prior spatial information. To estimate the number of clusters (K), we ran 10 independent runs for each K value varying between 1 and 13. A burn-in period of 1 000 000 followed by 1 000 000 Markov Chain Monte Carlo (MCMC) permutations was run to allow statistical parameters to reach stability and gave consistent results over the 10 independent runs. To determine the true number of populations (K) in the dataset,

the method described by Evanno et al. (2005) [52] was used. The STRUCTURE output was visualized using DISTRUCT v1.1 [53]. Thirdly, TESS v2.3 [54,55], which has an option to implement an admixture model that uses spatial coordinates of populations as prior information, was used to detect spatial genetic structure. Results from STRUCTURE and TESS were concordant; therefore we only discuss results from STRUCTURE (for

Table 2. Sampling locations of *Ceratitis capitata* in Ceres, the Western Cape and South Africa; and genetic diversity indices for the mitochondrial COI gene with N, sample size; N_h, the number of haplotypes; h, the haplotype diversity; π, nucleotide diversity, Fu's F-statistic with corresponding significance value (p) ($\pm$ = standard error) and the identities of the haplotypes (Genbank accession numbers: JX855840- JX855921).

Location	ID	N	N_h	h	π	Fu's FS	p	Haplotypes
Regional scale (Western Cape)								
Barrydale	BE	10	8	0.956±0.059	0.007±0.004	−1.760	0.130	2Cc1, Cc2, 2Cc3, Cc4, Cc5, Cc6, Cc7, Cc8
Calitzdorp	CD	10	10	1.000±0.045	0.007±0.004	−5.364	0.001	Cc5, Cc7, Cc11, Cc14, Cc15, Cc16, Cc17, Cc18, Cc19, Cc20
Ceres	CT	9	8	0.933±0.077	0.007±0.004	−1.595	0.165	Cc1, Cc7, Cc29, Cc38, Cc47, 2Cc48, Cc49, Cc50
Citrusdal	CR	10	8	0.933±0.077	0.009±0.005	−10181	0.227	Cc7, Cc9, Cc38, Cc42, 3Cc43, Cc44, Cc45, Cc46
Clanwilliam	CA	10	7	0.867±0.107	0.004±0.002	−2.178	0.065	Cc3, Cc7, 4Cc9, Cc10, Cc11, Cc12, Cc13
Ladismith	LS	9	9	1.000±0.052	0.008±0.005	−4.279	0.009	Cc11, Cc26, Cc31, Cc40, Cc56, Cc57, Cc58, Cc59, Cc60
Lutzville	CV	10	8	0.956±0.060	0.007±0.004	−1.770	0.138	Cc9, Cc29, Cc44, Cc51, 2Cc52, 2Cc53, Cc54, Cc55
Porterville	CE	10	10	1.000±0.045	0.007±0.004	−5.520	0.003	Cc9, Cc11, Cc21, Cc22, Cc23, Cc24, Cc25, Cc26, Cc27, Cc28
Robertson	RN	10	9	0.978±0.054	0.006±0.004	−3.604	0.023	Cc7, Cc8, Cc11, 2Cc12, Cc15, Cc38, Cc61, Cc62, Cc63
Simondium	CJ	10	9	0.978±0.054	0.008±0.004	−3.006	0.047	Cc7, Cc9, Cc11, Cc15, Cc26, 2Cc29, Cc30, Cc31, Cc32
Stellenbosch	CL	10	9	0.978±0.054	0.006±0.003	−4.031	0.004	Cc9, 2Cc11, Cc12, Cc25, Cc33, Cc34, Cc35, Cc36, Cc37
Tulbagh	TH	8	7	0.964±0.077	0.004±0.003	−2.928	0.020	Cc3, Cc7, Cc9, Cc59, Cc64, 2Cc65, Cc66, Cc67
Wellington	CN	9	8	0.972±0.064	0.005±0.003	−3.427	0.016	2Cc7, Cc9, Cc12, Cc29, Cc38, Cc39, Cc40, Cc41
Total		125	67	0.980±0.005	0.007±0.004	−25.310	0	
Broad scale (South Africa)								
Calitzdorp	CD	10	10	1.000±0.045	0.007±0.004	−5.364	0.004	Cc5, Cc7, Cc11, Cc14, Cc15, Cc16, Cc17, Cc18, Cc19, Cc20
Ceres	CT	9	8	0.933±0.077	0.007±0.004	−1.595	0.166	Cc1, Cc7, Cc29, Cc38, Cc47, 2Cc48, Cc49, Cc50
Levubu	LV	10	9	0.978±0.054	0.008±0.005	−2.751	0.057	Cc12, Cc56, Cc70, Cc76, Cc77, 2Cc78, Cc79, Cc80, Cc81
Makhado	LT	10	8	0.956±0.059	0.007±0.004	−1.665	0.146	Cc14, 2Cc26, 2Cc38, Cc71, Cc72, Cc73, Cc74, Cc75
Lutzville	CV	10	8	0.956±0.059	0.007±0.004	−1.848	0.125	Cc9, Cc29, Cc44, Cc51, 2Cc52, 2Cc53, Cc54, Cc55
Port Elizabeth	PE	5	4	0.900±0.161	0.007±0.004	0.552	0.529	Cc22, Cc54, 2Cc56, Cc82
Stellenbosch	CL	10	9	0.978±0.054	0.006±0.003	−4.067	0.010	Cc9, 2Cc11, Cc12, Cc25, Cc33, Cc34, Cc35, Cc36, Cc37
Upington	UP	10	6	0.889±0.075	0.007±0.004	0.491	0.581	Cc7, 2Cc55, 2Cc68, 3Cc69, Cc70
Total		74	50	0.990±0.004	0.007±0.004	−25.342	0	

TESS results see Fig. S2). Spatial genetic structure was investigated using two different approaches. First, a Mantel test was used to assess the significance of the association between geographic and genetic distance (IBD) (implemented in ARLEQUIN v3.5.1.2). Secondly, in SPAGEDI v1.3 [56] we characterised the spatial genetic structure of sampled locations using global *F*-statistics. The standard error of the F_{ST} computation was estimated using 10 000 permutations. The results were visualized by plotting the $F_{ST}/(1-F_{ST})$ values against the geographic distance between sampled locations.

Results

Mitochondrial DNA Analysis

Genetic structure within the Western Cape (regional spatial scale). We identified 67 distinct haplotypes for the 125 individuals included at the regional spatial scale. Haplotype diversity was notably high with an average value for the region of 0.980 (Table 2). Overall nucleotide diversity was 0.007 (which is comparable to other *Ceratitis* spp.; see e.g. [57]) (Table 2). Fu's F-statistic was negative and significant (FS = −25.310; p<0.001) indicating a deviation from equilibrium.

The haplotype network showed no explicit spatial pattern of genetic variation across the Western Cape (Fig. 2). Rather, the spatial population structure appeared almost random. Although variation was overwhelmingly partitioned within sampling locations (only 2.9% of the variation was accounted for by the between-site component), we found that the overall Φ_{ST} value was significant ($\Phi_{ST} = 0.03$, p = 0.01). Pairwise Φ_{ST} comparisons between sampling sites were mostly non-significant except for the Tulbagh sampling location which was significantly differentiated from four other sampling localities (Simondium, Citrusdal, Ladismith and Robertson) (see Table S2). A closer inspection revealed the presence of four unique haplotypes in this sampling location. Excluding the Tulbagh sampling location from the AMOVA returned a marginally non-significant partitioning of genetic variation ($\Phi_{ST} = 0.022$, p = 0.051). No correlation between genetic and geographic distances were found (regression coefficient = −0.000086, p = 0.927) indicating the absence of isolation by distance.

Genetic structure across South Africa (broad spatial scale). We identified 50 haplotypes for the 75 individuals included from across South Africa. Haplotype diversity was high with an average value for South Africa of 0.990 (Table 2) with nucleotide diversity being 0.007 (comparable to other *Ceratitis* spp.; see e.g. [57]) (Table 2). Fu's F-statistic was highly negative and significant (FS = −25.342; p<0.001) indicating a deviation from equilibrium.

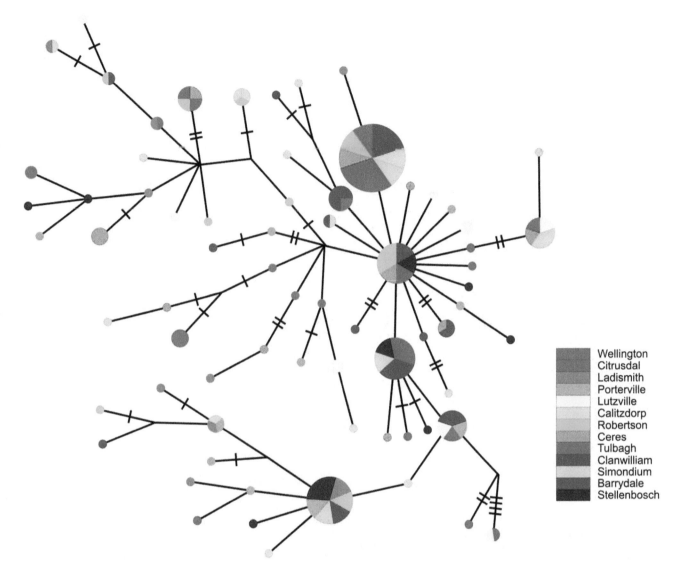

Figure 2. The parsimony haplotype network for *Ceratitis capitata* **in the Western Cape (regional scale).** The size of the pie charts is representative of the number of individuals that possess that haplotype. The small pie charts show haplotypes with a frequency of one individual. Every connecting line represents a mutational step of one between the different haplotypes. The perpendicular lines indicate additional mutational steps.

The haplotype network showed no clear spatial pattern of genetic variation in South Africa (Fig. 3). Variation was overwhelmingly partitioned within localities (Φ_{ST} = 0.028; p = 0.04). Pairwise Φ_{ST} values between sampling sites were small and non-significant. The only significant comparison was between Stellenbosch and Makhado, situated more than 1 600 km apart (Table S3). No correlation between genetic and geographic distances was found (regression coefficient = −0.00001, p = 0.789) indicating the absence of isolation by distance.

Microsatellite DNA Analysis

No linkage disequilibrium was observed among the 11 polymorphic microsatellite markers. All sampling locations deviated from HWE (genotype frequencies differ from ideal population which are characterized by random mating, no drift, mutation or migration) with relatively high levels of inbreeding indicative of non-random mating within sampling locations (Table 1).

Genetic structure within the Ceres Valley (fine spatial scale). From the Ceres Valley an average of 10.657 alleles were found with all of the sampling localities having private alleles (N_{AP}) except for Ceres 7 (Table 1). Genetic diversity, as indicated by mean expected heterozygosity (H_E), was 0.805 (Table 1). Allelic richness (A_R) ranged from 3.991 (Ceres 5) to 4.147 (Ceres 6) (mean = 4.082, based on minimum of 3 individuals; Table 1) and the inbreeding coefficient (F_{IS}) ranged between 0.185 (Ceres 1) and 0.256 (Ceres 12) (Table 1).

The overall F_{ST} value was 0.004 (p = 0.182), indicating no significant population differentiation. All of the pairwise F_{ST} values (quantification of genetic structure between populations) for sampling localities in the Ceres valley were not significant after corrections for multiple testing using False Discovery Rates. Both the Mantel test and the global F-statistics did not indicate any pattern of isolation by distance (r = 0.000002, p = 0.45; Fig. 4A).

The investigation of population differentiation in the Ceres valley using a Bayesian clustering approach in STRUCTURE

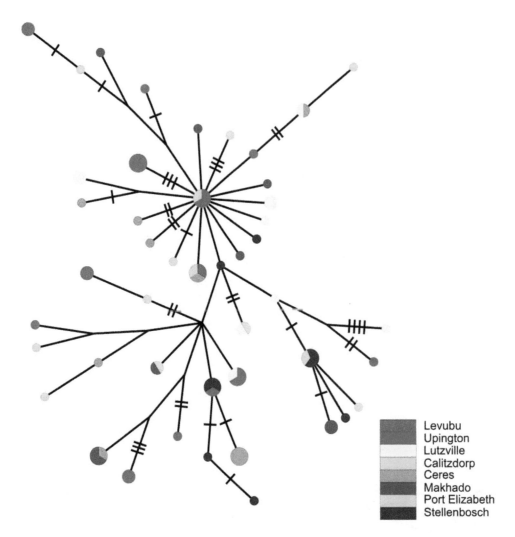

Levubu
Upington
Lutzville
Calitzdorp
Ceres
Makhado
Port Elizabeth
Stellenbosch

Figure 3. The parsimony haplotype network for *Ceratitis capitata* **in South Africa (broad scale).** The size of the pie charts is representative of the number of individuals that possess that haplotype. The small pie charts show haplotypes with a frequency of one individual. Every connecting line represents a mutational step of one between the different haplotypes. The perpendicular lines indicate additional mutational steps.

(Fig. S1) showed a lack of population differentiation. Evanno's method for estimating the optimal number of clusters (K) cannot calculate a ΔK value at $K = 1$, as it uses the second order rate of change. Examination of the log of the posterior probability of the data [ln P(D)] for each K value revealed the highest ln P(D) value at $K = 1$, an indication of the lack of population differentiation.

Genetic structure within the Western Cape (regional spatial scale). An average of 10.566 alleles was detected with all of the sampling localities having private alleles (N_{AP}) except for Lutzville and Citrusdal (Table 1). Average genetic diversity (H_E) for *C. capitata* in the Western Cape was high (0.803) (Table 1). Allelic richness (A_R) ranged from 3.950 (Lutzville) to 4.210 (Calitzdorp) (mean = 4.050, based on a minimum of 3 individuals; Table 1) and the inbreeding coefficient (F_{IS}) ranged between 0.138 (Lutzville) and 0.264 (Citrusdal) (Table 1).

Similar to the mtDNA findings, almost all of the pairwise F_{ST} comparisons were not significant after False Discovery Rate corrections (Table S4). The highest level of genetic differentiation was between Tulbagh and Citrusdal ($F_{ST} = 0.019$). The overall F_{ST} value of 0.006 (p = 0.002) indicated weak but significant population differentiation possibly due to the few significant pairwise comparisons. Results from STRUCTURE for the Western Cape

(Fig. S1) indicated no population differentiation and no isolation by distance (Mantel test: r = 0.000011, p = 0.128; see also Fig. 4B for global F-statistics).

We identified an average of 9.636 alleles with all of the sampling locations having private alleles (N_{AP}) (Table 1). Genetic diversity (H_E) for *C. capitata* in South Africa was high (0.778) (Table 1) and the mean value of Allelic richness (A_R) was 3.973 (based on a minimum of 3 individuals; Table 1). The inbreeding coefficient (F_{IS}) ranged between 0.138 (Lutzville) and 0.302 (Makhado) (Table 1).

Pairwise F_{ST} values were used to quantify the genetic structure between sampling locations and some of the comparisons were not significant after FDR corrections (Table S5). The highest level of genetic differentiation ($F_{ST} = 0.083$) was between Makhado and Port Elizabeth a distance of approximately 1300 km). The overall F_{ST} value was 0.021 (p = 0.0001), which indicated weak but significant population differentiation possibly due to the significant pairwise comparisons between a few localities. STRUCTURE results (Fig. S1) showed a lack of population differentiation. Neither the Mantel test nor global F-statistics indicated any pattern of isolation by distance (r = 0.000002, p = 0.269; Fig. 4C).

(A)

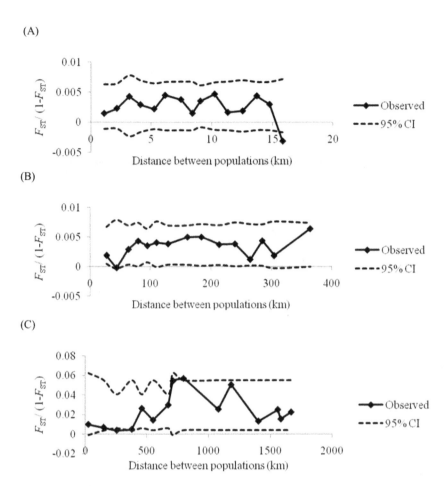

(B)

(C)

Figure 4. The spatial genetic structure of *Ceratitis capitata* **in (A) the Ceres valley, (B) the Western Cape and (C) South Africa.** The solid line represents the mean of the multilocus pairwise $F_{ST}/(1-F_{ST})$ values within each distance class and the dashed lines represent the 95% confidence intervals of the null distributions obtained from 1000 random permutations.

Discussion

In this study we characterized the genetic diversity, population genetic structure and population connectivity of a major international pest of agriculture, *C. capitata*, at three different spatial scales within South Africa. Several important and perhaps unexpected findings emerged from our results, with significant implications for other regions where this pest occurs.

Genetic Diversity

Exceptionally high levels of genetic diversity characterize medfly, from the scale of a single fruit-growing valley (Ceres) to its entire range across South Africa. The values reported here are notably higher than those reported for other *Ceratitis* species (*Ceratitis rosa* and *Ceratitis fasciventris*, [58]), but similar to those reported from other countries in Africa ($H_E = 0.750$ [27]; $H_E = 0.896$ [30]). Furthermore, the diversity values found for *C. capitata* in South Africa are higher than those reported for this species from other invaded regions of the world (Réunion island: $H_E = 0.660$; [27]; Australia: $H_E = 0.238–0.606$; [30]; the Mediterranean Basin: $H_E = 0.484–0.630$; [30]).

Three possible scenarios may account for these higher levels of genetic diversity in medfly. First, it may reflect the importance of propagule pressure and the effect of multiple introductions on their genetic diversity [1,59]. For example, medfly may spread relatively easily from Kenya to other African countries, including South

Africa, perhaps through human-mediated dispersal, thereby invading these countries multiple times. In contrast, the number of introductions to more distant countries such as Australia and Hawaii may be limited because of stricter quarantine control coupled with relative geographic isolation, and as such, result in lower genetic diversity values [60]. Second, although the native range for medfly is considered to be Afrotropical, this is contentious and the natural range may indeed have been broader [6]. Alien/invasive species typically have lower genetic diversity in their introduced ranges compared to their native ranges [60] and, in this respect, may indicate that South Africa falls within the native range. Third, the reported high levels of genetic diversity may be linked with exceptionally large population sizes in South Africa, among other factors (e.g. high mutation rate, see discussions in [61,62]). This high genetic diversity may contribute to their ability to colonize novel habitats through increased evolutionary potential [63,64].

Despite high genetic diversity and proposed large population sizes, all locations sampled from South Africa showed slight levels of inbreeding (F_{IS} (Ceres) ranged between 0.185–0.260; F_{IS} (Western Cape) ranged between 0.138–0.260; F_{IS} (South Africa) ranged between 0.138 and 0.302). Microsatellite estimates of F_{IS} in medfly have only been reported once previously [17], based on only 36 individuals randomly sampled on the Island of Chios, Greece. It is therefore difficult to place these findings in a global perspective, or evaluate the expected range of F_{IS} in medfly.

Although inbreeding seems unlikely given the large effective population sizes we report and that our results suggest that populations in South Africa may be expanding (Fu's F-statistic) (possibly due to an increase in the area under fruit production), this possibility cannot be excluded as medlfy demonstrates lekking behaviour [61]. Specifically, a small percentage of males can account for the majority of matings [65] suggesting non-random gene combination may be characteristic of the species.

Genetic Structure

In addition to high levels of genetic diversity, we find limited genetic differentiation among sampling locations across all spatial scales in South Africa. Given that the natural dispersal distances of C. capitata rarely exceeds 10 km [66], this was unexpected. Although some sampling locations are genetically significantly differentiated, these are mostly confined to single localities (such as Tulbagh) although the majority of sampling localities exhibit private alleles. Bonizzoni et al. (2004) [30] reported genetic homogeneity and a lack of spatial population differentiation in C. capitata populations in the coastal regions of Australia from Perth northwards. By contrast, Alaoui et al. (2010) [31] found population genetic structure in Moroccan C. capitata populations in endemic Argan forests, predominantly driven by their occurrence at different altitudes. We suggest that the structure of medfly in South Africa may be the result of complex interactions among factors at play at local scales (such as limited dispersal ability and adaptation -genetic and/or physiological- to local environments) and broad spatial scales, such as human-mediated or other forms of long distance dispersal. Indeed in South Africa large parts of the country are essentially climatically unsuitable for medfly and lacking in host plants; however, there are active commercial trade routes linking fruit distribution among these areas.

Dispersal

Owing to its global pest status, many investigations have focused on quantifying the dispersal ability of C. capitata. Although limited genetic structure and higher levels of population connectivity may have been expected for the Ceres valley (maximum distance between sampling localities is 17 km), this was an unexpected finding for our regional and broad scale investigation as sampling sites are geographically more distant, being up to 1600 km apart. Sharp and Chambers (1976) [67] showed that C. capitata can fly a maximum distance of 7–8 km, whereas most individuals only flew between 1–3 km within 2–3 hours. Similarly, Meats and Smallridge (2007) [66] indicated dispersal in C. capitata of between 0.5–9.5 km through natural dispersal (mainly flight), but that only a very small percentage of individuals are likely to do so (90% of released individuals remain within 400–700 m from release point). Therefore, given their limited dispersal ability based on direct estimates, the gene flow estimates are perhaps higher than might be expected. However, direct estimates are typically measured under natural or semi-natural conditions at a specific time point and may exclude the movement of individuals under extreme conditions, over longer time scales or multiple generations [68]. Therefore, although natural long-distance dispersal events by active flight or passive dispersal on wind are rare, they might nevertheless be frequent enough to cause genetic homogeneity among populations [68,69].

The distribution of suitable host plants (both wild and cultivated) in South Africa is probably able to facilitate the movement of C. capitata using their natural dispersal ability, although these are interspersed with large areas of unfavourable habitat. Therefore, these long distance dispersal events might occur in parallel or in series with human-mediated dispersal which

can take place over even longer distances. The most likely vectors of human-mediated dispersal include fruit consignments, the movement of nursery material (ornamental plants, e.g. in the Rosaceae and Liliaceae families) and also the movement of fruit between different locations throughout the country [19,70,71]. Regardless of the precise mode of movement, the overall result is potentially increased genetic homogenization.

Implications for Pest Management Strategies

If these results are assumed to be a reliable indication of movement patterns and population structure in South Africa, one potential implication of this work is that it suggests area-wide pest management should perhaps be undertaken at a broad scale, rather than on a fine scale (eg. farm or valley basis), as is presently the case [72]. Our results therefore imply that the whole of South Africa should potentially be considered a management unit, although this result should be verified further. While managing the whole of South Africa as a single unit may be considered unfeasible, it raises a number of practical issues for current control practise. First, if the whole country is not simultaneously targeted for control, any small-scale, localized control effort is likely to fail owing to high connectivity. It may therefore be worthwhile restricting the movement of fruit produce within the country, or allowing trade movement only after ensuring it is pest-free. Second, it may be important to consider high-risk routes of fruit movement and simply screen or quarantine those consignments most likely to move C. capitata around the country in order to eliminate this invasion pathway. Finally, it also suggests that import trade could be contributing to the high gene flow patterns detected (i.e. through propagule pressure). In this respect, it would be worthwhile to assess gene flow into South Africa from other adjacent countries or revisit global estimates of gene flow in medfly to determine if the global pattern previously documented [3,27] is changing, as this could help identify the reason for such low genetic structure in South Africa. It is increasingly clear, however, that the appropriate management unit should encompass wild host areas, home gardens and multi-owner fruit orchards to prevent C. capitata from taking refuge and recolonizing agricultural areas under control [73]. Furthermore, results of the current study can be used in future as a reference point to assess the success of SIT in South Africa. The expectation would be that genetic diversity would decrease as a result of decreasing population size if SIT is successful in suppressing the pest population. The present study is thus useful for better understanding the population structure of C. capitata, and in turn, could facilitate improved area-wide pest management programs for sustainable crop production in these and other geographic regions.

Supporting Information

Figure S1 Analysis of Ceratitis capitata individuals from (A) the Ceres valley, (B) the Western Cape and (C) South Africa using the Bayesian based method implemented in the program STRUCTURE. Each individual is indicated with a vertical line, the different shades of grey represent the individual's estimated percentage membership to the K clusters. Genetic population structure is shown for K = 2.

Figure S2 Analysis of Ceratitis capitata individuals from (A) the Ceres valley, (B) the Western Cape and (C) South Africa using the Bayesian based method implemented in the program TESS. Each individual is indicated with a vertical line, the different shades of grey represent the individual's estimated percentage membership to the K clusters. Genetic population structure is shown for $K = 2$.

Acknowledgments

Sequence and microsatellite data were genotyped at the Central Analytical Facility, Stellenbosch University. We are grateful to farmers and landowners in the Western Cape for access and sample collection. We are grateful for constructive comments and discussion of this work by the anonymous referees, Nadia Singh, Chris Weldon, Jaco le Roux and Aruna Manrakhan which greatly improved the manuscript.

References

1. Wilson JRU, Dormontt EE, Prentis PJ, Lowe AJ, Richardson DM (2009) Something in the way you move: dispersal pathways affect invasion success. Trends Ecol Evol 24: 136–144.
2. Blackburn TM, Pyšek P, Bacher S, Carlton JT, Duncan RP, et al. (2011) A proposed unified framework for biological invasions. Trends Ecol Evol 26: 333–339.
3. Malacrida AR, Gomulski LM, Bonizzoni M, Bertin S, Gasperi G, et al. (2007) Globalization and fruit fly invasion and expansion: the medfly paradigm. Genetica 131: 1–9.
4. De Meyer M, Robertson MP, Peterson AT, Mansell MW (2008) Ecological niches and potential geographical distributions of Mediterranean fruit fly (Ceratitis capitata) and Natal fruit fly (Ceratitis rosa). J Biogeogr 35: 270–281.
5. Villablanca FX, Roderick GK, Palumbi SR (1998) Invasion genetics of the Mediterranean fruit fly: variation in multiple nuclear introns. Mol Ecol 7: 547–560.
6. De Meyer M, Copeland RS, Wharton RA, McPheron BA (2002) On the geographic origin of the Medfly Ceratitis capitata (Weidemann) (Diptera: Tephritidae). Proceedings of the 6th International Fruit Fly Symposium, Stellenbosch, South Africa, 45–53.
7. White IM, Elson-Harris MM (1992) Fruit flies of economic significance: Their identification and bionomics. ACIAR, CAB International, Wallingford Oxon.
8. Annecke DP, Moran VC (1982) Insects and mites of cultivated plants in South Africa. Sigma Press (Pty) Ltd, Pretoria.
9. De Meyer M (2000) Systematic revision of the subgenus Ceratitis MacLeay s.s. (Diptera, Tephritidae). Zool J Linn Soc 128: 439–467.
10. Bedford ECG, Van den Berg MA, De Villiers EA (1998) Citrus pests in the Republic of South Africa. Institute for Tropical and Subtropical Crops, Nelspruit.
11. Malacrida AR, Guglielmino CR, Gasperi G, Baruffi L, Milani R (1992) Spatial and temporal differentiation in colonizing populations of Ceratitis capitata. Heredity 69: 101–111.
12. Lance D, McInnis D (2005) Biological basis of the sterile insect technique. In: Dyck VA, Hendrichs J, Robinson AS (eds) Sterile insect technique. Principles and Practice in Area-Wide Integrated Pest Management, Springer.
13. Duyck PE, David P, Quilici S (2006) Climatic niche partitioning following successive invasions by fruit flies in La Réunion. J Anim Ecol 75: 518–526.
14. Nyamukondiwa C, Terblanche JS (2009) Thermal tolerance in adult Mediterranean and Natal fruit flies (Ceratitis capitata and Ceratitis rosa): Effects of age, gender and feeding status. J Therm Biol 34: 406–414.
15. Ekesi S, Billah MK (2007) A field guide to the management of economically important Tephritid fruit flies in Africa. Icipe Science press second edition, Kenya.
16. Hendrichs J, Robinson AS, Cayol JP, Enkerlin W (2002) Medfly area wide sterile insect technique programmes for prevention, suppression or eradication: The importance of mating behaviour studies. Fla Entomol 85: 1–13.
17. Bonizzoni M, Katsoyannos BI, Marguerie R, Guglielmino CR, Gasperi G, et al. (2002) Microsatellite analysis reveals remating by wild Mediterranean fruit fly females, Ceratitis capitata. Mol Ecol 11: 1915–1921.
18. Barnes B, Rosenberg S, Arnolds L, Johnson J (2007) Production and quality assurance in the SIT Africa Mediterranean fruit fly (Diptera: Tephritidae) rearing facility in South Africa. Fla Entomol 90: 41–52.
19. Estoup A, Guillemaud T (2010) Reconstructing routes of invasion using genetic data: why, how and so what? Mol Ecol 19: 4113–4130.
20. Kourti A (2004) Estimates of gene flow from rare alleles in natural populations of medfly Ceratitis capitata (Diptera: Tephritidae). Bull Entomol Res 94: 449–456.
21. Katzner TE, Ivy JAR, Bragin EA, Milner-Gulland EJ, DeWoody JA (2011) Conservation implications of inaccurate estimation of cryptic population size. Anim Conserv 14: 328–332.
22. Barclay HJ, Matlock R, Gilchrist S, Suckling DM, Reyes J, et al. (2011) A conceptual model for assessing the minimum size area for an area-wide

integrated pest management program. Int J Agron. Available: http://dx.doi.org/10.1155/2011/409328.
23. Roderick GK (1996) Geographic structure of insect populations: gene flow, phylogeography and their uses. Annu Rev Entomol 41: 325–352.
24. Bohonak AJ (1999) Dispersal, gene flow, and population structure. Q Rev Biol 74: 21–45.
25. Broquet T, Petit EJ (2009) Molecular estimation of dispersal for ecology and population genetics. Annu Rev Ecol Evol Syst 40: 193–216.
26. Gasperi G, Guglielmino CR, Malacrida AR, Milani R (1991) Genetic variability and gene flow in geographical populations of Ceratitis capitata (Wied.) (medfly). Heredity 67: 347–356.
27. Gasperi G, Bonizzoni M, Gomulski LM, Murelli V, Torti C, et al. (2002) Genetic differentiation, gene flow and the origin of infestations of the medfly, Ceratitis capitata. Genetica 116: 125–135.
28. Malacrida AR, Marinoni F, Torti C, Gomulski LM, Sebastiani F, et al. (1998) Genetic aspects of the worldwide colonization process of Ceratitis capitata. Heredity 89: 501–507.
29. Bonizzoni M, Zheng L, Guglielmino CR, Haymer DS, Gasperi G, et al. (2001) Microsatellite analysis of medfly bioinfestations in California. Mol Ecol 10: 2515–2524.
30. Bonizzoni M, Guglielmino CR, Smallridge CJ, Gomulski M, Malacrida AR, et al. (2004) On the origins of medfly invasion and expansion in Australia. Mol Ecol 13: 3845–3855.
31. Alaoui A, Imoulan A, Alaoui-Talibi Z El, Meziane A El (2010) Genetic structure of Mediterranean fruit fly (Ceratitis capitata) populations from Moroccan Endemic Forest of Argania spinosa. Int J Agric Bio 12: 291–298.
32. Wang IJ (2010) Recognizing the temporal distinctions between landscape genetics and phylogeography. Mol Ecol 19: 2605–2608.
33. Simon C, Frati F, Beckenbach A, Crespi B, Liu H, et al. (1994) Evolution, weighting, and phylogenetic utility of mitochondrial gene sequences and a compilation of conserved polymerase chain reaction primers. Ann Entomol Soc Am 87: 651–701.
34. Barr NB, McPheron BA (2006) Phylogenetics of the genus Ceratitis (Diptera: Tephritidae). Mol Phylogenet Evol 38: 216–230.
35. Nei M (1987) Molecular Evolutionary Genetics. Columbia University Press, New York.
36. Excoffier L, Lischer HEL (2010) Arlequin suite version 3.5: a new series of programs to perform population genetics analyses under Linux and Windows. Mol Ecol Resour 10: 564–567.
37. Fu YX (1997) Statistical tests of neutrality of mutations against population growth, hitchhiking and background selection. Genetics 147: 915–925.
38. Clement M, Posada D, Crandall KA (2000) TCS: a computer program to estimate gene genealogies. Mol Ecol 9: 1657–1660.
39. Mantel N (1967) The detection of disease clustering and a general regression approach. Cancer Res 27: 209–220.
40. Bonizzoni M, Malacrida AR, Guglielmino CR, Gomulski LM, Gasperi G, et al. (2000) Microsatellite polymorphism in the Mediterranean fruit fly, Ceratitis capitata. Insect Mol Biol 9: 251–261.
41. Casey DG, Burnell AM (2001) The isolation of microsatellite loci in the Mediterranean fruit fly Ceratitis capitata (Diptera: Tephritidae) using a biotin/streptavidin enrichment technique. Mol Ecol Notes 1: 120–122.
42. Stratikopoulos EE, Augustinos AA, Petalas YG, Vrahatis MN, Mintzas A, et al. (2008) An integrated genetic and cytogenetic map for the Mediterranean fruit fly, Ceratitis capitata, based on microsatellite and morphological markers. Genetica 133: 147–157.
43. Raymond M, Rousset F (1995) An exact test for population differentiation. Evolution 49: 1280–1283.
44. Rousset F (2008) GENEPOP'007: a complete re-implementation of the GENEPOP software for Windows and Linux. Mol Ecol Resour 8: 103–106.
45. Ihaka R, Gentleman R (1996) R: a language for data analysis and graphics. J Comput Graph Stat 5: 299–314.

Author Contributions

Conceived and designed the experiments: JST BJVV MK AB. Performed the experiments: MK. Analyzed the data: MK AB BJVV JST. Contributed reagents/materials/analysis tools: JST BJVV. Wrote the paper: MK AB BJVV JST.

46. Storey JD (2002) A direct approach to false discovery rates. J R Stat Soc Series B Stat Methodol 64: 479–498.

47. Belkhir K, Borsa P, Chikhi L, Raufaste N, Bonhomme F (1996–2004).GE-NETIX 4.05, *logiciel sous Windows TM pour la génétique des populations*. Montpellier: Laboratoire Génome, Populations, Interactions, CNRS UMR 5000, Université de Montpellier II.

48. Peakall R, Smouse PE (2006) GENALEX 6: genetic analysis in Excel. Population genetic software for teaching and research. Mol Ecol Notes 6: 288–295.

49. Goudet J (2002) FSTAT, A program to estimate and test gene diversities and fixation indices (version 2.9.3.2). Available from: <http://www.unil.ch/popgen>.

50. Pritchard JK, Stephens M, Donnelly P (2000) Inference of population structure using multilocus genotype data. Genetics 155: 945–959.

51. Falush D, Stephens M, Pritchard JK (2003) Inference of population structure using multilocus genotype data: linked loci and correlated allele frequencies. Genetics 164: 1567–1587.

52. Evanno G, Regnaut S, Goudet J (2005) Detecting the number of clusters of individuals using the software STRUCTURE: a simulation study. Mol Ecol 14: 2611–2620.

53. Rosenberg NA (2004) DISTRUCT: a program for the graphical display of population structure. Mol Ecol Notes 4: 137–138.

54. Chen C, Durand E, Forbes F, François O (2007) Bayesian clustering algorithms ascertaining spatial population structure: a new computer program and a comparison study. Mol Ecol Notes 7: 747–756.

55. Durand E, Jay F, Gaggiotti OE, François O (2009) Spatial inference of admixture proportions and secondary contact zones. Mol Biol Evol 26: 1963–1973.

56. Hardy OJ, Vekemans X (2002) SPAGeDi: a versatile computer program to analyse spatial genetic structure at the individual or population levels. Mol Ecol Notes 2: 618–620.

57. Virgilio M, Backeljau T, Barr N, De Meyer M (2008) Molecular evaluation of nominal species in the *Ceratitis fasciventris*, *C. anonae*, *C. rosa* complex (Diptera: Tephritidae). Mol Phylogenet Evol 48: 270–280.

58. Baliraine FN, Bonizzoni M, Guglielmino CR, Osir EO, Lux SA, et al. (2004) Population genetics of the potentially invasive African fruit fly species, *Ceratitis rosa* and *Ceratitis fasciventris* (Diptera: Tephritidae). Mol Ecol 13: 683–695.

59. Facon B, Pointier JP, Jarne P, Sarda V, David P (2008) High genetic variance in life-history strategies within invasive populations by way of multiple introductions. Curr Biol 18: 363–367.

60. Lockwood JL, Cassey P, Blackburn T (2005) The role of propagule pressure in explaining species invasions. Trends Ecol Evol 20: 223–228.

61. Yuval B, Hendrichs J (2000) Behavior of flies in the genus *Ceratitis (Dacinae: Ceratitidini)*. In: Aluja M, Norrbom AL (eds) Fruit flies (Tephritidae): Phylogeny and evolution of behavior, 429–456. CRC Press, Boca Raton, Florida.

62. Chapuis MP, Popple JAM, Berthier K, Simpson SJ, Deveson E, et al. (2011) Challenges to assessing connectivity between massive populations of the Australian plague locust. Proc R Soc B 278: 3152–3160.

63. Lavergne S, Molofsky J (2007) Increased genetic variation and evolutionary potential drive the success of an invasive grass. Proc Natl Acad Sci U S A 104: 3883–3888.

64. Novak SJ (2007) The role of evolution in the invasion process. Proc Natl Acad Sci U S A 104: 3671–3672.

65. Arita LH, Kaneshiro KY (1985) The dynamics of the lek system and mating success in males of the Mediterranean fruit fly, *Ceratitis capitata* (Weidemann). Proc Hawaii Entomol Soc 25: 39–48.

66. Meats A, Smallridge CJ (2007) Short- and long-range dispersal of medfly, *Ceratitis capitata* (Dipt., Tephritidae), and its invasive potential. J Appl Entomol 131: 518–523.

67. Sharp JL, Chambers DL (1976) Gamma irradiation effect on the flight mill performance of *Dacus dorsalis* and *Ceratitis capitata*. Proc Hawaii Entomol Soc 22: 335–344.

68. Slatkin M (1985) Gene flow in natural populations. Annu Rev Ecol Syst 16: 393–430.

69. Coyne JA, Milstead B (1987) Long-distance migration of *Drosophila*. 3. Dispersal of *D. melanogaster* alleles from a Maryland orchard. Am Nat 130: 70–82.

70. Tolley KA, Davies SJ, Chown SL (2008) Deconstructing a controversial local range expansion: conservation biogeography of the painted reed frog (*Hyperolius marmoratus*) in South Africa. Divers Distrib 14: 400–411.

71. Hulme PE (2009) Trade, transport and trouble: managing invasive species pathways in an era of globalization. J Appl Ecol 46: 10–18.

72. Hendrichs J, Kenmore P, Robinson As, Vreysen MJB (2007) Area-wide integrated pest management (AW-IPM): principles. Practice and prospects. In: Vreysen MJB, Robinson AS, Hendrichs J (eds) Area-wide control of insect pest. From research to field implementation, Springer.

73. Manrakhan A, Addison P (2007) Monitoring Mediterranean fruit fly and Natal fruit fly in the Western Cape, South Africa. South African Fruit Journal 6: 16–18.

18

Evaluating the Effectiveness of an Ultrasonic Acoustic Deterrent for Reducing Bat Fatalities at Wind Turbines

Edward B. Arnett[1]*[¤], Cris D. Hein[1], Michael R. Schirmacher[1], Manuela M. P. Huso[2], Joseph M. Szewczak[3]

1 Bat Conservation International, Austin, Texas, United States of America, 2 Forest and Range Experiment Station, United States Geological Survey, Corvallis, Oregon, United States of America, 3 Department of Biological Sciences, Humboldt State University, Arcata, California, United States of America

Abstract

Large numbers of bats are killed by wind turbines worldwide and minimizing fatalities is critically important to bat conservation and acceptance of wind energy development. We implemented a 2-year study testing the effectiveness of an ultrasonic acoustic deterrent for reducing bat fatalities at a wind energy facility in Pennsylvania. We randomly selected control and treatment turbines that were searched daily in summer and fall 2009 and 2010. Estimates of fatality, corrected for field biases, were compared between treatment and control turbines. In 2009, we estimated 21–51% fewer bats were killed per treatment turbine than per control turbine. In 2010, we determined an approximate 9% inherent difference between treatment and control turbines and when factored into our analysis, variation increased and between 2% more and 64% fewer bats were killed per treatment turbine relative to control turbines. We estimated twice as many hoary bats were killed per control turbine than treatment turbine, and nearly twice as many silver-haired bats in 2009. In 2010, although we estimated nearly twice as many hoary bats and nearly 4 times as many silver-haired bats killed per control turbine than at treatment turbines during the treatment period, these only represented an approximate 20% increase in fatality relative to the pre-treatment period for these species when accounting for inherent differences between turbine sets. Our findings suggest broadband ultrasound broadcasts may reduce bat fatalities by discouraging bats from approaching sound sources. However, effectiveness of ultrasonic deterrents is limited by distance and area ultrasound can be broadcast, in part due to rapid attenuation in humid conditions. We caution that an operational deterrent device is not yet available and further modifications and experimentation are needed. Future efforts must also evaluate cost-effectiveness of deterrents in relation to curtailment strategies to allow a cost-benefit analysis for mitigating bat fatalities.

Editor: Danilo Russo, Università degli Studi di Napoli Federico II, Italy

Funding: This study was funded by Iberdrola Renewables, U.S. Department of Energy (DOE), National Renewable Energy Lab, Pennsylvania Game Commission, Bat Conservation International, American Wind and Wildlife Institute, and member companies of the American Wind Energy Association. The funders had no role in study design, data collection and analysis, decision to publish, or preparation of the manuscript.

Competing Interests: The authors have declared that no competing interests exist.

* E-mail: earnett@trcp.org

¤ Current address: Theodore Roosevelt Conservation Partnership, Loveland, Colorado, United States of America

Introduction

As wind energy production has steadily increased worldwide, bat fatalities have been reported at wind facilities worldwide [1,2,3,4] in a wide range of landscapes. A recent synthesis reported that approximately 650,000 to more than 1,300,000 bats have been estimated to have been killed from 2000–2011 in the U.S. and Canada [5]. Given these fatality rates, accelerating growth of the wind industry [6], and suspected and known population declines in many species of bats [7,8,9], it is imperative to develop and implement solutions to reduce future bat fatalities at wind facilities.

Prior studies have demonstrated that a substantial portion of bat fatalities consistently occur during relatively low-wind conditions over a relatively short period of time during the summer-fall bat migration period [2,4]. Curtailment of turbine operations under these conditions and during this period has been proposed as a possible means of reducing impacts to bats [1,2,10]. Indeed, recent studies in Canada [11] and the U.S. [12] indicate that increasing turbine "cut-in speed" (i.e., wind speed at which wind-generated electricity enters the power grid) from the manufactured speed (usually 3.5–4.0 m/s for modern turbines) to between 5.0 and 6.5 m/s resulted in at least a 50% reduction in bat fatalities (and as high as 93%) compared to normally operating turbines [12]. While costs of lost power from curtailment can be factored into the economics and financing and power purchase agreements of new projects, altering turbine operations even on a partial, limited-term basis potentially poses operational and financial difficulties for existing projects, so there is considerable interest in developing other solutions to reduce bat fatalities that do not involve turbine shutdowns. Also, changing turbine cut-in speed may not be effective in other regions that experience bat fatalities although this strategy may ultimately prove sufficiently feasible and economical for reducing bat fatalities. Thus, research on alternative mitigation strategies and their associated costs are warranted.

Studies in Scotland suggest that bat activity may be deterred by electromagnetic signals from small, portable radar units [13]. This study reported that bat activity and foraging effort per unit time were significantly reduced during experimental trials when their radar antenna was fixed to produce a unidirectional signal that maximized exposure of foraging bats to their radar beam. The effectiveness of radar as a potential deterrent has not been tested at an operating wind facility to determine if bat fatalities could be significantly reduced by these means. Moreover, the effective range of electromagnetic signals as well as the number of radar units needed to affect the most airspace near individual turbines would need to be determined to fully evaluate effectiveness and to allow some cost-benefit analysis relative to other potential deterrents or curtailment [11,12].

Echolocating bats produce high frequency vocal signals and perceive their surroundings by listening to features of echoes reflecting from targets in the path of the sound beam [14]. Thus, bats that use echolocation depend heavily on auditory function for orientation, prey capture, communication, and obstacle avoidance. Bats of some species avoid certain territorial social calls emitted by conspecifics [15] and are deterred by "clicks" emitted by noxious moths [16]. Because echolocating bats depend upon sensitive ultrasonic hearing, we hypothesized that broadcasting ultrasound from wind turbines may disrupt or "jam" their perception of echoes and serve as a deterrent. Such masking of echo perception, or simply broadcasting high intensity sounds at a frequency range to which bats are most sensitive, could create an uncomfortable or disorienting airspace that bats may prefer to avoid.

Few studies have investigated the influence of ultrasound broadcast on bat behavior and activity, particularly under field conditions. Broadband random ultrasonic noise may mask bat echolocation somewhat, but not completely [17]. Ultrasound broadcasts can reduce bat activity, perhaps due to greater difficulty in the bats hearing echoes of insects and thus reduced feeding efficiency [18]. A laboratory test of the response of big brown bats (*Eptesicus fuscus*) to a prototype eight speaker deterrent device emitting broadband white noise at frequencies ranging from 12.5–112.5 kHz in the laboratory and found that during non-feeding trials, bats landed in a quadrant containing the device significantly less when it was broadcasting broadband noise (J. Spanjer, University of Maryland and E. Arnett, Bat Conservation International, unpublished data). During feeding trials in this experiment, bats never successfully captured a tethered mealworm when the device broadcasted sound but captured mealworms near the device in about 1/3 of trials when it was silent. Field tests of the same acoustic deterrent found that when placed by the edge of a small pond, where nightly bat activity was consistent, nightly activity decreased significantly on nights when the deterrent was activated (J. Szewczak, Humboldt State University and E. Arnett, Bat Conservation International, unpublished data).

Our goal was to improve deterrent devices we previously developed and tested by increasing the effective area of ultrasonic emissions from the nacelle of wind turbines, and to test their effectiveness on reducing bat fatalities. The objectives of this study were 1) to conduct carcass searches and field bias trials (searcher efficiency and carcass removal) to determine rate of bat fatalities at treatment (those with deterrent devices) and control turbines; and 2) compare bat fatality rates at turbines treatment and control turbines to determine effectiveness. We successfully tested our ultrasonic deterrent device at an operating wind facility and offer suggestions for future efforts regarding this potential mitigation strategy to reduce bat fatalities at wind facilities.

Methods

Study Area

The study was conducted at the Locust Ridge Wind Project located near the towns of Shenandoah, Mahanoy City, and Brandonville in Columbia and Schuylkill Counties, Pennsylvania, and consisted of two different facilities. The Locust Ridge I (LRI)

Wind Farm has 13 Gamesa G87 2.0 MW turbines, each on 80 m monopoles with a rotor diameter of 87 m and a swept area of 5,945 m^2. There were 51 Gamesa G83 2.0 MW turbines, each on 80 m monopoles with a rotor diameter of 83 m and a swept area of rotor-swept area of 5,411 m^2, at the Locust Ridge II (LRII) Wind Farm. The facilities lie within the Appalachian mixed mesophytic forests ecoregion and the moist broadleaf forests that cover the plateaus and rolling hills west of the Appalachian Mountains [19,20]. Elevations along ridges where turbines are located range from 530–596 m. All turbines were located along a ridge in deciduous forest, with some species of evergreen trees interspersed. Vegetation across the area included thickets of scrub oak (*Quercus berberidifolia*) interspersed with chestnut oak (*Quercus prinus*) and gray birch (*Betula populifolia*), and mature hardwood forests of red oak (*Quercus rubra*), red maple (*Acer rubrum*), yellow birch (*Betula alleghaniensis*), American beech (*Fagus grandifolia*) and scrub oak.

Turbine Selection and Deterrent Installation

We randomly selected 15 of the 51 turbines located at LR II to be searched as part of a separate study to determine post-construction fatality rates and to meet permitting requirements of the Pennsylvania Game Commission's voluntary agreement for wind energy [21]. These 15 turbines were our control turbines for comparing with treatment turbines, those fitted with deterrent devices. In 2009, unforeseen mechanical and safety issues arose at the LRII site and many of these turbines had to be excluded from our potential treatment group due to potential safety hazards. Thus, we included all 13 turbines at LRI as well as the remaining available turbines at LRII (n = 36) when randomly selecting our 10 turbines to be fitted with deterrent devices; 3 turbines were randomly selected from the 13 available at the LRI site and 7 of 36 available at LRII.

We did not assess whether there were any potential inherent differences between the two types of turbines in 2009 and for this year assumed there were no confounding differences in our findings. However, in 2010, we attempted to assess inherent differences between control and treatment turbines by modifying our design and analysis to reflect a Before-After Control-Impact (BACI) design. The same sets of control and treatment turbines were monitored for a period of time prior to implementation of the deterrent treatment (1 May to 26 July 2010), then again during the deterrent implementation period (31 July through 9 October 2010). This design allowed for incorporating initial inherent differences between the two experimental treatment sets prior to implementation of the treatment as a reference for interpreting any differences detected during implementation of the treatment.

The deterrent devices used in our study consisted of a waterproof box (~45×45 cm, ~0.9 kg) that housed 16 transducers that emitted continuous broadband ultrasound from 20–100 kHz (manufactured by Deaton Engineering, Georgetown, Texas; Figure S1). We did not test other types of emissions (e.g., short pulses) concurrent with broadband emission because more devices and sample turbines would be required, thus resulting in cost and sample size constraints. Transducers we used had an optimum transmission level at their resonant frequency of 50 kHz transmission and reduced transmit levels at higher and lower frequencies over a broadband range of 20–100 kHz. This frequency range overlaps that of all bats known in the study area. Three factors influence the predicted effective transmitted power at a given distance: 1) the original transmitted power (sound pressure level; SPL); 2) attenuation with distance due to the wave front spreading (inversely proportional to the square of the distance, frequency independent); and 3) attenuation (absorption) in air of the sound wave (dependent on frequency, humidity and distance; Tables S1 and S2). The following discussion describes our estimation to base the target signal level of our deterrent:

A typical bat emits calls at about 110 dB sound pressure level (SPL) at 10 cm [22]. During search phase flight a typical North American species of bat emits about 12 calls per second, each about 5 milliseconds in duration [23,24]. Given the speed of sound at 340 m/sec and duration of an open air call, the bat's own call will theoretically mask echoes returning from objects within about 1.5 m (i.e., the bat cannot hear early return echoes while vocalizing). An echo from a target about 1.5 m away will return about 45 dB less than the original 110 dB signal, or at about 65 dB. The bat's next call would mask echoes returning from about 25 m away. By this first order estimation, a bat would theoretically perceive information from returning echoes with amplitudes of $\leq$65 dB over a range from about 1.5–25 m. Thus, we estimated that a broadband signal of $\geq$65 dB would begin jamming or masking most bat's echo perception from targets beyond about a 1.5 m range.

We attached 8 individual deterrent devices to the nacelle of each of 10 sample turbines. Three devices on each side of the nacelle were evenly spaced and pointed downward with one aimed into the rotor-swept area, one parallel with the monopole, and one aimed toward the back of the nacelle (Figures S2 and S3). Additionally, two devices were aimed at reflector plates; one that projected emissions into the upper part of the rotor-swept area, and one toward the rear of the nacelle. All devices connected to control boxes that were powered from outlets located in the nacelle and each was set on a timer to operate from ½ hour before sunset to ½ hour after sunrise each night of the study.

Delineation of Carcass Search Plots and Habitat Mapping

We delineated a rectangular plot 126 m north-south by 120 m east-west (60 m radius from the turbine mast in any direction; 15,120 m^2 total area) centered on each turbine sampled; this area represents the maximum possible search area for this study. Transects were set 6 m apart within each plot and in an east-west direction, due to the topography and layout of turbines at this facility. However, dense vegetation and the area cleared of forest at this facility was highly varied and, thus, we eliminated unsearchable habitat (e.g., forest) and usually did not search the entire possible maximum area. We used a Trimble global positioning system (GPS) to map the actual area searched at each turbine. The density-weighted area searched was used to standardize results and adjust fatality estimates (see statistical methods). The habitat visibility classes within each plot were also mapped using a GPS unit. We recorded the percent ground cover, height of ground cover (low [<11 cm], medium [11–50 cm], high [>50 cm]), type of habitat (vegetation, brush pile, boulder, etc), and the presence of extreme slope and collapsed these habitat characteristics into visibility classes that reflect their combined influence on carcass detectability (Table S3) [21].

Fatality Searches

We conducted daily searches at 15 control turbines and 10 treatment turbines from 15 August to 10 October 2009 and 1 May to 26 July and 31 July to 9 October 2010. Each searcher completed 5–7 turbine plots each day during the study. Searchers walked at a rate of approximately 10–20 m/min along each transect searching out to 3 m on each side for fatalities. Searches were abandoned only if severe or otherwise unsafe weather (e.g., heavy rain, lightning) conditions were present and searches resumed that day if weather conditions permitted. Searches

commenced at sunrise and all turbines were searched within 8 hr after sunrise.

We recorded date, start time, end time, observer, and weather data for each search at turbines. When a dead bat or bird was found, the searcher placed a flag near the carcass and continued the search. After searching the entire plot, the searcher returned to each carcass and recorded information on date, time found, species, sex and age (where possible), observer name, identification number of carcass, turbine number, perpendicular distance from the transect line to the carcass, distance from turbine, azimuth from turbine, habitat surrounding carcass, condition of carcass (entire, partial, scavenged), and estimated time of death (e.g., $\leq$1 day, 2 days, etc.). A field crew leader confirmed all species identifications at the end of each day. Disposable nitrile gloves were used to handle all carcasses to reduce possible human scent bias for carcasses later used in scavenger removal trials. Each carcass was placed into a separate plastic bag and labeled. Fresh carcasses, those determined to have been killed the night immediately before a search, were redistributed at random points on the same day for searcher efficiency and scavenging trials.

Ethics Statement

This study was conducted on private property and authorized by the operator Iberdrola Renewables Locust Ridge Wind LLC. All downed bats were euthanized, even if no physical injury was observed due to the possibility of barotraumas, following requirements by the Pennsylvania Game Commission protocol [21] and using acceptable methods suggested by the American Society for Mammalogists [25]; because sedation or anesthesia was not used in our study, we employed cervical dislocation. Our work did not require approval by an Institutional Animal Care and Use Committee and all aspects of the field work and permission to collect carcasses found during our study was conducted under the auspices of permits issued each year by the state of Pennsylvania, Pennsylvania Game Commission, and the U.S. Fish and Wildlife Service.

Field Bias Trials

Searcher efficiency and removal of carcasses by scavengers (herein referred to as carcass persistence) was quantified to adjust estimates of total bat and bird fatalities for detection bias. We conducted bias trials throughout the entire study period and searchers were never aware which turbines were used or the number of carcasses placed beneath those turbines during trials. Prior to the study's inception, we generated a list of random turbine numbers and random azimuths and distances (m) from turbines for placement of each bat used in bias trials.

We used only fresh killed bats for searcher efficiency and carcass removal trials during the study. At the end of each day's search, a field crew leader gathered all carcasses from searchers and then redistributed fresh bats at predetermined random points within any given turbine plot's searchable area. Data recorded for each trial carcass prior to placement included date of placement, species, turbine number, distance and direction from turbine, and visibility class surrounding the carcass. We attempted to distribute trial bats equally among different visibility classes throughout the study period and succeeded in distributing roughly one-third of all trial bats in each visibility class (easy, moderate, and difficult; difficult and very difficult were combined). We attempted to avoid "over-seeding" any one turbine with carcasses by placing no more than 4 carcasses at any one time at a given turbine. Because we used fresh bats for searcher efficiency trials and carcass removal trials simultaneously, we did not mark bats with tape or some other previously used methods [26] that could impart human or other scents on trial bat carcasses. Rather, we used trial bat placement details (i.e. azimuth, distance, sex, species) and signatures from hair and tissue samples (i.e., hair removed between the scapulae and wing punches) to distinguish them from other fatalities landing nearby. Each trial bat was left in place and checked daily by the field crew leader or a searcher not involved with the bias trials at turbines where carcasses were placed. Thus, trial bats were available to be found by searchers on consecutive days during daily searches unless removed by a scavenger. We recorded the day that each bat was found by a searcher, at which time the carcass remained in the scavenger removal trial. If, however, a scavenger removed a carcass before detection it was removed from the searcher efficiency trial and used only in the removal data set. When a bat carcass was found, the searcher determined if a bias trial carcass had been found by looking for markings described above and contacting the crew leader to determine if the location (direction and distance) matched any possible trial bats. All trial bats were left in place for the carcass removal trial. Carcasses were left in place until removed by a scavenger or they decayed and disintegrated to a point beyond recognition. Carcass condition was recorded daily up to 20 days, as present and observable or missing or no longer observable.

Statistical Methods

Carcass persistence/removal. Estimates of the probability that a bat carcass was not removed in the interval between searches were used to adjust carcass counts for removal bias. Removal included scavenging, wind or water, or decomposition beyond recognition. In most fatality monitoring efforts, it is assumed that carcass removal occurs at a constant rate that is not dependent on the time since death; this simplifying assumption allows us to estimate fatality when search intervals exceed one day. The length of time a carcass remains on the study area before it is removed is typically modeled as an exponentially distributed random variable. The probability that a carcass is not removed during an interval of length I can be approximated as the average probability of persisting given its death might have occurred at any time during the interval:

$$\hat{r}_{jk} = \hat{\bar{t}}_{jk}(1 - \exp(-I_{ij}/\hat{\bar{t}}_{jk}))/I_{ij}$$

where:

$\hat{r}_{jk}$ is the estimated probability that a carcass in the kth visibility class that died during the interval preceding the jth search will not be removed by scavengers;

$\hat{\bar{t}}_{jk}$ is the estimated average persistence time of a carcass in the kth visibility class that died during the interval preceding the jth search;

I_{ij} is the length of the effective interval preceding the jth search at the ith turbine;

Data from 351 and 408 bat carcasses in 2009 and 2010, respectively, were used in our analysis, with carcass persistence time modeled as a function of visibility class. We fit carcass persistence/removal data for bats to an interval-censored parametric failure time model, with carcass persistence time modeled as a function of size and/or visibility class. We used a relatively liberal alpha of 0.15 to identify factors (e.g., carcass size, visibility classes) that influence bias parameter values (i.e., searcher efficiency and carcass persistence) for removal of bat carcasses.

Searcher efficiency. Estimates of the probability that an observer will visually detect a carcass during a search were used to adjust carcass counts for observer bias. Failure of an observer to detect a carcass on the search plot may be due to its size, color, or

time since death, as well as conditions in its immediate vicinity (e.g., vegetation density, shade). In most fatality monitoring efforts, because we cannot measure time since death, it is assumed that a carcass' observability is constant over the period of study, which it likely is not. In this study, searches were conducted daily and carcass persistence times were long, providing an opportunity for a searcher to detect a carcass that was missed on a previous search. We used a newly derived estimator [27] that assumes a carcass missed on a previous search will not be observed on a subsequent search (i.e., there are inherent environmental conditions that make the carcass unobservable like heavy foliage, terrain). If this assumption is not met, it can lead to overestimates of fatality. Other estimators [26] assume that a carcass missed on a previous search has the same probability of being observed as it had on the first search (i.e., there is nothing inherent in the environment surrounding the carcass that makes it unobservable), missing it is purely a chance event and that if the carcass is not removed by predators and enough searches are conducted, it will eventually be observed. If this assumption is not met, it can lead to underestimates of fatality. It is likely that neither assumption is appropriate in all cases.

Searcher efficiency trial carcasses were placed on search plots and monitored for 20 days. The day on which a bat carcass was either observed or removed by a scavenger was noted. In these trial data, if a carcass had not been found within the first 8 searches it had essentially no chance of being found. This lends empirical support to the idea that there are some environmental conditions surrounding the carcass that determine its probability of being found. However, several carcasses missed on the first search were found on subsequent searches, lending support to the idea that at least for some carcasses, the probability of missing them is purely a chance event. To allow for some possibility of observing a carcass once having missed it, the set of trial carcasses comprised those found or still observable but not found within the first 8 searches. After accounting for carcasses removed before a searcher had the chance of observing them, we fit data from 139 (2009) and 169 (2010) bat carcasses to a logistic regression model, with odds of observing a carcass given that it persisted, modeled as a function of visibility class. Again, we used a relatively liberal alpha of 0.15 to determine if a significant effect among visibility classes existed. Because we found no bats in the very difficult visibility class, SE was not modeled for this class.

Density of carcasses and proportion of area surveyed. Density of carcasses is known to diminish with increasing distance from the turbine [26], so a simple adjustment to fatality based on area surveyed would likely lead to overestimates, because unsearched areas tend to be farthest from turbines where carcass density is lowest. The calculated function (see below) relating density to distance from a turbine was used to weight each square meter in the plot. The density-weighted fraction of each plot that was actually searched was used as an area adjustment to per-turbine fatality estimates rather than using a simple proportion.

The density of bat carcasses (number of carcasses/m^2) was modeled as a function of distance (m) from the turbine. Because searcher efficiency and visibility class are confounded with distance, only fresh bat carcasses found in Easy visibility class were used for this analysis and all non-incidental data from all searched turbines were used, yielding a total of 172 fresh bat carcasses. We assumed that the carcass persistence time and searcher efficiency would be equal for all carcasses within this class and would not change as a function of distance from the turbine. We also assumed that no bat carcasses killed by turbine blades would fall >200 m from the turbine. Carcasses were "binned" into 2 m rings extending from the turbine edge out to the theoretical maximum plot distance (Figure S4). We determined the total area among all search plots that was in the easy visibility class (m^2) in each ring and calculated carcass density (number of carcasses/m^2) in each ring. Density was modeled as a conditional cubic polynomial function of distance (dist):

If distance ≤50 m, then density = exp (−1.77328+0.0346454*dist −0.00271076* dist^2+0.0000229885* dist^3) − 0.01, else density = 0.009363847*exp (−0.05*(distance-50)).

Relative density was derived by dividing the predicted density of each m^2 unit by the total predicted density within 200 m of a turbine, providing a density-weight for each m^2 unit. The density weighted area (DWA) of a plot was calculated as the sum of the density weights for all m^2 units within the searchable area. If no portion of a designated plot was unsearchable, the density weight for the plot would be 1. The physical area surveyed within a plot differed among turbines and ranged from 20–47% of the delineated theoretical maximum search plot, with an average of 31% whereas the weighted density area of plots averaged 62% (range: 44–78%). In addition, using this density weight, we estimated 7.2% of the carcasses killed at a turbine would be found beyond the boundaries of the designated search plot.

Fatality estimates. We adjusted the number of bat fatalities found by searchers by estimates of searcher efficiency and by the proportion of carcasses expected to persist unscavenged during each interval using the following equation:

$$\hat{f}_{ijk} = \frac{c_{ijk}}{\hat{a}_i \hat{p}_{jk} \hat{r}_{jk} \hat{e}_{jk}}$$

where:

$\hat{f}_{ijk}$ is the estimated fatality in the kth visibility class that occurred at the ith turbine during the jth search;

c_{ijk} is the observed number of carcasses in the kth visibility class at the ith turbine during the jth search;

$\hat{a}_i$ is the density-weighted proportion of the area of the ith turbine that was searched;

$\hat{p}_{jk}$ is the estimated probability that a carcass in the kth visibility class that is on the ground during the jth search will actually be seen by the observer;

$\hat{r}_{jk}$ is the probability than an individual bird or bat that died in the kth visibility class during the interval preceding the jth search will not be removed by scavengers; and

$\hat{e}_{jk}$ is the effective interval adjustment (i.e., the ratio of the length of time before 99% of carcasses can be expected to be removed to the search interval) associated with a carcass in the kth visibility class that died during the interval preceding the jth search.

The value for $\hat{p}_{jk}$ was estimated through searcher efficiency trials with estimates given above; $\hat{r}_{jk}$ is a function of the average carcass persistence rate and the length of the interval preceding the jth search; and $\hat{r}_{jk}$, $\hat{e}_{jk}$ and $\hat{p}_{jk}$ are assumed not to differ among turbines, but differ with search interval (j) and visibility class (k).

The estimated annual per turbine fatality for bats and birds was calculated using a newly derived estimator [27] and the equation is:

$$\hat{f} = \frac{\sum_{i=1}^{10} \sum_{j=1}^{n_i} \sum_{k=1}^{3} \hat{f}_{ijk}}{10}$$

where n_i is the number of searches carried out at turbine i, 1 = 1, ..., 10, and $\hat{f}_{ijk}$ is defined above. The per turbine estimate and confidence limits were multiplied by 64, the total number of

turbines, and divided by 0.9279 to adjust for actual density-weighted area searched to give total annual fatality estimates [28]. This estimate assumes that no fatalities occurred during the winter, i.e. prior to April and after November. No closed form solution is yet available for the variance of this estimator, so 95% confidence intervals of this estimate were calculated by bootstrapping [29]. Searcher efficiency was estimated from a bootstrap sample (with replacement) of searcher efficiency data, carcass persistence estimated from a bootstrap sample of carcass persistence data, and these values were applied to the carcass data from a bootstrap sample of turbines to estimate average fatality per turbine. This process was repeated 1000 times. The 2.5^{th} and 97.5^{th} quantiles from the 1,000 bootstrapped estimates formed the 95% confidence limits of the estimated fatality [27].

Comparison between treatment and control turbines. In 2009, we compared average fatality at control with treatment turbines for all bats and for each species using one-way analysis of variance with each turbine as the experimental unit and log$_e$ transformed estimated total fatalities as the response. In 2010, estimated average bat fatality per turbine at control and treatment turbines, during the treatment phase and the period immediately preceding it (pre-treatment phase) was analyzed using a BACI approach [26,30,31], employing ANOVA repeated measures with the turbine as the experimental unit, repeatedly measured twice. Our approach determined whether the ratio of average per-turbine fatality at control turbines (n = 15) to treatment turbines (n = 10) during implementation of the deterrents was significantly greater than it was in the period immediately preceding implementation of the treatments. In both years, the fatality data were log transformed to satisfy assumptions of normality and homogeneity of variance [32].

Results

In 2009, we searched 15 control turbines and 10 treatment turbines each day between 15 August and 10 October, and did not assess inherent variability among turbines. We found 194 carcasses (135 at control, 59 at treatment) of 6 species and two carcasses were not identifiable to species in 2009 (Table 1). During the pre-treatment period between 1 May and 26 July 2010, we searched 15 control turbines daily for all but 2 days (16 May and 2 June) and 10 Deterrent turbines daily for all but 4 days (9, 20, 24 25 July 2010) due to heavy rain, or facility maintenance. During the treatment period between 1 August and 15 October, we searched 15 control turbines daily for all but 4 days (26 August; 22, 29, 30 September 2010) and 10 Deterrent turbines daily for all but 3 days (19 August; 9, 30 September 2010) due to heavy rain or facility maintenance. During the pre-treatment period from 1 May to 26 July 2010, we found 59 carcasses comprising 6 species of bats (37 at control, 22 at treatment; Table 2). During the treatment period, we found 223 carcasses comprising 6 species of bats (162 at control, 61 at Deterrent; Table 3). Fatalities were found at all 25 turbines searched and time required to search each plot ranged from 12–100 minutes in both years of the study. Based on data from turbines not equipped with deterrents, the estimated fatality rate for this site ranged from 16–29.3 bats/turbine/year (8–14.7/ MW/year) from 2009–2010.

Fatality Estimates in 2009

A total of 278 trial carcasses were used to estimate searcher efficiency in this study. One hundred thirty-nine of the 145 (96%) carcasses in the easy class that persisted >7 days were found by searchers, while 105 of the 123 (85%) carcasses in the moderate class that persisted long enough to be observed were found. Eight

Table 1. Number of bats by species and age/sex class found under turbines at the Locust Ridge Wind Project, Columbia and Schuylkill Counties, Pennsylvania, 1 April–15 November 2009.

2009	Adult male	Adult female	Juvenile male	Juvenile female	Unknown	Total
Control						
Big brown	3	–	2	3	2	10
Eastern red	6	2	1	–	4	13
Hoary	11	8	2	3	6	30
Little brown	12	2	6	2	2	24
Silver-haired	12	8	3	2	1	26
Tri-colored	12	2	8	5	4	31
Unknown	–	–	–	–	1	1
Sub-total	56	22	22	15	20	135
Treatment						
Big brown	1	–	2	–	1	4
Eastern red	2	3	1	2	1	9
Hoary	6	1	–	1	2	10
Little brown	9	2	1	–	1	13
Silver-haired	1	1	–	1	5	8
Tri-colored	3	2	2	4	2	13
Unknown	–	–	–	–	2	2
Sub-total	22	9	6	8	14	59
Total	78	31	28	23	34	194

Table 2. Number of bats by species and age/sex class found under turbines at the Locust Ridge Wind Project, Columbia and Schuylkill Counties, Pennsylvania, 1 May–26 July (Pre-experiment phase).

2010 Pre-treatment period (1 May–26 July)	Adult male	Adult female	Juvenile male	Juvenile female	Unknown	Total
Control						
Big brown	5	1	–	–	2	8
Eastern red	4	7	–	–	–	11
Hoary	6	4	–	–	1	11
Little brown	1	2	–	–	–	3
Silver-haired	1	1	–	–	–	2
Tri-colored	2	–	–	–	–	2
Unknown	–	–	–	–	–	–
Sub-total	19	15	–	–	3	37
Treatment						
Big brown	5	1	–	–	–	6
Eastern red	6	1	–	–	–	7
Hoary	4	1	–	1	1	7
Little brown	–	–	–	–	–	–
Silver-haired	–	–	–	–	–	–
Tri-colored	2	–	–	–	–	2
Unknown	–	–	–	–	–	–
Sub-total	17	3	–	1	1	22
Total	36	18	0	1	4	59

of 10 (80%) carcasses in the difficult class were found. A logistic regression model of the odds of detection given persistence as a function of visibility classes was fit to the data and there was strong evidence of a difference in searcher efficiency among the visibility classes ($\chi_2^2 = 10.32$, $p = 0.006$). Data from 351 scavenger removal trial carcasses were fit to an interval-censored parametric failure time model. Average carcass persistence time was found to be strongly related to visibility classes ($\chi_2^2 = 6.58$, $p = 0.037$). Average persistence time was estimated to be 9.4 days (95% CI: 7.7, 11.7 days), 13.9 days (95% CI: 10.8, 18.3 days) and 8.7 days (95% CI: treatment 4.6, 16.1 days) in easy, moderate and difficult visibility classes respectively. Estimates of the probability of a bat carcass persisting for 1 day (r) were 0.948 (95% CI: 0.938, 0.958), 0.964 (95% CI: 0.955, 0.973) and 0.942 (95% CI: 0.900, 0.970), respectively.

The average per-turbine fatality rate at treatment turbines was significantly less than at control turbines ($F_{1,23} = 14.7$, $p < 0.001$). We estimated an average of 11.6 bats (95% CI: 9.4, 14.1) were killed per turbine at treatment turbines during this period, compared to 18.4 bats (95% CI: 16.0, 21.3) killed per turbine at control turbines. We estimated 60% higher fatality (95% CI: 26%, 104%) per control turbine than per treatment turbine from 15 August to 10 October 2009, or conversely, 21–51% estimated fewer bats were killed per treatment turbine than per control turbine.

We estimated twice as many hoary bats ($\bar{x} = 2.09$, 95% CI = 1.18, 4.04) killed per control turbine than treatment turbine,

and nearly twice as many silver-haired bats ($\bar{x}1.88$, 95% CI = 0.92, 5.14), although the estimated effect was not significant for this species (Tables 3 and 4). Results for other species were highly variable with no statistically significant difference between turbine groups (Tables 4 and 5).

Fatality Estimates in 2010

A total of 169 bat carcasses were used to estimate searcher efficiency in this study. Eighty three of 86 (97%) carcasses in the easy class that persisted >7 days were found by searchers, while 59 of 70 (84%) carcasses in the moderate class that persisted long enough to be observed were found. Eight of 13 (62%) carcasses in the difficult class were found. Because no fatalities were found in the very difficult class, we removed the 6 bats placed in this class from our analysis. A logistic regression model of the odds of detection given persistence was fit to the visibility classes and there was strong evidence of a difference in searcher efficiency among the visibility classes ($\chi_2^2 = 14.59$, $p = 0.007$).

Data from 408 scavenger removal trial carcasses were fit to an interval-censored parametric failure time model. Average carcass persistence time was found not to be related to visibility class ($\chi_2^2 = 0.56$, $p = 0.907$), but there was moderate evidence that average persistence time was longer before the treatment period than during the treatment period ($\chi_2^2 = 4.27$, $p = 0.12$). Average persistence time was estimated to be 7.8 days (95% CI: 6.4, 9.6

Table 3. Number of bats by species and age/sex class found under turbines at the Locust Ridge Wind Project, Columbia and Schuylkill Counties, Pennsylvania, 31 July–9 October (experiment phase) 2010.

2010 Treatment period (31 July–9 August)						
	Adult	Adult		Juvenile		
Juvenile						
	male	female	male	female		
Unknown	Total					
Control						
Big brown	2	4	2	1	–	9
Eastern red	28	19	–	–	3	50
Hoary	32	10	4	4	11	61
Little brown	6	–	–	–	–	6
Silver-haired	9	10	–	–	1	20
Tri-colored	8	2	1	1	4	16
Unknown	–	–	–	–	–	–
Sub-total	85	45	7	6	19	162
Treatment						
Big brown	1	–	–	–	–	1
Eastern red	9	10	–	–	3	22
Hoary	11	6	–	2	3	22
Little brown	1	1	–	–	1	3
Silver-haired	1	1	1	–	2	5
Tri-colored	2	2	1	–	3	8
Unknown	–	–	–	–	–	–
Sub-total	25	20	2	2	12	61
Total	**110**	**65**	**9**	**8**	**31**	**223**

period and 0.923 (95% CI: 0.912, 0.933) during the treatment period.

Bat fatality data from the pre-treatment period were used to evaluate if there were inherent difference between control and treatment turbines. We determined there was marginal evidence that the ratio of control:treatment fatalities was greater during the treatment period than in the pre-treatment period ($F_{1,23} = 3.9$, p = 0.061). During the pre-treatment period, prior to implementation of the deterrents, fatality per control turbine was estimated to be 1.09 times greater than per treatment turbine (95% CI: 0.74–1.61). We determined the initial inherent difference was about 9% in the fatality rate between the two sets and, while this was not statistically significant, we chose to adjust our comparison of fatalities between control and treatment turbines accordingly.

During the treatment period, we estimated an average of 12.8 bats (95% CI: 9.5, 17.2) were killed per turbine at treatment turbines compared to 22.9 bats (95% CI: 18.0, 29.3) killed per turbine at control turbines. Bat fatalities per control turbine was estimated to be 1.8 times greater than per treatment turbine (95% CI: 1.22–2.64); in other words, 18–62% fewer bats killed per treatment turbines relative to control turbines during the treatment period. As stated above, however, we determined an approximate 9% inherent difference between treatment and control turbines and fatality per control turbine was estimated to be 1.09 times greater than per treatment turbine (95% CI: 0.74–1.61) prior to implementation of the treatment. Thus, the ratio of fatality per control turbine relative to treatment turbines after implementing the treatment was estimated to be 1.64 times greater than the pre-treatment period ratio (95% CI: 0.98, 2.76). In other words, between 2% more and 64% fewer bats were killed per treatment turbine relative to control turbines after accounting for inherent turbine differences prior to treatment implementation.

In 2010, prior to implementation of the deterrent treatment, we estimated 1.47 times as many hoary bats (95% CI = 0.39, 3.42) and 1.32 times as many silver-haired bats (95% CI = 0.47, 3.27) killed per control turbine than treatment turbine. Although we estimated nearly twice as many hoary bats ($\bar{x} = 1.88$, 95% CI = 1.19, 2.82) and nearly 4 times as many silver-haired bats ($\bar{x} = 3.78$, 95% CI = 1.12, 12.82; Tables 5 and 6) killed per control turbine than treatment turbine during the treatment period, these represented only about a 20% increase in fatality relative to the pre-treatment period. High variation among turbines, small numbers of carcasses found and frequent zero-counts of these and other species at each turbine prevented formal statistical tests of these ratios using the BACI design (Tables 6 and 7).

days) prior to implementation of the treatments and 6.2 days (95% CI: 5.4, 7.1 days) during the implementation of the treatments. This slight difference in average persistence time had little effect on the probability of a carcass persisting through the search interval. The estimated probability of a bat carcass persisting for 1 day (r) was 0.939 (95% CI: 0.926, 0.950) prior to the treatment

Table 4. Number of each species found (N) and the estimated bat fatalities/turbine (mean and 95% confidence intervals [CI]) for each species of bat per turbine, adjusted for searcher efficiency, carcass removal, and area, at control and treatment turbines at the Locust Ridge Wind Project in Columbia and Schuylkill Counties, Pennsylvania, 15 August–10 October 2009.

Species	N	Control Turbines			N	Treatment Turbines		
		Mean	Lower 95% CI	Upper 95% CI		Mean	Lower 95% CI	Upper 95% CI
Big brown bat	10	1.34	0.35	2.59	4	0.78	0.20	1.36
Eastern red bat	13	1.81	0.95	2.83	9	1.73	0.73	2.73
Hoary bat	30	4.14	3.13	5.19	10	1.98	1.12	3.22
Little brown bat	24	3.36	2.14	5.05	13	2.66	1.57	3.82
Silver-haired bat	26	3.51	2.08	4.98	9	1.85	0.75	3.27
Tri-colored bat	31	4.15	2.36	6.20	13	2.47	1.29	3.99
Unknown bat	1	0.12	0.10	0.48	1	0.17	0.16	0.51

Table 5. Ratio between bat fatalities per control turbine relative to treatment turbines (mean and 95% confidence intervals [CI]) for each species of bat from the Locust Ridge Wind Project in Columbia and Schuylkill Counties, Pennsylvania, 15 August–10 October 2009.

Species	Mean Ratio control:treatment	Lower 95% CI	Upper 95% CI
Big brown bat	1.74	0.41	6.13
Eastern red bat	1.06	0.44	2.75
Hoary bat*	2.09	1.18	4.04
Little brown bat	1.27	0.71	2.36
Silver-haired bat	1.88	0.92	5.14
Tri-colored bat	1.68	0.80	3.58
Unknown bat	0.12	0.00	2.28

Confidence intervals that do not include 1.0 are considered statistically significant (*).

Discussion

Previous research has indicated difficulty when attempting to mask or "jam" bats' echolocation except under specific conditions [e.g., 17, 33]. Indeed, bats can actually adjust their echolocation under jamming conditions [34,35]. Bats are, however, likely "uncomfortable" when broadband ultrasound is present because it forces them to shift their call frequencies to avoid overlap, which in turn will lead to suboptimal use of echolocation or they may not echolocate at all [14,34].

In contrast to previously tested acoustic "repellers" [36], the device we have developed and tested shows some promise for deterring bats from the surrounding airspace near wind turbines. This study represents the first field test of a deterrent device to reduce bat fatalities at wind turbines by comparing fatalities at treated and untreated turbines. Our findings generally corroborate our previous conclusions from unpublished laboratory and field experiments that a regime of presumably uncomfortable or disorienting ultrasound may deter bats from occupying such a treated airspace. While the treatment response we observed generally falls within the range of variation of fatalities among turbines we studied, nothing in the statistical evaluation of our data suggested that our random selection of the 10 treatment turbines somehow skewed mortality rates among the turbines we chose. We acknowledge that 3 of our treatment turbines had to be located on the Locust Ridge I portion of the facility where no control turbines were selected. While this could have influenced the results, we noted in 2009 that two of these three turbines had

fewer mean fatalities relative to the overall mean for deterrent turbines, while in 2010, the mean fatalities of all three of these turbines generally were equal to or greater than the overall mean for treatment turbines. Fatalities at other turbines in both the control and treatment sets also varied from one year to the next and we do not believe data from the three turbines from LR I biased our findings.

In 2010, we examined potential inherent difference between the two sets of turbines and our findings suggested a marginal difference existed in fatalities between control and treatment turbines prior to implementation of the treatment. However, we caution that data from our pre-treatment period in 2010 was collected prior to migration of migratory tree roosting species and the ratio of migrant to non-migrant species was different between these two periods in our study. Thus, different levels of fatality, different species composition, and possibly different behaviors of the bats during the two phases may have influenced our findings regarding inherent differences between control and treatment turbines. Future field tests of deterrent devices should better account for potential differences in fatalities among different species when determining inherent variation among sample turbines.

The effectiveness of ultrasonic deterrents as a means to prevent bat fatalities at wind turbines is limited by the distance and area that ultrasound can be broadcast. Unfortunately, rapid attenuation of ultrasound in air, which is heavily influenced by humidity (Table S1), limits the effective range of broadcasts. Nightly humidity in this region of Pennsylvania averaged 86.5% in August

Table 6. Estimated bat fatalities/turbine (mean and 95% confidence intervals [CI]) for each species of bat per turbine, adjusted for searcher efficiency, carcass removal, and area, at control and treatment turbines at the Locust Ridge Wind Project in Columbia and Schuylkill Counties, Pennsylvania, 31 July–9 October 2010.

Species	Control Turbines				Treatment Turbines			
	N	Mean	Lower 95% CI	Upper 95% CI	N	Mean	Lower 95% CI	Upper 95% CI
Big brown bat	9	1.19	0.39	2.12	2	0.38	0.23	0.85
Eastern red bat	50	7.16	5.32	9.27	22	4.77	2.70	6.92
Hoary bat	61	9.12	7.08	11.70	22	5.02	3.37	7.31
Little brown bat	6	0.87	0.39	1.38	3	0.65	0.20	1.27
Silver-haired bat	20	2.87	1.48	4.47	5	1.00	0.18	2.03
Tri-colored bat	16	2.32	1.37	3.38	8	1.55	0.91	2.23

Table 7. Ratio between bat fatalities per control turbine relative to treatment turbines (mean and 95% confidence intervals [CI]) for each species of bat from the Locust Ridge Wind Project in Columbia and Schuylkill Counties, Pennsylvania, 31 July–9 October 2010.

Species	Mean Ratio control:deterrent	Lower 95% CI	Upper 95% CI
Big brown bat	3.72	0.70	7.87
Eastern red bat	1.59	0.93	2.78
Hoary bat*	1.88	1.19	2.82
Little brown bat	1.72	0.43	5.22
Silver-haired bat*	3.78	1.12	12.82
Tri-colored bat	1.59	0.84	2.96

Confidence intervals that do not include 1.0 are considered statistically significant (*).

2009, 84.8% in September 2009, 80% in August 2010, and 76.8% in September 2010 (source http://climate.met.psu.edu/www_prod/). Assuming a constant temperature of 20°C and air pressure of 101.325 kPa and 80% humidity, the theoretical distance to "jam" bats at the assumed 65 dB level only extends to 20 m for the 20–30 kHz range, and declines to only 5–10 m for the upper frequency ranges of broadcast (70–100 kHz;). Ultrasound emission in the perpendicular plane of the rotor-swept area may be adequate to affect approaching bats, particularly those species influenced at the lower frequencies. However, it is clear that effective emissions in the parallel plane of the rotor-swept area will be difficult if not impossible to achieve based on sound attenuation in humid environments. The effective airspace would be different and larger in more arid environments, however (Table S1). We also note that some devices were not operating all the time during our study, due to malfunctions. Although we were unable to account for this factor in our analysis, clearly the affected airspace was reduced when some devices were inactive, which further influenced our findings.

We assume that when bats encounter a gradient of increasingly strong emissions as they approach the deterrent device, they will respond by flying opposite to that gradient to escape the effect of emissions. However, at present we know little about the general responses that various species of bats have upon entering a field of ultrasound emissions. It is therefore important to consider our assumptions when interpreting results of this study. Although our acoustic deterrent device could only generate a limited effective volume of uncomfortable airspace, bats could have detected the presence of such airspace from a greater range, possibly beyond the rotor swept area. Bats previously experiencing the discomfort of ultrasound broadcast may avoid approaching other treated towers, which they could detect as treated from beyond the zone of discomfort. In this way, ultrasound broadcast may effectively serve as acoustic beacons to direct bats away from wind turbines. Over time, bats may learn to avoid all turbines from their experience with those equipped with deterrents, similar to documented behavior of bats encountering other discomforting experiences such as mist nets. Just as bat capture success in mist nets declines on successive nights as bats apparently learn to detect the presence of nets and thereafter avoid them [37,38,39,40], we speculate that after experiencing a disagreeable encounter with ultrasound treated airspace bats may opt to subsequently avoid it. Other lines of evidence indicate that bats learn and remember spatial locations or stimuli associated with obstacles or threats. A study that modified experiments conducted by Griffin [14] challenged bats to maneuver through vertical wires, and they did so by tilting and scrunching their wings; the same bats continued these

maneuvers at the locations of wires even after they were removed [41]. In practice, the actual decline of activity at any treated site will likely depend upon immigration of naïve bats into the area. We did not monitor bat activity with night vision or thermal imaging cameras [42] and, thus, were unable to assess activity patterns of bats simultaneous with fatality searches. It is also possible that insects preyed on by bats in this region were deterred from the turbines, which could represent the ultimate cause of avoiding treated turbines. Indeed, studies have demonstrated that ultrasound can repel insects [43] and influence their reproduction [44]. However, we did not assess insect abundance and suggest future studies should attempt to address causal factors of avoidance including effect on insect prey.

Conversely, bats may habituate to the presence of ultrasound emissions and acoustic deterrents may actually lose their effectiveness over time. However, in prior field tests of deterrents, bats did not appear to habituate or accommodate to the presence of ultrasound emitted from a previous prototype deterrent at least over short periods of 5–7 days (J. Szewczak, Humboldt State University and E. Arnett, Bat Conservation International, unpublished data). Habituation to deterrents should, however, continue to be investigated in future studies.

The effectiveness of acoustic deterrents will likely vary among different species of bats. Hoary bats, for example, employ the lowest frequency range of the species we studied (~20–25 kHz) and may be affected more so than other species that use higher frequencies and perhaps fly at further distances from the device. Hoary bats had significantly fewer fatalities at turbines with deterrents relative to those without them in both years, and silver-haired bats also had fewer fatalities at turbines with deterrents in 2010. In 2010, however, after accounting for inherent differences between turbine sets prior to treatment, hoary and silver-haired bats killed per control turbine relative to treatment turbines during the treatment period represented only a 20% increase in fatality over the pre-treatment period. Species-specific effectiveness warrants further investigation in a study with more power to detect differences among species. Future studies hopefully will also elucidate whether deterrents can eventually serve as a mitigation tool for minimizing or eliminating take of threatened or endangered species such as the Indiana bat (*Myotis sodalis*). The limited range of ultrasound broadcast from a wind turbine tower or nacelle might have only a moderate contribution toward reducing impacts of bats randomly flying through the rotor-swept area. However, for bats that may be drawn to and approach turbine towers as potential roosts or gathering sites [1,45], the combination of effective range and learned avoidance response to ultrasound broadcast may have longer term effects in reducing bat mortality at wind turbines. We also note that we only tested broadband ultrasound emission (20–100 kHz) and short pulses mimicking echoes of insects [16], for example, could prove to be more effective for some species and should be tested in future studies.

Introducing ultrasound emissions into the environment could potentially yield negative environmental effects on other species of wildlife, but we do not feel this is of concern because the device we tested only had a limited effective range because of rapid attenuation of ultrasound with distance from its source. Within the effective range of the treated airspace, emissions could affect ultrasound-sensitive) insects and disperse them, providing less reason for bats to occupy that airspace, assuming food sources attract bats to turbines [1,10]; and 2) passerines that may be deterred from the airspace, thus reducing strikes. If ultrasound could reach the ground from locations where deterrent devices are mounted, which was not the case in our study, ultrasound sensitive small mammals could be dispersed away from the base of the turbines, but this would likely be a positive impact that could reduce strikes of stooping raptors.

Conclusions

This study, and previous experiments with earlier prototypes, revealed that broadband ultrasound broadcasts may affect bat behavior directly by discouraging them from approaching the sound source, or indirectly by reducing the time bats spend foraging near a turbine if insects are repelled by ultrasound [e.g., 42, 43; also recognizing not all insects have ears to detect ultrasound] and ultimately reduce bat fatalities at wind turbines. However, variation among turbines yielded inconclusive evidence of a strong effect of deterrents on bat fatality and while the approach may hold some promise, further refinement and investigation is needed. We did experience technical issues in both years of the study, including water leakage that rendered some deterrents inoperable during portions of the study period which clearly influenced our findings. Thus, results from this study may reflect a more conservative estimate of potential fatality reduction achievable through application of the deterrent device we tested. Still, we caution that the response estimated in this study falls generally within the range of variation for bat fatalities among turbines in this and other studies in the region [2]. Additionally, deterrents resulted in lower reductions in bat fatality relative to curtailing turbine operations by increasing cut-in speeds (44–93%) [11,12]. We further caution that it would be premature and unwarranted to conclude or interpret from these initial results that this technology provides an operational deterrent device ready for broad-scale deployment at wind facilities. While we do not consider acoustic deterrents to be an acceptable mitigation strategy at this time, with further experimentation and modifications, this type of deterrent method may prove successful and broadly applicable for protecting bats from harmful encounters with wind turbine blades. Future research and development and field studies should attempt to improve the device and it's weatherproofing and emission performance, and optimizes the placement and number of devices on each turbine that would affect the greatest amount of airspace in the rotor-swept area to estimate potential maximum effectiveness of this tool to reduce bat fatalities. New studies also should test other emission types such as short ultrasonic pulses that mimic insects [16]. Finally, we did not attempt to develop comparative estimates of costs associated with our deterrent devices relative to lost revue of operational mitigation because current deterrent development costs are high and dynamic and operational costs to maintain

them over a period of time have not been established. Future efforts should determine production and maintenance costs of newly developed deterrents that can be factored into a cost-benefit analysis comparing different approaches for mitigating bat fatalities.

Supporting Information

Figure S1   An ultrasonic deterrent device used in this study (Photo by E. Arnett, Bat Conservation International).

Figure S2   Ultrasonic deterrent devices mounted on the side of the turbine nacelle (photo by E. Arnett, Bat Conservation International).

Figure S3   Depiction of acoustic deterrent placement on the nacelle of turbines and ultrasonic broadcast volume from devices (broadcast volume approximation of data from Senscorp beam pattern data, see supplemental material below).

Figure S4   Hypothetical carcass search plot for a wind turbine illustrating 2 m rings extending from the turbine edge out to the theoretical maximum plot distance and a depiction of "easy" searchable area (shaded area within line drawing) in the plot, used to develop weights for adjusting fatalities.

Table S1   Calculated decibel level at different distances and frequencies at two different levels of relative humidity (10 and 40%) for acoustic deterrent devices used in this study. Calculations assume ambient temperature of 20°C and air pressure of 101.325 kPa (kilopascal).

Table S2   The attenuation of sound in air due to viscous, thermal and rotational loss mechanisms is simply proportional to f^2. However, losses due to vibrational relaxation of oxygen molecules are generally much greater than those due to the classical processes, and the attenuation of sound varies significantly with temperature, water-vapor content and frequency. A method for calculating the absorption at a given temperature, humidity, and pressure can be found in ISO 9613-1 (1993). The table and figure below gives values of attenuation in dB m^{-1} for a temperature of 20°C and an air pressure of 101.325 kPa. The uncertainty is estimated to be ±10%.

Table S3   Habitat visibility classes used during this study, following Pennsylvania Game Commission Protocol [21]. Data for Classes 3 and 4 were combined during our final analyses.

Acknowledgments

We are indebted to R. Bergstresser, C. Bowersox, C. Martin, M. Dasilva, P.J. Falatek, B. Farless, P. Green, L. Heffernan, J. Kougher, S. Mitchell, S. O'Connor, S. Philibosian, M. VanderLinden, and J. Wilcox for conducting field surveys. We thank Iberdrola Renewables employees A. Linehan, L. Van Horn, D. DeCaro, D. Rogers, J. Roppe, and S. Webster for their logistical support. Z. Wilson conducted GIS analysis for the study. We also thank T. Allison, R. Barclay, P. Cryan, E. Gilliam, G. Jones, T. Kunz, R. Medellin, J. Rydell, and an anonymous reviewer for their critical reviews of this work. Finally, we wish to thank M. Jensen (Binary Acoustic Technology, Tucson, Arizona) for his initial work on the deterrent devices and Deaton Engineering (Georgetown, Texas) for their follow-up design work and manufacturing of the deterrents used in this study.

Author Contributions

Conceived and designed the experiments: EBA MMPH JMS. Performed the experiments: EBA MRS CDH. Analyzed the data: MMPH. Contributed reagents/materials/analysis tools: MMPH. Wrote the paper: EBA MMPH CDH JMS.

References

1. Kunz TH, Arnett EB, Erickson WP, Johnson GD, Larkin RP, et al. (2007) Ecological impacts of wind energy development on bats: questions, hypotheses, and research needs. Front Ecol Environ 5: 315–324.
2. Arnett EB, Brown K, Erickson WP, Fiedler J, Henry TH, et al. (2008) Patterns of fatality of bats at wind energy facilities in North America. J Wildl Manage 72: 61–78.
3. Baerwald EF, Barclay RMR (2009) Geographic variation in activity and fatality of migratory bats at wind energy facilities. J Mammal 90: 1341–1349.
4. Rydell J, Bach L, Dubourg-Savage M, Green M, Rodrigues L, et al. (2010) Bat mortality at wind turbines in northwestern Europe. Acta Chiropt 12: 261–274.
5. Arnett EB, Baerwald EF (2013) Impacts of wind energy development on bats: implications for conservation. In: Adams RA, Peterson SC, editors. Bat evolution, ecology, and conservation. New York: Springer Science Press (in press).
6. Energy Information Administration. (2010) Annual energy outlook 2010 with projections to 2035. U.S. Department of Energy, Energy Information Administration website. Available: http://www.eia.doe.gov/oiaf/ieo/world.html. Accessed 2011 December 15.
7. Racey PA, Entwistle AC (2003) Conservation ecology of bats. In: Kunz TH, Fenton MB, editors. Bat Ecology. Chicago: University of Chicago Press. 680–743.
8. Winhold L, Kurta A, Foster R (2008) Long-term change in an assemblage of North American bats: are eastern red bats declining? Acta Chiropta 10: 359–366).
9. Frick WF, Pollock JF, Hicks AC, Langwig KE, Reynolds DS, et al. (2010) An emerging disease causes regional population collapse of a common North American bat species. Science 329: 679–682.
10. Cryan PM, Barclay RMR (2009) Causes of bat fatalities at wind turbines: hypotheses and predictions. J Mammal 90: 1330–1340.
11. Baerwald EF, Edworthy J, Holder M, Barclay RMR (2009) A large-scale mitigation experiment to reduce bat fatalities at wind energy facilities. J Wildl Manage 73: 1077–1081.
12. Arnett EB, Huso MMP, Schirmacher MR, Hayes JP (2011) Changing wind turbine cut-in speed reduces bat fatalities at wind facilities. Front Ecol Environ 9: 209–214.
13. Nicholls B, Racey PA (2009) The aversive effect of electromagnetic radiation on foraging bats–a possible means of discouraging bats from approaching wind turbines. PLoS ONE 4(7): e6246. doi:10.1371/journal.pone.0006246.
14. Griffin DR (1958) Listening in the dark. New Haven: Yale University Press. 413 p.
15. Barlow KE, Jones G (1997) Function of pipistrelle social calls: field data and a playback experiment. Animal Behaviour 53: 991–999.
16. Hristov NI, Conner WE (2005) Sound strategy: acoustic aposematism in the bat-tiger moth arms race. Naturwissenschaften 92: 164–169.
17. Griffin DR, McCue JJG, Grinnell AD (1963) The resistance of bats to jamming. J Exp Zool 152: 229–250.
18. Mackey RL, Barclay RMR (1989) The influence of physical clutter and noise on the activity of bats over water. Can J Zool 67: 1167–1170.
19. Brown RG, Brown ML (1972) Woody Plants of Maryland. Baltimore: Port City Press 347 p.
20. Strausbaugh PD, Core EL (1978) Flora of West Virginia. Second edition. Grantsville: Seneca Books, 1079 p.
21. Pennsylvania Game Commission (2007) Pennsylvania Game Commission wind energy voluntary cooperation agreement. Pennsylvania Game Commission website. Available: http://www.portal.state.pa.us/portal/server.pt/gateway/PTARGS_0_0_114831_0_0_43/http;/pubcontent.state.pa.us/publishedcontent/publish/marketingsites/game_commission/content/wildlife/habitat_management/wind_energy/wind_energy_voluntary_cooperative_agreement_ci.html?qid=85232671&rank=6. Accessed 2013 April 24.
22. Surlykke A, Kalko EKV (2008) Echolocating bats cry out loud to detect their prey. PLoS ONE 3(4): e2036. doi:10.1371/journal.pone.0002036.
23. Fenton MB (2003) Eavesdropping on the echolocation and social calls of bats. Mammal Rev 33: 193–204.
24. Parson S, Szewczak JM (2009) Detecting, recording, and analyzing the vocalizations of bats. In: Kunz TH, Parsons S, editors. Ecological and behavioral methods for the study of bats, 2nd edition. Baltimore: Johns Hopkins University Press. 901 p.
25. Gannon WL, Sikes RS, and the Animal Care and Use Committee of the American Society of Mammalogists (2007) Guidelines of the American Society of Mammalogists for the use of wild mammals in research. J Mammal 88: 809–823.
26. Strickland MD, Arnett EB, Erickson WP, Johnson DH, Johnson GD, et al. (2011) Comprehensive guide to studying wind energy/wildlife interactions. National Wind Coordinating Collaborative website. Available: http://www.nationalwind.org/assets/publications/Comprehensive_Guide_to_Studying_Wind_Energy_Wildlife_Interactions_2011_Updated.pdf. Accessed 2013 April 24.
27. Huso MMP (2011) An estimator of wildlife fatality from observed carcasses. Environmetrics 22: 318–329.
28. Cochran WG (1977) Sampling techniques, 3rd edition. New York: John Wiley & Sons. 428 p.
29. Manly BFJ (1997) Randomization and Monte Carlo Methods in Biology, 2nd edition. New York: Chapman and Hall. 300 p.
30. Hurlbert SH (1984) Pseudoreplication and the design of ecological field experiments. Ecol Monogr 54: 187–211.
31. Hewitt JE, Thrush SE, Cummings VJ (2001) Assessing environmental impacts: effects of spatial and temporal variability at likely impact scales. Ecol Appl 11: 1502–1516.
32. Steele RGD, Torrie JH, Dickie DA (1997) Principles and procedures of statistics: a biometrical approach, 3rd edition. Boston: McGraw-Hill. 666 p.
33. Møhl B, Surlykke A (1989) Detection of sonar signals in the presence of pulses of masking noise by the echolocating bat, Eptesicus fuscus. J Comp Physiol 165: 119–194.
34. Ulanovsky N, Fenton MB, Tsoar A, Korine C (2004) Dynamics of jamming avoidance in echolocating bats. Proceed Royal Soc London B 271: 1467–1475.
35. Gillam EH, McCracken GF (2007) Variability in the echolocation of Tadarida brasiliensis: effects of geography and local acoustic environment. Animal Behav 74: 277–286.
36. Hurley S, Fenton MB (1980) Ineffectiveness of fenthion, zinc phosphide, DDT and two ultrasonic rodent repellers for control of populations of little brown bats (Myotis lucifugus). Bull Environ Contamin and Toxicol 25: 503–507.
37. Kunz TH, Hodgkison R, Weise C (2009) Methods for capturing and handling bats. In: Kunz TH, Parsons s, editors. Ecological and behavioral methods for the study of bats, 2nd edition. Baltimore: Johns Hopkins University Press. 1–35.
38. Larsen RJ, Boegler KA, Genoways HH, Masefield WP, Kirsch RA, et al. (2007). Mist netting bias, species accumulation curves, and the rediscovery of two bats on Montserrat (Lesser Antilles). Acta Chiropt 9: 423–435.
39. Robbins LW, Murray KL, McKenzie PM (2008). Evaluating the effectiveness of the standard mist-netting protocol for the endangered Indiana bat (Myotis sodalis). Northeast. Nat. 15: 275–282.
40. Winhold L., Kurta A. (2008). Netting surveys for bats in the northeast: differences associated with habitat, duration of netting, and use of consecutive nights. Northeast. Nat. 15: 263–274.
41. Pollak GD, Casseday JH (1989). The Neural Basis of Echolocation in Bats. Berlin: Springer-Verlag. 155 p.
42. Horn J, Arnett EB, Kunz TH (2008) Behavioral responses of bats to operating wind turbines. J Wildl Manage 72: 123–132.
43. Belton P, Kempster RH (1962) A field test on the use of sound to repel the European corn borer. Entomologia 5: 281–288.
44. Huang F, Subramanyam B, Taylor R (2011) Ultrasound affects spermatophore transfer, larval numbers, and larval weight of Plodia interpunctella (Hübner) (Lepidoptera: Pyralidae). J Stored Products Res 39: 413–422.
45. Cryan PM (2008) Mating behavior as a possible cause of bat fatalities at wind turbines. J Wildl Manage 72: 845–849.

Permissions

All chapters in this book were first published in PLOS ONE, by The Public Library of Science; hereby published with permission under the Creative Commons Attribution License or equivalent. Every chapter published in this book has been scrutinized by our experts. Their significance has been extensively debated. The topics covered herein carry significant findings which will fuel the growth of the discipline. They may even be implemented as practical applications or may be referred to as a beginning point for another development.

The contributors of this book come from diverse backgrounds, making this book a truly international effort. This book will bring forth new frontiers with its revolutionizing research information and detailed analysis of the nascent developments around the world.

We would like to thank all the contributing authors for lending their expertise to make the book truly unique. They have played a crucial role in the development of this book. Without their invaluable contributions this book wouldn't have been possible. They have made vital efforts to compile up to date information on the varied aspects of this subject to make this book a valuable addition to the collection of many professionals and students.

This book was conceptualized with the vision of imparting up-to-date information and advanced data in this field. To ensure the same, a matchless editorial board was set up. Every individual on the board went through rigorous rounds of assessment to prove their worth. After which they invested a large part of their time researching and compiling the most relevant data for our readers.

The editorial board has been involved in producing this book since its inception. They have spent rigorous hours researching and exploring the diverse topics which have resulted in the successful publishing of this book. They have passed on their knowledge of decades through this book. To expedite this challenging task, the publisher supported the team at every step. A small team of assistant editors was also appointed to further simplify the editing procedure and attain best results for the readers.

Apart from the editorial board, the designing team has also invested a significant amount of their time in understanding the subject and creating the most relevant covers. They scrutinized every image to scout for the most suitable representation of the subject and create an appropriate cover for the book.

The publishing team has been an ardent support to the editorial, designing and production team. Their endless efforts to recruit the best for this project, has resulted in the accomplishment of this book. They are a veteran in the field of academics and their pool of knowledge is as vast as their experience in printing. Their expertise and guidance has proved useful at every step. Their uncompromising quality standards have made this book an exceptional effort. Their encouragement from time to time has been an inspiration for everyone.

The publisher and the editorial board hope that this book will prove to be a valuable piece of knowledge for researchers, students, practitioners and scholars across the globe.

List of Contributors

Hiroki Obata and Aya Manabe
Faculty of Letters, Kumamoto University, Kumamoto City, Kumamoto Prefecture, Japan

Naoko Nakamura
Faculty of Law and Letters, Kagoshima University, Kagoshima City, Kagoshima Prefecture, Japan

Tomokazu Onishi
Faculty of Intercultural Studies, The International University of Kagoshima, Kagoshima City, Kagoshima Prefecture, Japan

Yasuko Senba
Project for Seed Impression Studies, Kumamoto University, Kumamoto City, Kumamoto Prefecture, Japan

Ilga Porth, Carol Ritland and Kermit Ritland
Department of Forest Sciences, University of British Columbia, Vancouver, British Columbia, Canada

Richard White
Department of Statistics, University of British Columbia, Vancouver, British Columbia, Canada

Barry Jaquish
Kalamalka Forestry Centre, British Columbia Ministry of Forests, Lands and Natural Resource Operations, Vernon, British Columbia, Canada

René Alfaro
Pacific Forestry Centre, Canadian Forest Service, Victoria, British Columbia, Canada

David L. Smith
Department of Epidemiology, Johns Hopkins Bloomberg School of Public Health, Baltimore, Maryland, United States of America
Malaria Research Institute, Johns Hopkins Bloomberg School of Public Health, Baltimore, Maryland, United States of America
Fogarty International Center, NIH, Bethesda, Maryland, United States of America

T. Alex Perkins and Thomas W. Scott
Fogarty International Center, NIH, Bethesda, Maryland, United States of America
Department of Entomology, University of California, Davis, California, United States of America,

Lucy S. Tusting
Department of Disease Control, London School of Hygiene and Tropical Medicine, London, United Kingdom

Steven W. Lindsay
Fogarty International Center, NIH, Bethesda, Maryland, United States of America
School of Biological and Biomedical Sciences, Durham University, Durham, United Kingdom

Soroush Parsa
CIAT (Centro Internacional de Agricultura Tropical), Cali, Colombia
Department of Entomology, University of California Davis, Davis, California, United States of America

Raúl Ccanto, Edgar Olivera and María Scurrah
Grupo Yanapai, Miraflores, Lima, Peru

Jesús Alcázar
CIP (Centro Internacional de la Papa), Lima, Peru

Jay A. Rosenheim
Department of Entomology, University of California Davis, Davis, California, United States of America

Xiongbing Tu
State Key Laboratory for Biology of Plant Diseases and Insect Pests, Institute of Plant Protection, Chinese Academy of Agricultural Sciences, Beijing, P. R. China
Department of Entomology, College of Agronomy and Biotechnology, China Agricultural University, Beijing, P.R. China

Zhihong Li
Department of Entomology, College of Agronomy and Biotechnology, China Agricultural University, Beijing, P.R. China

Jie Wang, Xunbing Huang, Huihui Wu and Zehua Zhang
State Key Laboratory for Biology of Plant Diseases and Insect Pests, Institute of Plant Protection, Chinese Academy of Agricultural Sciences, Beijing, P. R. China

Jiwen Yang and Chunbin Fan
Tianjin Binhai New Area of Dagang Agricultural Service Center, Tianjin, P.R. China

Qinglei Wang
Cangzhou Academy of Agricultural and Forestry Sciences of Hebei, Cangzhou, P.R. China

Nicole Wäschke, Kristin Hardge, Torsten Meinersa and Monika Hilker
Freie Universität Berlin, Institute of Biology, Applied Zoology/Animal Ecology, Berlin, Germany

Christine Hancock and Elisabeth Obermaier
University of Würzburg, Department of Animal Ecology and Tropical Biology, Würzburg, Germany

Michael Epelbaum
Independent Multidisciplinary Scientist, Nashville, Tennessee, United States of America

Mathieu Maheu-Giroux and Marcia C. Castro
Department of Global Health and Population, Harvard School of Public Health, Boston, Massachusetts, United States of America

Yan-Yan Chen
Ministry of Education Key Lab of Environment and Health, Department of Epidemiology and Biostatistics, School of Public Health, Tongji Medical College, Huazhong University of Science and Technology, Wuhan, China
Hubei Center for Disease Control and Prevention, Wuhan, China

Jian-Bing Liu, Xi-Bao Huang, Shun-Xiang Cai and Zheng-Ming Su
Hubei Center for Disease Control and Prevention, Wuhan, China

Rong Zhong, Li Zou and Xiao-Ping Miao
Ministry of Education Key Lab of Environment and Health, Department of Epidemiology and Biostatistics, School of Public Health, Tongji Medical College, Huazhong University of Science and Technology, Wuhan, China

Armel Djénontin
Faculté des Sciences et Techniques/MIVEGEC (IRD 224-CNRS 5290-UM1-UM2), Université d'Abomey Calavi/Centre de Recherche Entomologique de Cotonou (CREC), Cotonou, Bénin

Cédric Pennetier, Barnabas Zogo, Koffi Bhonna Soukou and Marina Ole-Sangba
MIVEGEC (IRD 224-CNRS 5290-UM1-UM2), Centre de Recherche Entomologique de Cotonou (CREC), Cotonou, Bénin

Martin Akogbéto
Faculté des Sciences et Techniques/Centre de Recherche Entomologique de Cotonou (CREC), Université d'Abomey Calavi/Centre de Recherche Entomologique de Cotonou (CREC), Cotonou, Bénin

Fabrice Chandre
MIVEGEC (IRD 224-CNRS 5290-UM1-UM2), Laboratoire de lutte contre les Insectes Nuisibles (LIN), Montpellier, France

Rajpal Yadav
Department of Control of Neglected Tropical Diseases, World Health Organization, Geneva, Switzerland

Vincent Corbel
MIVEGEC (IRD 224-CNRS 5290-UM1-UM2)/ Department of Entomology, Kasetsart University, Ladyaow Chatuchak Bangkok, Thailand

Lêda N. Regis, Cláudia M. F. Oliveira, Rosângela M. R. Barbosa, José Constantino Silveira Jr., Maria and Alice Varjal Melo-Santos
Departameto de Entomologia, Fundaûção Oswaldo Cruz-Fiocruz-Pe, Recife-PE, Brazil

Ridelane Veiga Acioli
Secretaria Estadual de Saúde, Recife-PE, Brazil

Wayner Vieira Souza
Departameto de Saúde Coletiva, Fundação Oswaldo Cruz-Fiofruz-PE, Recife-PE, Brazil

Cândida M. Nogueira Ribeiro
Secretaria Municipal de Saúde, Santa Cruz do Capibaribe-PE, Brazil

Juliana C. Serafim da Silva
Secretaria Municipal de Saúde, Ipojuca-PE, Brazil

Antonio Miguel Vieira Monteiro
Divisão de Processamento de Imagens, Instituto Nacional de Pesquisas Espaciais-INPE, São José dos Campos-SP, Brazil

Cynthia Braga
Departameto de Parasitologia, Fundação Oswaldo Cruz-Fiocruz-PE, Recife-PE, Brazil

Marco Aurélio Benedetti Rodrigues and Marilú Gomes N. M. Silva
Departameto de Eletrônica e Sistemas, Universidade Federal de Pernambuco, Recife-PE, Brazil

Paulo Justiniano Ribeiro Jr. and Wagner Hugo Bonat
Departameto de Estatística, Universidade Federal do Paraná, Curitiba-PR, Brazil

Liliam César de Castro Medeiros
Centro de Ciência do Sistema Terrestre, Instituto Nacional de Pesquisas Espaciais-INPE, São José dos Campos-SP, Brazil

Marilia Sa Carvalho
Fundação Oswaldo Cruz-Fiocruz, Rio de Janeiro-RJ, Brazil

André Freire Furtado
Departameto de Virologia, Fundação Oswaldo Cruz-Fiocruz-PE, Recife-PE, Brazil

Gislaine A. Carvalho, Andrea Oliveira B. Ribon and Luiz Orlando de Oliveira
Departamento de Bioquímica e Biologia Molecular, Universidade Federal de Viçosa, Viçosa, MG, Brazil

Juliana L. Vieira, Marcelo M. Haro and Raul Narciso C. Guedes
Departamento de Entomologia, Universidade Federal de Viçosa, Viçosa, MG, Brazil

Alberto S. Corrêa
Departamento de Entomologia e Acarologia, Escola Superior de Agricultura "Luiz de Queiroz", Universidade de São Paulo, Piracicaba, São Paulo, Brazil

Rory P. Wilson, Rebecca Richards, Angharad Hartnell, Andrew J. King, Justyna Piasecka, Yogendra K. Gaihre and Tariq Butt
Biosciences, College of Science, Swansea University, Singleton Park, Swansea, Wales, United Kingdom

Sébastien Marcombe and Dina M. Fonseca
Center for Vector Biology, Rutgers University, New Brunswick, New Jersey, United States of America,

Ary Farajollahi
Center for Vector Biology, Rutgers University, New Brunswick, New Jersey, United States of America, Mercer County Mosquito Control, West Trenton, New Jersey, United States of America,

Sean P. Healy
Monmouth County Mosquito Extermination Commission, Eatontown, New Jersey, United States of America

Gary G. Clark
Mosquito and Fly Research Unit, Agriculture Research Service, United States Department of Agriculture, Gainesville, Florida, United States of America

Benjamin Hornoy, Michéle Tarayre and Anne Atlan
Ecobio, Centre National de la Recherche Scientifique, Université de Rennes 1, Rennes, France

Maxime Hervé
BIO3P, Institut National de la Recherche Agronomique – Agrocampus Ouest, Université de Rennes 1, Rennes, France

Luc Gigord
Conservatoire Botanique National de Mascarin, Saint-Leu, La Réunion, France

Minette Karsten
Evolutionary Genomics Group, Department of Botany and Zoology, University of Stellenbosch, Matieland, South Africa
Centre for Invasion Biology, Department of Conservation Ecology and Entomology, Stellenbosch University, University of Stellenbosch, Matieland, South Africa

Bettine Jansen van Vuuren
Centre for Invasion Biology, Department of Zoology, University of Johannesburg, Auckland Park, South Africa

Adeline Barnaud
IRD, Montpellier, France

John S. Terblanche
Centre for Invasion Biology, Department of Conservation Ecology and Entomology, Stellenbosch University, University of Stellenbosch, Matieland, South Africa

Edward B. Arnett, Cris D. Hein and Michael R. Schirmacher
Bat Conservation International, Austin, Texas, United States of America,

Manuela M. P. Huso
Forest and Range Experiment Station, United States Geological Survey, Corvallis, Oregon, United States of America

Joseph M. Szewczak
Department of Biological Sciences, Humboldt State University, Arcata, California, United States of America

Index

9 781632 397812